D0893991

SIDEWAYS

SIDEWAYS

THE CITY GOOGLE COULDN'T BUY

JOSH O'KANE

RANDOM HOUSE CANADA

PUBLISHED BY RANDOM HOUSE CANADA

 Published in 2022 by Random House Canada, a division of Penguin Random House Canada Limited, Toronto. Distributed in Canada and the United States of America by Penguin Random House Canada Limited, Toronto.

www.penguinrandomhouse.ca

Library and Archives Canada Cataloguing in Publication

Title: Sideways : the city Google couldn't buy / Josh O'Kane.
Names: O'Kane, Josh, author.
Identifiers: Canadiana (print) 20220228183 | Canadiana (ebook) 20220228310 |
ISBN 9781039000780 (hardcover) | ISBN 9781039000797 (EPUB)
Subjects: LCSH: Google (Firm) | LCSH: Technology—Social aspects—Ontario—Toronto. | LCSH: Technology—Political aspects—Ontario—Toronto. | LCSH: Data privacy—Ontario—Toronto. | LCSH: Privacy, Right of—Ontario—Toronto. | LCSH: Waterfronts—Ontario—Toronto. | LCSH: City planning—Ontario—Toronto.
Classification: LCC T14.5 .O45 2022 | DDC 303.48/309713541—dc23

Text design: Matthew Flute
Jacket design: Matthew Flute
Image credits: (pixels) amtitus / Getty Images

Printed in the U. S. A.

2 4 6 8 9 7 5 3 1

For my parents, Joanne and Mick,
who left the Miramichi
and found a city

To approach a city, or even a city neighborhood, as if it were a larger architectural problem, capable of being given order by converting it into a disciplined work of art, is to make the mistake of attempting to substitute art for life.

The results of such profound confusion between art and life are neither life nor art. They are taxidermy.

—Jane Jacobs, *The Death and Life of Great American Cities*, 1961

LYLE LANLEY:	I swear it's Springfield's only choice. Throw up your hands and raise your voice!
TOWNSPEOPLE:	Monorail!
LYLE LANLEY:	What's it called?
TOWNSPEOPLE:	Monorail!
LYLE LANLEY:	Once again—
TOWNSPEOPLE:	Monorail!
MARGE SIMPSON:	But Main Street's still all cracked and broken.
BART SIMPSON:	Sorry Mom—the mob has spoken.
TOWNSPEOPLE:	Monorail! Monorail! Monoraaaaaaiiiil!

—"The Monorail Song," *The Simpsons*, 1993

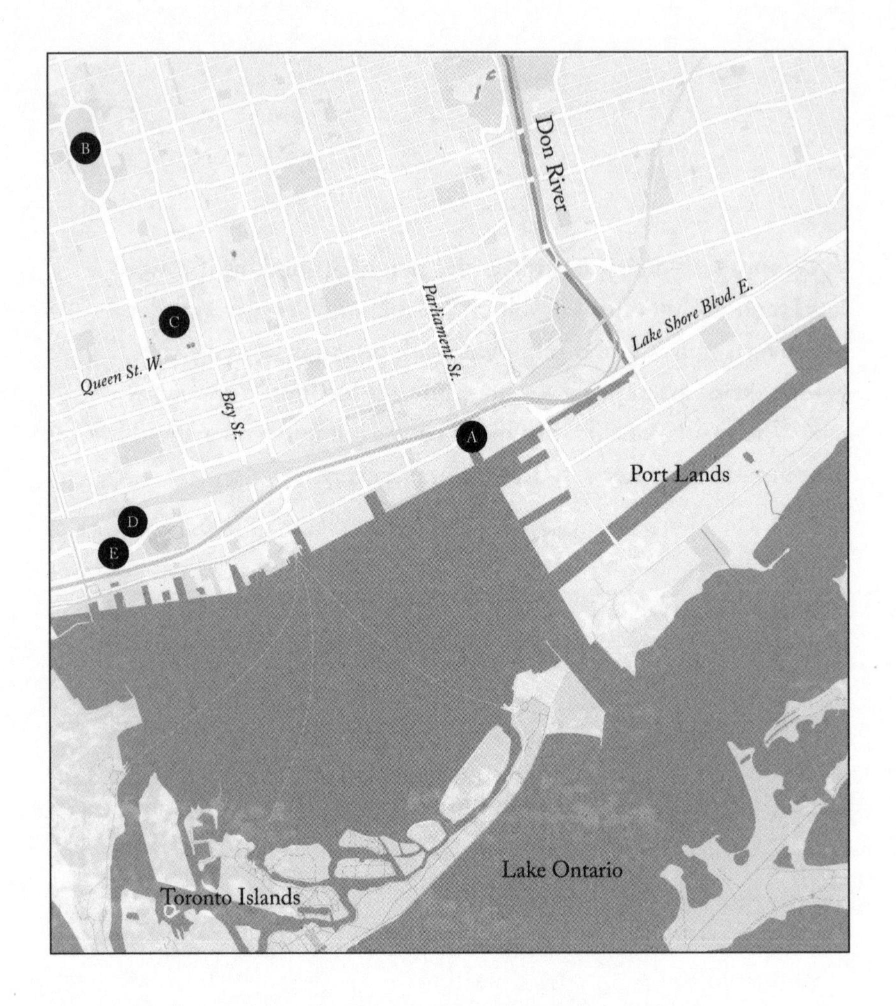

TORONTO'S DOWNTOWN CORE AND WATERFRONT

- A Quayside
- B Queen's Park
- C City Hall
- D CN Tower
- E Rogers Centre

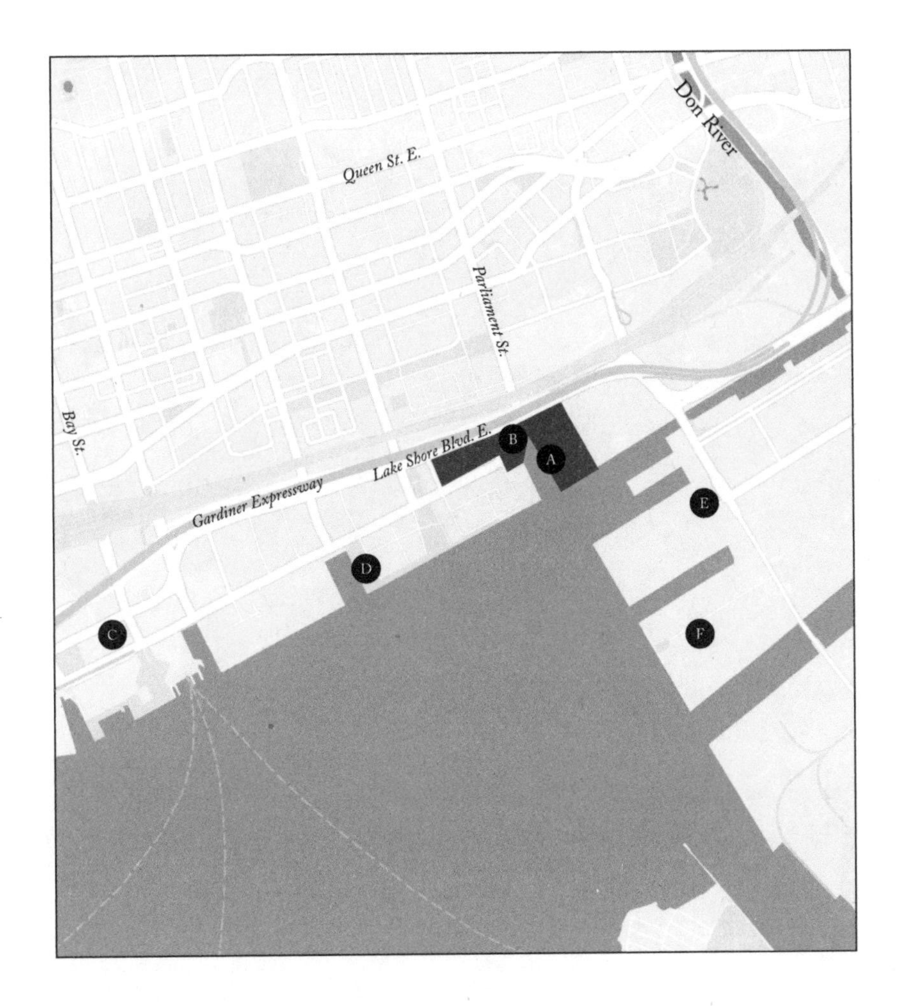

QUAYSIDE AND SURROUNDING AREA

- A Quayside (highlighted)
- B 307 Lake Shore Blvd. E. (Sidewalk Labs office)
- C 20 Bay St. (Waterfront Toronto office)
- D Sugar Beach
- E Future site of Villiers Island West
- F Polson Pier

Contents

PROLOGUE

After the Gold Rush

THE BERLINERS COULDN'T TELL whether to be excited or disappointed. It was just after noon on a late-October day in 2018, and their screens were lighting up in and around the neighbourhood of Kreuzberg, just east of the city's core. The loosely connected crew didn't all call themselves activists, but they sure acted like activists, and they were frantically texting and e-mailing each other, trying to figure out if they'd just won a two-year campaign against one of the world's most powerful companies. If they really had, they'd learned the news in the most mundane way possible: from a blog post.

"Campus in Kreuzberg substation becomes house for social engagement," the post began in German. Big changes were coming to the neighbourhood just east of Berlin's core. "Everything is different," the post continued. "Google is handing over the space to us as social experts to unleash the full power of social action."

Was the post some kind of hoax? The link the Berliners were bouncing around was from betterplace, a relatively unknown local online-donation platform, which would get access to a large space in a converted electrical substation instead of Google, the Californian corporate giant that had become a household name by making the world's information easier to find. It was Google that had the lease on the space. Giving it away would mean the company was abandoning a long-planned incubator-style "campus" for start-ups that had faced two years of pushback from digital rights campaigners, anti-gentrification advocates and plain-old frustrated neighbours from Kreuzberg and Neukölln—the communities that met where the building sat on the Landwehr Canal.

The allies were unlikely: young and old, graffiti taggers and lawyers, recent immigrants and long-time Berliners. But their concerns about Google's plans for Kreuzberg, first announced in late 2016, were all variations on the same themes. After watching Big Tech spend two decades concentrating power in the digital world, they had noticed its power spilling over into the physical world. A tech company's data collection can give it market power, market power can become purchasing power, and purchasing power can price people out of neighbourhoods. And affordability was only part of their concern: unregulated data collection can morph into surveillance, surveillance can chip away at the right to privacy, and the decline of privacy can constrain other rights and freedoms. Google's enemies in Kreuzberg and across the rest of Berlin weren't just challenging an annoying new neighbour—they were contesting the way the world's richest companies were reshaping the world's cities and democracies.

Kreuzberg's anti-gentrification factions had a history that stretched back to before the fall of the Berlin Wall. But against Google, ordinary neighbours with no history of protest found themselves suddenly working alongside anarchists to defend their neighbourhood against an encroaching giant with unfathomable resources. Their flyers and pamphlets were augmented by a loose crew of self-described nerds who took the battle into territory that Google had long ago claimed—the internet—to bring global attention to their campaign. Some of those opponents branded their battle as a movement with a name few would forget: Fuck Off Google.

Tens of thousands of stickers bearing their campaign's name were slapped across Berlin's courtyard walls, train station halls and bathroom stalls. Just a month before the blog post was published, dozens of the movement's supporters had occupied the old electrical substation, with a handful getting taken away by the police.

Their tactics had worked. Or, well, *maybe* they had.

Some of the Fuck Off Google crew weren't sure they could believe the post. They traded messages furiously. There was a poorly staged photo attached to the announcement, of a Google executive in a grey sweater handing the keys over to the new tenants; one member of the crew zoomed in as close as possible to see if the image had been photoshopped. "Doesn't look fake," the Berliner e-mailed in English to the rest of the group. "Have we won already?? We were only starting having fun!!"

Eventually, headlines began proving that the news was real. In Europe, the *Guardian*, BBC and *Le Monde* covered the end of the Google Kreuzberg campus. So did Bloomberg, the *New York Times*, CNN and the *Wall Street Journal*. The coverage came with new details: Google would spend €14 million to cover five years of rent, utilities and retrofitting costs so that betterplace and a social co-operative could take over the company's space. For the Berliners who'd fought the campus, this carried a subtext that seemed much more consequential than corporate benevolence. It looked like Google was willing to spend €14 million just to get rid of a nuisance—to rid itself of *them*.

Soon, night had fallen in Berlin, and some of the Fuck Off Google crew decided they should mark the occasion. A few dozen of them gathered in front of the substation, cheering and popping bottles of cheap sparkling East German wine named after Little Red Riding Hood—Rotkäppchen—as the canal glistened in the dark behind them.

But the celebration was muted. The day had at first seemed cathartic, but some of the crew began worrying about what came next. The new tenants had the space for only five years—just a third of the length of Google's lease at the substation. The California company could easily outwait the opposition and come back to Kreuzberg. And Fuck Off Google's battle wasn't just about Berlin. For many of them, shutting down the start-up campus dealt with only a symptom of what they felt was wrong with the company, not the cause. Google was still everywhere.

Ten or fifteen people were still hanging around when one of them pulled a bedsheet out of their bag and laid it on the grey cobblestones in front of the substation. As empty bottles of Rotkäppchen clinked and rolled around next to the sheet, someone pulled out a can of spray paint. The dwindling crowd began to call out the names of distant cities where like-minded counterparts were still fighting Google.

The painter lifted the can from the bedsheet, now a makeshift banner. *Solidarity from Berlin to SF, to San Jose, to Rennes*, it read.

The group gathered around for a picture—but then someone remembered another city that shared in their fight. It wasn't with Google, but with a lesser-known sister company funded by Google's money: one facing accusations of sidestepping democratic oversight and plotting a sensor-filled neighbourhood that some worried would be rife with surveillance.

The painter lowered the spray can back to the sheet, filling out the banner's bottom-right corner: *to Toronto.*

They hoisted the banner and threw their fists in the air.

One year and eight days earlier, a Toronto real estate executive named Julie Di Lorenzo was in a teal-and-orange cab heading south through the city's core, racing to the office of a public land-development agency called Waterfront Toronto. She was deeply, deeply worried. She was supposed to be doing due diligence with a corporate lawyer, not rushing to a board meeting that had been rescheduled to help accommodate the schedules of some of the world's most influential men. Worse, the Waterfront board was meeting to vote *about* the project she wanted to talk through with the corporate lawyer: a partnership with a Google affiliate called Sidewalk Labs to develop an ambitious, sustainable, high-tech community on the shore of Lake Ontario in Canada's biggest city.

It was a little after 8 a.m. on Monday, October 16, 2017. A veteran property developer with an anxious attention to detail, Di Lorenzo was on the phone with a different lawyer—one she'd hired to advise her on corporate governance—just as she had been until 10 p.m. the night before. She was trying to figure out if she should vote against the project, abstain or leave the Waterfront board altogether.

Taken at face value, the partnership meant Toronto would gain access to both brand affiliation with Google and the deep pockets of its parent company, Alphabet, to build a first-of-its-kind neighbourhood. The company promised to "set a new standard for downtown communities and help relieve the pressures of the city's remarkable growth." Sidewalk had put enormous effort into its bid for the project, promising energy-efficient building designs, self-driving taxis, garbage-moving underground robots and sensors that would use artificial-intelligence software to track traffic flows. Modular building components could be assembled quickly, and Sidewalk promised that those buildings could have a "radical" mix of retail, office, residential and gathering spaces. Clever construction and financing ideas could even make the cost of living more affordable, meaning Sidewalk could potentially solve some of the problems that the Berliners were battling, though through a more overtly capitalist approach than they may have liked.

Much of Waterfront's management was smitten with Sidewalk's pitch to plan out this neighbourhood on 12 acres of underused lakeshore property. The agency had tentatively agreed on the partnership a month earlier. As a partner, Sidewalk would bring more than just Google's clout; it was ready to cut a $50 million cheque to help with planning and consultations.

Waterfront had never taken on a project like this. The agency was launched by the federal, provincial and Toronto governments a decade and a half earlier to make the city's once-industrialized slice of Great Lakes shoreline less barren. Each of the three governments appointed directors like Di Lorenzo to make sure the agency was following this mandate. And now, after weeks spent finalizing the deal, Waterfront was asking those directors to vote on it four days earlier than planned, in large part because prime minister Justin Trudeau, Alphabet chair Eric Schmidt and Sidewalk CEO Dan Doctoroff had a narrow window of opportunity to be in the same place for a celebratory announcement.

That opportunity was on Tuesday. Several armies of schedulers were trying to make that event happen, and all that was left was for

Waterfront's directors to sign off on the project. But Di Lorenzo didn't want to. The developer saw herself as a shrewd, professional dealmaker, one both experienced and fortunate enough in life to dedicate her spare time to public service. On this October day, she felt she was being forced to vote on a project without enough time to make a fair decision on the public's behalf.

Di Lorenzo chaired a Waterfront board committee that was supposed to parse these kinds of deals for the rest of the directors, using her decades of experience in development to guide the board's decisions, and something about the deal she was being rushed to vote on didn't feel right. Waterfront had a duty to ensure that its projects were in the best interests of Torontonians, of Ontarians, of Canadians, yet in her mind the agreement had raised more questions than answers. It wasn't quite clear who would be accountable if something went wrong, and the company had no real track record. Though Sidewalk executives had development and planning experience in New York, the company itself hadn't built anything before. New York had a very different political climate than Toronto's, too, and a mayor endowed with greater powers. Pushing such a project through would be harder here: as happy as Toronto would be for a Google co-sign, caution was built into the fabric of its politics.

Sidewalk's wildly rich parent company did have a track record, however, in building technology. It had changed the world through internet searches, advertising auctions, mobile computing and data collection, and at the time was worth nearly $700 billion—more than double the Canadian government's budget for the year. Waterfront had no experience in going into business with a company with that kind of money and technical savvy. The power imbalance would be profound.

Di Lorenzo's cab slipped down Bay Street, past the curving twin towers of City Hall and into the business district, as her lawyer, Carol Hansell, coached her on her options for the vote. Hansell was a well-regarded governance expert who'd been a director at Canada's central bank and on countless other boards. She was just as baffled as Di

Lorenzo: forcing a board to vote early on a deal of this magnitude so the prime minister and corporate executives could announce it the next day was no way to reach a sound decision.

Resigning was an option, Hansell told Di Lorenzo. But Di Lorenzo didn't want to do that. Even if the project went forward, she wanted to be in a position to fix problems, or at least warn the public of what might go wrong. And merely abstaining from the vote wouldn't send a clear message. So Di Lorenzo asked about dissenting. Given everything she was worried about, would it be appropriate?

She had every reason in the world to dissent, Hansell said, and she could make sure to ask that her position be recorded.

They got off the phone, and Di Lorenzo's cab arrived at the foot of Bay Street just before 9 a.m. She stepped out and into the cloudy, chilly morning and walked into the office tower.

She took the elevator to Waterfront Toronto's offices, turned to the south door and stepped into its boardroom, where a thirteenth-floor view of Lake Ontario was blocked by a generic-looking office tower that Waterfront's design team probably would have frowned upon. The other directors and staff in the room seemed calm. Di Lorenzo wasn't. As she took her seat, she was shaking. Her mind was racing. Did Waterfront have enough savvy to make sure they weren't getting screwed by one of the world's most powerful tech companies?

Over the next two and a half years, the deal that terrified Julie Di Lorenzo would explode into a battle for the future of cities, privacy, wealth and democratic decision-making. For all of Sidewalk and Waterfront Toronto's vaunting talk of innovation as a vessel to make cities better—and for all the accusations of embracing surveillance that they were about to face—technology and data were almost afterthoughts in the calamity that unfolded.

The story of Sidewalk's Toronto project is also the story of the failed ambitions of powerful men who deployed the language of collective progress as they tried to build their personal legacies and influence. Once Sidewalk walked into Toronto, those men largely stepped back and let others fight their battles for them.

Left to manage a project that routinely spun out of control, Waterfront and Sidewalk's staff spent years trying to convince the public that the neighbourhood they were planning really had the public in mind. Working against them were droves of unexpected allies who used whatever means they had to call out what could go wrong. Digital activists found themselves working alongside tenured professors, patent lawyers, start-up executives, an erstwhile smartphone magnate, a backroom-loving venture capitalist, a former attorney general whose life had nearly collapsed, an auditor general with a flair for drama, an ex-punk parliamentarian, and a mayor and a provincial leader who might have wanted each other's jobs. Some of these people fought against what Sidewalk wanted to build in Toronto. Some of them fought against each other. And some of them simply wanted to make the project better. But they each managed to tell the world what could happen when one of the world's biggest companies tried to reinvent cities in its own image.

Nothing went according to plan. And over the course of four years, I spoke with nearly everyone involved as I tried to figure out why. What started as a series of stories in 2018 looking at Sidewalk Labs and the economic consequences of data collection for the *Globe and Mail*—Canada's most widely read national newspaper—led me to dozens of investigations and news stories about the company that reached into every corner of Canadian business and politics, into the great American centres of influence, and into the long shadow of an old electrical substation across the Atlantic Ocean. When Sidewalk pulled out of Toronto in May 2020, I realized I'd witnessed something that said more about how power works in the twenty-first century than a sensor-filled 12-acre neighbourhood flooded with garbage-hauling robots ever could.

This book is based on interviews with more than 150 people, conducted in Toronto, New York and Berlin—and over pandemic phone and video calls—as well as thousands of pages of documents, some of which Sidewalk and government officials spent years trying to shield from the public. It's a story about technology, ego, influence and failure, and how the rest of the world began asking the same questions as Julie Di Lorenzo was asking herself that October morning.

CHAPTER 1

Everything Now

BEFORE LARRY PAGE CHANGED how the world found information, he wanted to change the way University of Michigan students got around Ann Arbor. He wanted to build a monorail.

The eventual Google co-founder went to a summer leadership camp for Michigan students in the early nineties and came to admire its slogan—to have a "healthy disregard for the impossible." He happened to hate the local transit network; he'd spent many evenings in bus shelters in the frigid Midwest cold while commuters whizzed by in cozy cars. With the leadership camp's encouragement, Page began to research how to build out the single-rail transportation system. He eventually proposed a design that looked less like a train and more like a tiny car, with a series of small, two-person pods zipping along the fixed rail—an individualist solution to a collective struggle.

"It was a futuristic way of solving our transportation problem," he told Michigan's graduating class in a 2009 speech, a decade and a half after he'd graduated. To Page, it made little sense that the university didn't drop everything to build his transit network. "You're still working on that, I hear," he told the Michigan crowd jokingly, deviating from both his usual robotic cadence and the prepared remarks that his company, Google, posted on its website. He rarely enjoyed speaking in public—some people saw him as shy, others as dismissive—and didn't always engage well with people. In Ann

Arbor, he clung to his notes, leaning on bad jokes to carry the speech. But the transit matter was only half a joke. He'd brought the idea up again with the university's president in the months leading up to his commencement appearance.

Page had built one of the fastest-growing companies in history, but couldn't figure out why a university and its home city didn't invest in a one-of-a-kind monorail proposed by an undergrad. He'd mastered the business of the internet, but couldn't seem to wrap his head around the messy constraints of city life—the budgets, the consultations, the studies, the infrastructure, the slog of process.

"I still think a lot about transportation," he told the Michigan grads. "You never lose a dream—it just incubates as a hobby. Many things people labour hard to do now, like cooking, cleaning and driving, will require much less human time in the future." That can only happen, he said, if people really do disregard the impossible—"and actually build the solutions."

Larry Page grew up in Michigan, socially awkward but brimming with dreams, idolizing the brilliance of Nikola Tesla but fearing the obscurity in which he died. Page wanted his ideas to change the world. To do that, he had to do what Tesla didn't: build a successful business behind them.

He was born in 1973. His father, Carl, was an early artificial-intelligence researcher and computer-science professor at Michigan State who bequeathed Page with an argumentative streak; his mother, Gloria, taught programming at the same school. His early love of computing came alongside an intense interest in how things worked. By age nine, he was using screwdrivers to take apart all the power tools in the house, learning their intricacies along the way. He reportedly didn't bother putting them back together.

When Page left Michigan for Stanford University's graduate computer science program in Silicon Valley, he was still dreaming up new forms of transportation. In a plant-filled office in the William

Gates Computer Science Building, he'd rant to his officemates about a future filled with automated, robotic, low-cost taxis that would roam cities picking people up before packing together on hyper-efficient roads and freeways. His attention became more focused, however, after he made friends with a National Science Foundation fellow named Sergey Brin who, like him, needed a dissertation topic.

Their first interactions were more arguments than conversations, often about city planning and zoning regulations. But a much different dissertation opportunity quickly presented itself: It being the mid-nineties, the internet was confusing to navigate. Page and Brin thought about building a system for publicly annotating websites, but realized that some of those annotations would need to take precedence over others, depending on the authority of the person writing them. So they figured they could rank the comments using complex cocktails of coded instructions—algorithms—to make the decisions for them, minimizing human effort. Given that the trust in and importance of a piece of academic research was often measured by how frequently it had been cited, Page realized he could translate that trust not just to annotations, but to websites themselves—by counting how many times a page had been linked to by other sites. By using this to develop a hierarchy of trustworthiness, Page's system became capable of ranking every site on the internet by how much the rest of the world already valued it.

Doing this, it so happened, would also make searching for reliable information much, much easier. This idea trumped the more cumbersome, less precise technology that backed other early web-search leaders and popular portals like AltaVista and Yahoo. And as Page and Brin refined their budding search engine technology, it kept delivering more accurate results, benefitting from the growth of the web while others tripped over the constant flow of new information. Becoming less interested in academia and keen to avoid Tesla-like obscurity, Page and Brin set out to turn their web-searching idea into a real business. As they sought a name for it, Page's roommate offered up the word *googol*—a number beginning with one followed by a hundred zeroes. When Page jotted the idea down, he wrote *Google* instead. The typo stuck.

Google's first outside funding was a $100,000 cheque from Sun Microsystems co-founder Andy Bechtolsheim; Page and Brin marked the milestone with a meal at Burger King. Other investors quickly lined up, and soon they had a million dollars to play with. The new company's central values—a pledge of service to its users and ceaseless belief in the power of algorithms—sent its growth skyward. By the day in 2000 when the company signed on to handle Yahoo's search traffic, Google said it had indexed at least a billion pages. More searching gave Google more data to analyze, which helped it continuously improve the quality of its search results, in turn generating even more data.

Data became core to the way Google thought about its products: by amassing logs of user activity, from the keywords people searched to the links they eventually clicked, the company could learn not just what people were looking for, but reverse-engineer how they made decisions. This data could help engineers understand a user's intentions when they typed keywords that could be construed as different things—a search for "Gettysburg Address," for instance, did not usually mean someone was trying to find out where an office was in southern Pennsylvania—and tailor results in the right direction. Measuring user behaviour became central to the evolution of Google Search, setting the stage for a broader shift in consumer research.

Suspicious of legacy power structures, Page and Brin strong-armed two of Silicon Valley's most influential tech investors in 1999, forcing them to co-invest at half the stake they each wanted, thereby minimizing the control of outside shareholders. When those venture investors signed on, though, they did manage to find a way to exert *some* control. They made the Google co-founders find an experienced CEO to run the shop. The company settled on former Novell and Sun Microsystems executive Eric Schmidt, a bespectacled, calm and confident speaker far more likely than Page or Brin to charm a room full of suits. What he had in common with the co-founders, though, was a willingness to keep secrets from the public to protect the magic of Google's technology. He called it the "hiding strategy."

Soon, Google was hiding how well it had been shaping the economics of the internet. Its AdWords platform let people and businesses pay to have ads show up when Google users punched in the keywords of their choice—like placing an ad in the Yellow Pages in real time. Through auctions and algorithms, Google built a system that drew millions of advertisers to bid on ad spaces that ran alongside Google's billions of search results. In 2002, Google brought in $347 million in revenue and made $185 million in profit. By 2004, as Google prepared to list as a public company, Page and Brin didn't want to make the type of financial and strategic disclosures required of one. Regardless, Google was beholden to regulators that required significant disclosures from companies with as many shareholders as it had. So the co-founders gave in, going public that August. It worked out well. Page and Brin kept a controlling interest and became billionaires. A few days later, they went to Burning Man, the hedonistic Nevada desert celebration beloved by the Silicon Valley crowd that tries to be a "temporary autonomous zone"—a self-sustaining city that writes its own rules.

Search was only Google's first frontier. The company released free, web-based e-mail through Gmail, offering a gigabyte of storage when its opponents offered next to nothing. Hoping to monetize this costly approach to e-mail from the jump, engineers built it to run ads alongside messages, analyzing text for keywords and serving up ads to match them. The co-founders saw e-mail as another important aspect of life that people would want to search, and didn't think analyzing such correspondence seemed intrusive. "It never occurred to me as a privacy thing," Brin said in 2011.

As Google grew in the 2000s, so too did the ways it collected data. Buying YouTube for $1.65 billion gave it another mountain of search information and a driver's-seat view of the world's shifting entertainment habits. Acquiring the ad company DoubleClick for $3.1 billion in 2007 gave Google an even greater ability to track people's behaviour. Its Maps software evolved from a fine-grained lens through which people could gawk at the world to a ruthlessly efficient route plotter that knew where many of the world's people were travelling at all times. With each passing year, Google could measure more of the world's behaviour in an increasing number of ways.

Some of Google's growing ventures were less about expanding what people could search and more about drawing additional users to its existing business lines—like when it acquired a mobile-software development shop called Android in 2005 for a reported $50 million. More and more data were being sent over wireless communications networks, and Page thought it would be smart to buy into the trend. Google wouldn't race to build phones: it would focus on powering them. It made its Android operating system free and open to running a wide range of applications—the opposite strategy to Apple's closed iPhone operating system, in which software and hardware were tightly intertwined and controlled.

Page and Brin liked this version of innovation. By letting others build on top of what Google had started, they ensured that people would come to the Android ecosystem in droves and use the services it did monetize. By 2017, Android had beat Windows to become the most popular operating system in the world, with more than two billion devices using it monthly. An enormous and ever-increasing proportion of the world's people saw the digital world through Google's eyes.

In the 2010s, Google became more like the gargantuan Microsoft of the nineties than its founders might want to admit. It was difficult to compete with on numerous fronts and commanded an enormous influence over computing, but because it was a generation younger than Bill Gates's behemoth, Google extended its influence much further into the networked future rather than the personal computer past. Its long-held slogan, "don't be evil," was routinely criticized and eventually abandoned. Google's quest to catalogue the world's data resulted in frequent market dominance, industry upheaval and privacy breaches, including a late-2000s attack that gave hackers access to its servers for at least a year.

At the turn of the decade, Larry Page was frustrated with Google's growing pains. He was worried the company was losing its knack for innovation as it aged and settled into routines. Page had reshaped the world once, and he wanted to do it again. He soon summoned Brin and Schmidt to discuss the company's future. He made it clear he wanted more control. After a long discussion, Schmidt agreed to step down as CEO. He became executive chairman weeks later, and Page ascended to the helm of the company he had co-founded.

Under Page's watch, the company put greater emphasis on developing "moonshots"—ideas that might at first seem audacious but had a chance to be as globally transformative as Google Search. He became obsessed with the "gospel of 10x": developing entirely new products and services at least ten times better than anything on the market. The company had even launched a moonshot division, called X, in 2010, just before Page's ascension to CEO. Google became a teenager that decade, Page argued, and needed to find new ways to grow.

Some of X's early projects showed standard Silicon Valley ambition. Developing self-driving cars, for instance, became a race between Google and transportation start-ups like Uber. Some of the moonshots failed spectacularly: its "smart" glasses, Google Glass, both looked ridiculous on those who wore them and, because they could discreetly record video, became a privacy nightmare. Google pulled Glass from the consumer market in 2015, after even its legions of "glassholes" gave up on the product.

X worked on nanoparticles that could float through bodies to monitor for diseases; electricity-generating kites; energy-storing salt; and a network of balloons beaming internet access to parts of the world without much wired infrastructure. The head of Google X, Astro Teller, once described Page's moonshot ambitions by theorizing what would happen if he brought a time machine into Page's office and plugged it in: Page, Teller said, would wonder why the machine needed to plug in at all. Page himself put it this way: "A big part of my job is to get people focused on things that are not just incremental." Inventing the tools of the future took time and money, which Google had plenty of. "One thing we're doing is providing long-term, patient capital," he told the *Financial Times*.

Soon, X wasn't the only Google division aiming sky-high. Page planned to pour a nine-figure sum into Calico, a biotech company that would try to cure death. Google also launched Fiber, a division that promised to wire fast and affordable broadband networks into cities. But Fiber failed very publicly in Louisville, Kentucky, because one of its core features—the service's cheapness—resulted in too many cut corners. It buried some of its network cables just two inches deep on roadsides. Within months, cables became exposed.

After a few years, Page's colleagues were getting tired of all the we'll-save-the-world projects. One former executive later put it harshly: "No one wants to face the reality that this is an advertising company with a bunch of hobbies."

There was no single event that pushed Larry Page to become more deeply fixated on moonshots, but the obsession appeared to accelerate after Sergey Brin, who was nominally in charge of X, got caught in a messy extramarital affair with an X staffer in 2013. Page was furious. He hated conflict and tried to avoid people who frustrated him, and appeared to stop talking to Brin for awhile, according to a *Vanity Fair* exposé of the affair. At around the same time, Page summoned about a dozen senior executives for a new project called Google 2.0. They were tasked with finding new ways for the company to change the world—new projects with smaller, nimbler teams. Page envisioned Google changing everything from finance to education. Air travel, for one, wasn't efficient enough for him. He didn't understand some transportation experts' love for high-speed trains, because he felt that planes could be just as efficient—if he could somehow disrupt the concept of airports.

Soon, Google's day-to-day fell to the wayside for the CEO. "Today's big bets won't seem so wild in a few years' time," Page wrote in 2014, the same year he promoted Sundar Pichai, then Google's Android and Chrome head, to oversee all of Google's products, just as he himself once had. This freed Page up to focus on new ideas. In doing so, he sometimes returned to Google's obsession with personal data. He wanted to know where someone was, down to the inch, ostensibly to give people information about the aisle they were shopping in or who else was in a store with them, even though such detail could easily be used or manipulated to surveil someone. Some of the Google 2.0 ideas were even "creepier," one Google staffer told the tech publication The Information, without going into further detail.

As Page's deputies hashed out ideas, Adrian Aoun, one of Google's artificial-intelligence leaders, helped convince him to set up a new, more secretive moonshot group, separate from the dramas at X and with the potential to spin out successful concepts as whole new companies. But some people who knew Page in this era grew concerned about his single-mindedness. One close friend, Elon Musk, wondered if Page was considering the consequences of his broadening work, especially as he put increased faith in the power of artificial intelligence. "He could produce something evil by accident," Musk told a biographer in the mid-2010s. Musk felt Page meant well, but might one day create something with the power to end all humanity. "I'm really worried about this," he added.

Still, Page and Aoun forged ahead with the new division, which they called Javelin. By then, Page was so intent on controlling information that he'd gotten guests at his own wedding to sign non-disclosure agreements. The Javelin team, with Aoun as director of "special projects," operated with the same level of secrecy. It would, they felt, give them freedom to find new ways to change the world for the better—a mission that inspired some of the people who worked with Page at the time.

Some of the ideas Javelin explored didn't get very far. The team brainstormed developing the kind of plant-based, meat-like food that other companies would popularize a few years later. They prototyped a tube for biking in, filled with the same helium-oxygen mixture that scuba divers breathe; its resistance is less than air in the atmosphere, which could help cyclists travel faster. Another project was more ambitious. As Page considered how Google could make a moonshot out of the world's physical spaces, he set his sights on something much bigger than a monorail in Ann Arbor or self-driving taxis flooding the Bay Area.

Cities everywhere have spent decades watching their budgets dwindle, and their governments tend to love it when someone else is willing to spend money to make them look good. So tech companies spent the first decades of this century viewing cities as their next multi-billion-dollar frontier, only to find them harder to navigate than the internet. Everyone ostensibly gets a say in how cities are run, necessitating endless meetings and layers of bureaucracy to make decisions. Fast-moving companies tend to hate regulation, and cities are full of it. But like Larry Page, many of the executives behind what was eventually called the smart city movement viewed cities' inefficiencies as problems to solve for profit.

Founded in 1911 to make recording and organizing information easier, IBM had always sold patterns to businesses. A hundred years later, it decided to sell patterns to cities. The pitch: "Intelligent Operations Centers" that would swallow up information from municipal services, public safety teams, transportation networks and more, then analyze it all and hand over results. Eventually, the company found a high-profile partner. Rio de Janeiro bought into the dream after devastating floods displaced thousands of people in 2010, six years before it was scheduled to host the summer Olympic Games. The city asked IBM to build it a disaster management system that could warn of future catastrophes and get information to the right people in Rio's government. What it got was a full-fledged operations centre, for which data was gathered from hundreds of cameras and thirty agencies to coordinate daily life. Accidents and responses were itemized. Riots could be classified and dealt with; floods could be predicted. The operations centre was costly and imperfect, however. Some crises saw swift, coordinated responses; others didn't. Rio said the centre cost it only $14 million, but some residents wondered whether the investment could have gone directly to infrastructure upgrades, such as those that might have prevented a 2011 trolley derailment that killed five people—not to mention the inequities that ruled the city's social dynamics. IBM made the city look smart, but it didn't fix its problems.

The Munich multinational Siemens got its start in the nineteenth century, wiring telegraph lines across Europe and lighting up its streets. By 2010, it had a hand in power grids the world over and was investing heavily in making those grids more efficient—making them smart. Within four years, the company hoped, it would earn €6 billion from smart-grid technology alone, helping not just individual customers but whole jurisdictions monitor and lower power consumption. Siemens called the Canadian province of New Brunswick a "rebel with a cause" in 2019 for teaming up on a pilot project with the local power utility, which remotely preheated houses before customers woke up, in an attempt to reduce the strain on grids. By then, Siemens had expanded its smart-city offerings. Singapore was piloting an operating system developed by Siemens that claimed to "harness big data from billions of intelligent devices" to help manage everything from energy usage to transit passenger experiences as part of the city's "Smart Nation" strategy—one that has been accused of massive surveillance and hacking vulnerabilities.

South Korea's Songdo International Business District, at the edge of Seoul's sprawl, was a purpose-built creature of the digital era. The master-planned community near Incheon International Airport branded itself as the world's "smartest" city. Plans for Songdo first hatched in 2001: it would be an energy-efficient business centre with airport access, low taxes and minimal regulation, filled with shiny new technologies that could compete with Asian commerce hubs like Shanghai and Hong Kong. Water was recycled; buildings minimized energy; sensors were everywhere. It was often billed as the world's biggest-ever private real estate development, and may have cost north of $40 billion—an investment that would prove well worthwhile for the real estate, construction and technology firms that built Songdo were it replicated around the world. For a while, this seemed likely, especially given the way building functions like heating, lighting and ventilation could be centralized to both keep costs low and keep residents comfortable. Some called Songdo a "city in a box," a model that could be exported broadly.

In 2011, San Jose's Cisco Systems promised to spend $47 million to wire Songdo's many sensors together to help officials conserve energy and make decisions. Traffic patterns could be tracked and predicted, a smart grid could manage energy loads, and street lights could be turned off until pedestrians approached them. "The city aims to do nothing less than banish the problems created by modern urban life," *The Independent* declared in 2009. And if Songdo could be replicated, Cisco saw a $30 billion business opportunity.

But problems piled up. By 2019, Songdo was still struggling to attract people and businesses to its vision of a dream city. Only about 100,000 people were believed to live there, falling short of the 300,000-person goal. Few businesses set up in the district: less than half of the planned office space actually got built, and nearly a third of the offices that were built were vacant at the time. Cisco, however, still made a mint there, walking away with at least $1 billion from video-conferencing system sales alone.

At the same moment that Larry Page thought Google's innovative spirit was being stifled, it was becoming clear that cities were welcoming the warm embrace of massive multinational technology companies. And so it was little surprise that Page returned to an old obsession—the same one that had riled him up at the Michigan commencement speech. He saw, in the world's rapidly growing yet bureaucracy-stifled metropolises, a new collection of problems that could be solved and a multi-billion-dollar market waiting on the other side: how people moved around, how infrastructure was designed, how community services were rolled out. Google didn't need to build an operations centre like IBM; it could build the city itself, wire it to harvest data about how residents use the city, and then study the patterns in that data for ideas that could become technologies to make urban life easier. Then those technologies could be sold at a premium to governments around the world.

Google's first steps into cities were small: wiring them with Fiber and building self-driving cars to fill their streets with the project it would later call Waymo. But under the auspices of Javelin, Page's team wanted to approach city problems from entirely new angles, much as his old colleague Sebastian Thrun was trying to build flying cars while the rest of Silicon Valley was building self-driving vehicles that were stuck on the ground. Page's team was urged to think about what constraints they could remove that would make cities better. Flying cars themselves were a good start to this thought experiment; like the monorail Page had proposed for Ann Arbor, they wouldn't get stuck in traffic. Massive stretches of highway in and around cities sit empty most of the day. What could cities look like if flying cars made urban design more efficient? You could live 100 miles from work if your car could fly you there in just a few minutes. The land used for highways could be repurposed. And if you could live in a city 100 miles from work, what would you want that city to look like?

The ideas evolved with enormous ambition, says Isaac Taylor, a long-time Google special projects adviser who had worked on Google Glass before taking a leadership role on the Javelin team. What if Javelin built a city at sea, free from any city or country's regulations and restrictions?

As a general rule, Silicon Valley billionaires like Page had long felt that laws and regulations held them back from building a world that looked more like *Star Trek*. "Instead of Captain Kirk and the *USS Enterprise*, we got the Priceline Negotiator and a cheap flight to Cabo," wrote one of the partners at Founders Fund, the venture capital firm led by the self-styled contrarian tech investor Peter Thiel, in 2011. It so happened that Thiel had also funnelled half a million dollars into the Seasteading Institute, an organization dedicated to "building floating societies with significant political autonomy."

Javelin's urban thought experiments were headed in the same direction. Things like building-height stipulations, speed limits and zoning restrictions get in the way of ambitious companies on a good day. When someone with the unbound imagination of Larry Page tries to reimagine how cities could be built better, rules can seem even more pesky. "From the very beginning, there's this fascination with data-driven government that isn't going to be messed up by bureaucrats and politicians—it's going to be rational because it was made by software developers," Taylor says of Google's then-burgeoning interest in cities.

The people who took part in this research or watched it unfold held differing views of its intentions. Some were inspired by the spirit of Page's vision and saw the experiment as just that—the spirit of an idea. Some were surprised at how deep the rabbit hole needed to be dug in order to please Page. And some were caught somewhere in the middle, pleased to be part of a mission that was both energizing and exhausting. At one point, a group of them took a boat out into San Francisco Bay. They'd hoped to build something there, on the water. As the waves lapped around them, Taylor shared some bad news with a superior who had a direct line to Page: after consulting with lawyers and lobbyists, his team had learned that, as much as they felt they could engineer a floating city, it turned out that the bay was beholden to regulations, too.

The full moon was visible in the daylight, and Taylor pointed up to it. "It would be easier for Larry to build his city on the moon than right here in San Francisco Bay," Taylor said.

His superior looked at him sternly. "Don't joke about that," he told Taylor. "Larry wants to build on the moon, too."

When Page said he wanted moonshots, he meant it.

Though it was prevented from building a city that would be a cross between Burning Man and Atlantis, Javelin had developed a handful of ideas in its research that would work on land. The biggest was a dome—a towering, air-locked bubble of carbon fibre tendons connecting pieces of transparent film. If they could imagine a dome sheltering a city from the sea's might, researchers figured, they could engineer a dome able to shelter a regular city from the earth's increasingly hostile climate. Staff went to great pains to study how to build and sustain a massive dome. They searched around the world to explore what was possible, speaking with architects and designers who'd worked with dome-like structures. They studied designs that might work in extreme conditions: in the world's hottest deserts, or in the Canadian Arctic—"preparing for the gnawing catastrophes of climate chaos," Taylor says. They ran scenarios to see if a dome could remain intact if a plane crashed into it, too. The team even crunched rough numbers for a business model: if it cost $5 billion to build a Google-designed dome to cover a $50 billion development, the developer could easily shave 10 percent off the cost of building the neighbourhood inside, because many of its structures wouldn't need walls or mechanisms for heating and cooling. Life would be easier, and thanks to those savings, the dome would pay for itself. Some of the team thought Google could even build a factory to manufacture the dome's carbon fibre and film pieces, becoming a world leader in materials science along the way.

Javelin staff searched rural communities in northern California, and even explored teaming up with sovereign Indigenous Nations, Taylor says, as they sought regulatory freedom to enable their visions. The intense, earnest thought experiment eventually came crashing down to reality, however, and the Javelin team began to draw up a few different plans that could work with existing cities. In one, they would partner with landowners in a handful of locations to construct dome-covered mixed-use neighbourhoods filled with modular building units, flexible home financing models, self-driving transit vehicles and delivery drones. This, according to an internal document, would help Alphabet build excitement and "gauge city willingness to release regulatory control for innovative developments demanded by residents." In another plan, they would build out a network of suburbs, luring applicants with a bidding process—one that would ask municipalities to minimize zoning restrictions and building codes, create dedicated airspace for delivery drones, allow Google to handle building permits and strip away landlord-tenant rules.

At another point, the Javelin team drafted up a loose vision for what a Google community or city might look like, framing its benefits around the economic fallout of the Great Recession and the rising cost of gas. Infrastructure should adapt to life, not the other way around, they argued. They proposed an individualist approach to transportation, in the spirit of the monorail pods Larry Page had proposed for Ann Arbor in the nineties, with autonomous vehicles that would shuttle people around at their whims.

Javelin didn't let prescribed notions of how the world worked stop it from thinking big. The ideas weren't just utopian, they were Googley, as the company liked to describe the way it imprinted its proprietary blend of engineering and whimsy on the world. And the ideas were becoming focused enough that the company needed to find someone to steer them into reality. Page was busy, and not particularly suited to the work. "Larry's the worst person you want designing your product," a Google product manager once said. "He's very smart but not your average user."

The Javelin team built a spreadsheet of potential hires to lead the new company. At one point, Taylor suggested Angela Merkel. The German chancellor was a scientist who had been born in the communist East and had come to power as a politician after reunification, during which she'd watched the government systems of her youth be systematically replaced. She was a rational leader who knew how to steer through change. But Taylor added another name to the list that caught the eye of Eric Schmidt and Larry Page. It was someone who understood the power of technology but had proven he could navigate the messy intricacies of cities. As it happened, he was a special kind of leader that Schmidt admired: a diva. "As long as their contributions match their outlandish egos, divas should be tolerated and even protected," Schmidt once argued. He believed that if you could look past the self-centred nature that made them frustrating—and past the collateral damage that often accrued behind them—divas were good at getting things done. So to build Google's dream city, Page and Schmidt reached out to the diva. His name was Dan Doctoroff.

CHAPTER 2

Big Rings

MICHAEL BLOOMBERG HAD spent his career as a rational number cruncher, but the New York mayor had no problem ratcheting up the room's emotions as he made a final appeal for his city to host the Olympic Games. "Our city needs these Games in 2012," the mayor told the International Olympic Committee's hundred-plus members in Singapore in July 2005. "In our city's darkest hour, we asked ourselves: Can we recover? New Yorkers stood up then and said, 'Yes, we can recover, we will rebuild, and we must continue to welcome everyone.'"

He was onstage with then-senator Hillary Clinton and the team that built the city's bid for the 2012 Games, including his deputy mayor "for economic development and rebuilding," Daniel L. Doctoroff—a tall, athletic former private-equity boss whose title, like Bloomberg's pitch, strategically reframed 9/11 as an opportunity. Wearing matching blue-and-white Oscar de la Renta getups, they were making their final pitch after eleven years of work that Doctoroff had spearheaded. Since watching Italy oust Bulgaria 2–1 in a World Cup semifinal at New Jersey's suburban Giants Stadium in 1994, he'd been hell-bent on bringing an even bigger spectacle to New York City. As a kid, he'd loved the drama of the Olympic Games and the sense of global unity they'd conveyed in the midst of the Vietnam War. Doctoroff's hometown, Detroit, had lost its bid to host the 1968 Games, but he became obsessed with watching the broadcast from Mexico City on his black-

and-white TV. The Games showcased the world's size and diversity, and what screams big and diverse, he thought, like New York City?

In putting together the Olympic bid, Doctoroff had become an accidental city-builder. He had scouted New York's five boroughs for years, developing proposals for event sites. Then, as deputy mayor, he used his political influence to retool New York in such a way as to make it an excellent host for the Games. They needed a stadium, so he set about planning one above a rail yard on Manhattan's west side; even when that failed, he continued to help the site become one of the biggest real estate projects in U.S. history. Because he didn't think New Jersey would be helpful to the Olympic bid, he lavished attention on Brooklyn, removing roadblocks for the arena that would become the Barclays Center and investing in communities along the East River to prime them for sports venues. Obsessed with storytelling and self-mythology, he reimagined the way New York sold itself to the world. And he routinely sought private funding mechanisms to pay for public projects, granting the corporate world greater influence on one of the world's most influential cities.

Even in the final hours leading up to the Singapore presentation, Doctoroff had been tallying IOC delegate votes like a congressional whip, confident of the city's momentum. Though he'd barely slept, his piece of the presentation went off without a hitch. "Give us a chance," he told delegates with his folksy Michigan accent and grandfatherly cadence, "and we will make you proud."

His pride wouldn't last the day. New York was eliminated in the second round of voting—the victim, Doctoroff convinced himself, of double-crossing by Mexico and Russia, and certainly not because London had a better bid. Still, he was humiliated. He was terrified of failure, and the moment was seared in his brain. So he didn't let himself think of losing the Olympics as a failure. All that effort, he decided, had turned him into a city-builder. "The prospect of being shamed around the world," Doctoroff would later say, "is a powerful motivator."

Dan Doctoroff wasn't always a cities guy. He started out a suburban guy. He grew up in Birmingham, outside Detroit. It was a cushy life: his mother was a psychologist, his dad an FBI agent turned appeals court judge. His first visit to New York was a shock. It was a muggy August day in 1968. Through the window of his parents' 1965 Impala wagon, as the family approached the Bronx, the ten-year-old gawked at what seemed to him like an "alien city." The three dozen brown towers of Co-op City, he thought, were bleak. The boy declared to his parents that he would never live in New York.

He'd eat those words in his twenties. He moved to the city when his wife got a job with HBO. Staying at a friend's place in Gramercy Park, Doctoroff admired the finer touches of the brownstones, but worried that local dealers and sex workers made it feel run down. In 1993, the lifelong Democrat voted for Republican mayoral candidate Rudy Giuliani, who campaigned on quality-of-life issues, and came to admire the Giuliani administration's tough-on-crime policies. These included pioneering the "broken windows" approach to law enforcement—cracking down on misdemeanours to reduce overall crime—which many experts have come to view as a racist tactic that unfairly punishes vulnerable, often minority communities. This strategy can also pave the way for gentrification, pushing the vulnerable to the fringes. Giuliani's crime policies, Doctoroff thought, added "new hope" to the city.

Doctoroff viewed the world through the lens of money. He waited only until the third page of his 2017 memoir *Greater Than Ever: New York's Big Comeback* to point out that he made more than his affluent parents combined in his first year at Lehman Brothers. (The Upper West Side townhouse he later moved into would be worth $11.25 million by the time he sold it.) Spending money well became the defining drive of his professional life, too. Coming up in the private-equity game at Lehman, and later as a managing partner at Oak Hill Capital Partners, he came to see himself as a dealmaker adept at creative project financing. A year and a half after his Olympic crusade was born at the Giants Stadium World Cup game, he'd worked out plans to privately fund a New York Olympic Games with ticket sales, sponsorships and TV rights.

While seeking potential sites for venues, he became acutely aware of abandoned industrial slices of New York's 520 miles of waterfront. He learned to navigate dysfunctional bureaucracies and to make friends in high places who could help him stay in the know. Maintaining influence with officials was important; in the 2001 mayoral election, he donated to the top four candidates. By that point, he'd estimated the Games would cost $1.9 billion, but would bring New York $3.2 billion in revenue.

As he built out his Olympic plan, Doctoroff learned that changes to a city need to be politically acceptable, satisfying a maximum number of constituents in the face of opposition. It turned out that giving people a seat at the table tended to get them on your side. During the Olympic bid, Doctoroff used its board of directors as a sort of outreach program, throwing invitations to anyone who could be helpful in his campaign. One of those influential people was Michael Bloomberg, the business information magnate who was running to succeed Giuliani as mayor.

Bloomberg never showed up for Doctoroff's Olympic-bid board meetings, and the pair tended to struggle with small talk; Doctoroff didn't at first take his mayoral candidacy seriously. Or so he says—he also likes to tell stories to help humanize himself. Doctoroff claims it was his then-teenage son, Jacob, who convinced him to consider Bloomberg a serious contender.

Bloomberg won the job by a 2 percent margin. A week and a half later, one of Doctoroff's Olympic-bid allies called him up. Not even Bloomberg had expected Bloomberg to win, the colleague told Doctoroff, and now he needed to put together a slate of deputy mayors—a slate that could include Doctoroff, if he wanted. Doctoroff initially turned down the offer but agreed to take a meeting with Bloomberg. It was only after Bloomberg suggested that Doctoroff could use a perch at City Hall to advance his Olympic-bid campaign that he decided to take the job.

The newly minted political appointee stepped down from an official role with the bid to host the Games, but with Bloomberg's blessing spent the next six years intertwining his Olympic goals with the city's. His influence over the first half of the administration's three-term tenure reshaped New York both physically and philosophically. The urban-studies writer Scott Larson described the Bloomberg era as having "ever more tightly embedded the logic and assumptions of capital accumulation in the urban landscape."

Doctoroff gets a lot of comparisons to Robert Moses, the never-elected New York City official whose blind city-building ambition during the twentieth century ignored the plight of the communities he razed. If Moses's infamy was built on decades spent ruthlessly reorienting New York toward the car, Doctoroff and Bloomberg's legacy might be how they reimagined the city's built form as a product whose value should be maximized for shareholders. Those shareholders weren't just individual voters, but also the increasingly involved private sector, particularly as the Bloomberg administration reckoned with a budget shortfall in its early days. "There is an orientation here that says that government does not have to exercise control over every great idea," Doctoroff told the *New York Times* in 2002. "In many cases, you will look back and say, 'Yes the private sector played a significant role in creating things that would not have occurred otherwise.'"

Doctoroff tried to monetize everything New York could offer. Some ideas were simple, like using licences to squeeze money from well-known city trademarks like its police and fire department logos, and setting up an ad contract to fund thousands of bus shelters and hundreds of newsstands. In other cases, bringing in private interests to fund public services led to messy fights. These included an exclusive $166 million contract in 2003 to get Snapple drinks into public schools and government buildings. The money was supposed to help pay for after-school sports programs—with a promise to promote

New York in Snapple ads—and was touted as a way to get kids to drink something healthy. But health advocates railed against the promotion of sugar, while the city's comptroller, Bill Thompson, tried suing to cancel the agreement, arguing that it was a sweetheart deal that ignored potentially more profitable proposals. The deal wasn't even as lucrative as had been promised, and the city was forced to revise its revenue targets for the program to less than a third of what officials first expected. Taxpayers got a half-baked contract, but Snapple got endless press.

The blending of private and public interests took a step further with Doctoroff's ambitions for the Far West Side of Manhattan. He'd envisioned a whole new development above and around a rail yard along the Hudson River just west of Madison Square Garden. Tenements, warehouses and the land above idle trains could be repurposed into a whole new community with residential, commercial and public spaces. "We were asking to build the equivalent of Downtown Seattle or Downtown Minneapolis on a small, square patch of Manhattan," Doctoroff would later write. "A city within a city." The development would eventually be christened Hudson Yards. Not coincidentally, the land in question contained plenty of room for a stadium. The Olympics needed a stadium, which meant Doctoroff needed a stadium. So, too, did the National Football League's New York Jets, which in the early 2000s had new ownership that didn't want the team to keep playing on the Giants' field in New Jersey.

The business of sports cut through Doctoroff's life and helped engender a friendship that began in 1997 with the developer Stephen Ross, another scion of suburban Detroit, who soon became one of the Olympic bid's biggest fundraisers. Doctoroff and Ross each owned small stakes in the National Hockey League's New York Islanders in the late nineties, before the team was sold for a $40 million loss. When Doctoroff was hired as deputy mayor, Ross became responsible for a $3.2 million personal loan Doctoroff had made to New York's Olympic committee.

Ross came to be a significant source of advice for Doctoroff at about the same time that the developer's firm, Related Companies, was getting involved in city projects. It gained control of development around the South Street Seaport and won the right to redevelop the Bronx Terminal Market without a competitive process. "When you look at the pattern of behavior between the city and Related, one suspects that there is more going on than pure public policy considerations," one lobbyist told the *New York Times* after his clients had battled with both Doctoroff and Ross over another development.

Once New York's Conflicts of Interest Board decided not to block contact between the two, Doctoroff said he "set a standard that far exceeded what the conflicts of interest law required of me." That included not personally negotiating with Related or picking it for projects. After Doctoroff returned to the private sector a few years later, however, it was he who called Ross, after the first Hudson Yards developer backed out, to encourage him to raise his hand for the job. Ross won those rights in 2008. Later, he would offer to name Hudson Yards the Bloomberg Center as part of an effort to woo Bloomberg L.P. to move its offices there. At the time, Doctoroff was serving as Bloomberg's CEO and president.

Hudson Yards never got a stadium, despite a plan to have the New York Jets cover $800 million of the proposed $1.4 billion project for a space they would use for only a handful of home games each year. Many people got in Doctoroff's way, none more than New York state assembly speaker Sheldon Silver, who represented downtown Manhattan. A month before the final Olympic-bid vote in Singapore, Silver tried to force Doctoroff to build the stadium at Ground Zero, where the World Trade Center's towers had collapsed on 9/11. Doctoroff refused. Instead, with a tight deadline before the Olympic vote, he and his team brokered a deal in Queens with the New York Mets. Even when the Olympic loss came just weeks later, Citi Field would become a shining new home for New York's least-annoying professional baseball franchise. And even without a stadium on the Far West Side, Doctoroff likely experienced a financial windfall thanks to the towers built at Hudson Yards: he had a stake in some real estate in the area that would have seen extraordinary growth in value just from its proximity to the development.

As deputy mayor, Doctoroff liked to tell himself that much of his work was rooted in making New York better for everyone. He routinely visited far corners of the city to see what life was like on a local level, and he pushed for more funding for affordable housing units. And with more than a million new residents expected in the city by 2030, he helped build out a plan, called PlaNYC, that would ensure they'd all have housing and easy access to parks and transit. One of PlaNYC's key goals was a 30 percent reduction in carbon emissions, which would be achieved in part by charging drivers as much as eight bucks a pop for driving south of 60th Street in Manhattan during peak hours, a strategy called congestion pricing. Doctoroff admired his rivals in Olympic-bid-winning London for pulling off such a program; establishing a similar one in New York became a significant obsession in his waning days as a public servant. The federal government was even willing to pony up $354 million to set up fee collection and new bus routes. But the state government once again dealt the final blow, after politicians from the city's outer boroughs and suburbs argued that the comparatively wealthy residents of Manhattan would benefit most from the plan.

This was, after all, the Manhattan that Doctoroff and Bloomberg had relentlessly tried to reshape. When talking about his days as deputy mayor, Doctoroff likes to point out that he helped rezone 40 percent of New York to make better use of the city's space, including for affordable housing. That's true. But New York also underwent massive gentrification at the same time. One analysis found that the way the administration approached rezoning was likely to preserve places where wealthier homeowners lived, while less-wealthy, less-white neighbourhoods were more likely to be rezoned to allow denser development, threatening the homes of the more vulnerable people who already lived there.

Doctoroff even tried to take some credit for helping Brooklyn's Williamsburg and Greenpoint neighbourhoods become havens for artists and creative types—"perhaps the greatest concentration of hipsters in the world," he later wrote. The humblebrag, however, omits the most obvious reason why hipsters were bound for Brooklyn: many of them had been priced out of Manhattan. Rents soared during Doctoroff's time in City Hall, and from 2000 to 2014, roughly the duration of the entire Bloomberg administration, the proportion of New Yorkers who spent more than 30 percent of their income on rent shot above half.

Though Hudson Yards wasn't built until after Doctoroff had left his post as deputy mayor, it remained a product of his effort and his financial philosophy. The comparisons Doctoroff got to Robert Moses were usually kind only when they referred to his ambitions. They also tended to point out the New Yorkers left behind by his efforts. The development went on to become a shining—on a sunny day, literally gleaming—example of New York's transformation into a playground for the rich. Valued at nearly a billion dollars an acre, it was often described as the biggest private development project in American history. Residents of Harlem would later find out that Hudson Yards was built with at least $1.2 billion in financing from a cash-for-visas program that was supposed to be earmarked for reducing poverty in their neighbourhood. The Harlemites and the hipsters who had fled to Williamsburg probably couldn't afford to shop at the seven-floor Hudson Yards mall, filled with luxury-brand stores selling Fendi, Coach and Dior, but they were welcome to climb a piece of public art that looked like a pillar of shawarma meat—though, for awhile, if they posted a picture with the installation, Hudson Yards had a claim to some of the rights to those photos.

In 2007, Doctoroff left the Bloomberg administration to join Bloomberg L.P., the mayor's information and analytics company, once again eroding the distinction in Doctoroff's work between public and private interests. The mayor joked that he hoped his namesake company could give Doctoroff a raise from his dollar-a-year public service salary. Doctoroff would be able to say he earned it: under his watch as president and later as CEO, Bloomberg L.P.'s reputation and pocketbook ballooned. Revenue jumped 67 percent, to $9 billion, as forty-eight thousand more companies started using Bloomberg's eponymous information terminals, and its news division added more than five hundred staff, raising its prestige as a newswire service as it also brought the magazine *BusinessWeek* into the fold.

But Doctoroff liked control. And even though Bloomberg (the person) had said he didn't want to go back to Bloomberg (the company) when he finished his third mayoral term in 2013, he announced his return less than a year later. Publicly, Doctoroff said he left Bloomberg L.P. voluntarily, and that his friendship with the former mayor remained intact. But in fact, Bloomberg's return appeared to create a rift between them. Their visions for the money-losing news side of the company clashed. Michael Bloomberg was willing to keep bleeding cash for the sake of a quality product, but Doctoroff cared more about profit. By the end of 2014, people inside the company were leaking to other media that Bloomberg had edged Doctoroff out.

After two decades of non-stop work since he first dreamed up the idea of hosting the Olympics, Doctoroff suddenly faced the prospect of sitting idle. He wasn't accustomed to the feeling; he and Bloomberg used to talk about their usefulness in terms of their most recent job. "People only want you because of the role that you've had," the then-mayor had told him. Those words suddenly carried more weight.

But in November, an e-mail showed up in his inbox that would set him on a new path. It was from his friend Eric Schmidt, Google's executive chairman—another technocrat, just like Bloomberg. It was about rethinking cities.

CHAPTER 3

Building a Mystery

DAN DOCTOROFF WAS talking with Eric Schmidt within a day of getting the Google chairman's e-mail. Schmidt's socially awkward colleague, Larry Page, stayed out of the courtship at first, but Schmidt didn't hold back on his colleague's behalf. Page wanted to rethink everything people knew about cities, from the ways they moved around them to how they interacted with governments. The ask of Doctoroff was modest: the company wanted to tap him as an adviser for Page and the new team he was setting up. But the pitch for the project itself was astounding: Page wanted to build a new city of his own and fill it with experiments. Doctoroff was shocked.

At this stage in their careers, the two ex-Michiganders were worlds apart. It was November, and Doctoroff lived in dense, freezing New York, while Page had spent two decades enjoying California's sunny sprawl. The figurative distance was wider. Doctoroff had spent six years trying to reshape New York under Mayor Bloomberg, and he still dwelled on the many people and complex processes that had stopped him from getting things done. Page, meanwhile, wanted to build a new city from scratch. That wasn't just ambitious for a tech magnate—it was unheard of.

Doctoroff would later say he wasn't convinced Page's plan could actually work, but this was Google, which never met a roadblock it didn't try plowing through. The idea was audacious enough to be interesting. He soon made his way to the company's headquarters in Mountain View, California, to talk about it.

The first impressions at the Googleplex didn't all go smoothly. Some employees were confounded by the old government guy who dared wear a suit to their offices. And after meeting the Javelin team, Doctoroff still wasn't sure the ideas were solid. They hadn't yet figured out how to move from concept to reality in developing new processes for city-building. But Google's top executives admired Doctoroff's experience. And Doctoroff, now in his mid-fifties, had never forgotten a piece of advice from an old Lehman Brothers pal, Peter Solomon: *You don't want to go out on a bad note.*

So Doctoroff didn't just sign on as an adviser. He decided to partner with Google to start a whole new company. It would be called Sidewalk Labs, and Doctoroff would be its CEO.

Doctoroff came on board seeking independence and enough cash to make a real incursion into cities. He got one of those things. Javelin leader Isaac Taylor, who would leave for another project soon after, recalls Doctoroff pushing Larry Page for a budget of between $2 billion and $3 billion. This wasn't a huge ask when it came to the scale of cities; Songdo reportedly cost at least $40 billion, while Saudi Arabia's planned smart city of Neom had a $500 billion budget. In the end, however, Doctoroff and Page reached a figure for Sidewalk closer to $300 million. Page later called Alphabet's investment "relatively modest," but a nine-figure commitment would still help Doctoroff's new company fund operations, early project costs and eventually some in-house start-up incubation, giving Sidewalk a greater head start and more financial freedom than the vast majority of tech companies in human history.

Google itself was also going through big changes at the time. Just months after Doctoroff began speaking with Schmidt, Page and his team reorganized the entire company so that many of its side businesses could operate with similar independence to the new city-building subsidiary. They created a holding company, Alphabet, in 2015, which would oversee Google and allow for newer businesses to act as separate companies. In the years that followed, for example, both the self-driving division Waymo and the internet-beaming balloon project Loon became distinct companies under Alphabet. Page and Sergey Brin elevated their trusted deputy Sundar Pichai to take over Google.

One person close to Page described the Alphabet announcement as "Larry's retirement letter" to Google itself. With Pichai at the helm, Page was freer than ever to focus on moonshots. And by operating separately from the cash cow, sister companies like Sidewalk and thermostat maker Nest could also distance themselves from the increasing scrutiny over Google's hunger for data and market dominance.

Sidewalk kept a physical distance from Google, too, setting up shop on the other side of the country in New York. Some of Page's California corps stayed involved: Javelin's Adrian Aoun helped get Sidewalk off the ground before leaving to launch his own health-tech company, and an engineer and entrepreneur from the special projects team named Anand Babu joined Sidewalk as chief operating officer.

Sidewalk also began to borrow staff from Google's New York engineering office. They included the office's founder, Craig Nevill-Manning, a New Zealander who'd helped build the program that became Google Maps. Doctoroff also turned to friends and allies, bringing in some advisers from his City Hall days who specialized in complex negotiations and real estate. Rohit Aggarwala had worked with Doctoroff on the failed congestion pricing plan for Manhattan and had later joined Michael Bloomberg's namesake consultancy as a sustainability adviser; he had a Ph.D. in American history that focused on city development and was seen as a master of big-picture ideas. Doctoroff's former chief of staff Joshua Sirefman joined, too. Like Doctoroff, Sirefman took a public-private approach to working with cities—though it had once cost him $1,500 and some awkward headlines. (When he left his post at New York City's Economic

Development Corporation, he became a vice-president at the Toronto real estate firm Brookfield Properties, where he broke lobbying rules while working on a bid for the Hudson Yards site.)

The newly assembled technologists and political animals needed a third kind of person: big-picture urban thinkers. One of the first recruits was Anthony Townsend. He was a wavy-haired consultant and researcher with a doctorate in urban planning, a soft-spoken demeanour and an eye for the absurdities in life. Townsend had published a book called *Smart Cities: Big Data, Civic Hackers, and the Quest for a New Utopia* in 2013, and he'd built a reputation as a go-to thinker when it came to tech in cities. Though *Smart Cities* sought to champion citizen-led urban-technology initiatives around the world, it also critiqued the many ways IBM, Cisco, Siemens and others had failed at using technology to improve cities.

Around 2014, Babu had paid a visit to Townsend at New York University to get his take on urban tech. Townsend remembers Babu prying open his laptop and writing down Townsend's every word—"Silicon Valley–style sucking information out of me," he would later say. In 2015, Townsend got another visit, this time from Dan Doctoroff. He offered Townsend a consulting gig with Sidewalk. Townsend hadn't been a huge fan of the Bloomberg administration, but he did admire its ability to get things done. It turned out Doctoroff was a fan of Townsend's work, or at least what he'd heard of it. The timing was convenient: though Townsend remained skeptical of the work of past smart-city giants, he was warming up to the idea of consumer-facing tech companies trying their hand at city-building. He liked the team that Doctoroff had assembled, including Sirefman and Babu, and the approach they were pitching of creating value for citizens themselves, rather than optimizing city operations purely for profit or savings. "At that point, the benefit of the doubt was on Google's side that they were capable of doing no evil," Townsend says. So he signed on, joining what he saw as a chance to tear down the blueprints of cities as we know them without making the heavy-handed mistakes of earlier tech giants' efforts that he'd outlined in *Smart Cities*.

Time would prove him wrong. "If they had read the book," Townsend says, "they might not have hired me."

As Townsend and others began spending time at Sidewalk's early offices—first at Google's massive east coast headquarters in Chelsea, then in the neo-Gothic Woolworth Building downtown—they noticed a gap between how the company talked about itself and what it actually did. Sometimes the gap was so narrow it was nearly imperceptible, ensuring that the company incubated plenty of loyalists in its quest to remake cities. But sometimes, others noticed, you could drive a truck through the chasm between perception and reality.

"Many of you are reading this post while living in a city," Larry Page wrote on his company's widely panned, now-defunct social media platform Google+ in June 2015. "And you can probably think of a ton of ways you'd like your city to be better—more affordable housing, better public transport, less pollution, more parks and green spaces, safer biking paths, a shorter commute . . . the list goes on!" This was the formal announcement of the creation of Sidewalk Labs, half a year after Page had dispatched Eric Schmidt to recruit Dan Doctoroff. Bringing Silicon Valley's lord-and-saviour rhetoric to a realization that thousands of urban-planning undergrads had had before him, he pointed out that many of these issues were actually interrelated. "Availability of transportation affects where people choose to live, which affects housing prices, which affects quality of life," he continued. "So it helps to start from first principles and get a big-picture view of the many factors that affect city life."

Sidewalk began making some of its ideas public in February 2016, when chief operating officer Anand Babu and chief technology officer Craig Nevill-Manning presented early ideas at a Google event in New York. Nevill-Manning drew parallels to Bell Labs, the New York research and development company that had invented or helped develop technologies like the transistor, the laser, the Unix operating system and the C programming language. The long-time Googler was comfortable in front of the crowd, speaking in a calm New Zealand accent before passing the mic to Babu, a slight man who spoke more cautiously, leaning further into tech parlance. "We're calling this reimagining the city as a digital platform," he said. "When digital technologies become a peer with concrete and with laws and regulation and taxes, what happens? How does that change cities?" Nevill-Manning talked in relatable terms about how self-driving cars could free up parking spaces, while Babu homed in on the proposed "canopy"—the dome idea Sidewalk had inherited from Javelin that might cover some or all of the community—that could simplify building construction underneath. The pair positioned Sidewalk as an "equitable and human" company, distancing themselves from the heavy-handed work of companies like IBM, Cisco and Siemens, which had pioneered the so-called smart city movement.

This type of framing was key to Sidewalk's marketing in the early days. Urbanists would guide technologies through the intricacies of city-building to avoid the efficiency-focused sins of the past and build a human-centric city of the future. In blog posts and conversations, they regularly talked about bridging the "urban-tech divide." The bridge was, at best, only half-built. Real estate specialists scanned for potential development sites and ran numbers, while the insular, idealistic language of technology companies permeated the company's ideas.

By then, Javelin had wound down with the departure of Adrian Aoun, and many of its aspirations and ideas were foisted upon Sidewalk. The biggest of these was the dome, which became a source of great debate among Sidewalkers. Townsend played along with the idea, letting his imagination supersede an otherwise growing cynicism with the company. He asked his co-workers to think about the kind of ideas a Sidewalk city could incubate if they treated the first year under the dome as a fifty-two-week Burning Man–style free-for-all of innovation. Company leaders loved the idea of such concentrated experimentation.

People who were around Sidewalk in those days remember many conversations about making city life better for people, rather than optimizing cities for the sake of optimization. Townsend liked that. He walked his colleagues through modern urban-planning history, explaining ideas like the garden cities first proposed by British planner Ebenezer Howard at the turn of the twentieth century as an alternative to crowded industrial cities. The proposed small countryside cities had housing radiating from a central garden or civic complex, and were often rimmed at the edges with industry and agriculture. Some Sidewalk bosses loved the parallels between the garden city movement and their own dreams of reinventing cities to make daily life better. But they were particularly inspired by a much different twentieth-century city that was never built.

Walt Disney's Experimental Prototype Community of Tomorrow, or EPCOT, was meant to be more than just another theme park. The cartoon baron wanted to build a city. Disney even gave it the Googley-sounding code name Project X. Like Larry Page, Walt Disney didn't let real-world constraints get in his way. Like Dan Doctoroff, he had no problem partnering with the private-sector giants of his age to rethink how cities were built, working with the likes of General Electric and Ford to design exhibits for the 1964 World's Fair. And like the staff at Sidewalk Labs, Disney envisioned a place of constant innovation—a place that would "always be in a state of becoming." It would be filled with monorails and people movers. Garbage would be shuttled through a hidden network of pneumatic tubes. And at the centre of it all would be a dome—of course—to shelter residents and visitors from the elements.

Disney tried to obscure his grand vision from landowners and governments, registering innocently named businesses to buy up property near Orlando at bargain prices. He amassed acreage twice as large as Manhattan, then began pushing Florida lawmakers to give him as much control over it as possible, even securing the right to generate power with nuclear fission.

When Disney died in 1966 of complications following lung cancer, building a monorail-filled dome city with pneumatic garbage tubes suddenly seemed like a gargantuan task to the team he left behind. "There's a gigantic difference between the spark of a brilliant idea and the daily operation of an idea," one of his creative executives later told *Esquire*.

The EPCOT that opened in 1982 was just a theme park. But Sidewalk staff admired the story behind it—in particular, the way Disney and his namesake company had dealt with governments. In 1967, the Florida legislature created the Reedy Creek Improvement District, which would encompass much of Disney's property near Orlando and give it the authority to do many of the things governments are supposed to do, from running emergency services to maintaining roads to administering property taxes. It functioned like a private government, run by a board of local landowners. Two-thirds of the district's land belonged to the Disney company, ensuring that, in effect, the company governed itself.

Sidewalk didn't just want to build a city of the future. It wanted to run the city, too. The urbanist ideas and citizen-focused technologies would be there, but executives were slowly building out a vision to prototype a community with many functions run by a private company. Anthony Townsend watched Sidewalk's leadership—especially Doctoroff—embrace the wrong parts of the history lessons he'd been giving. "What he wanted was to run his own government," Townsend says. Sidewalk had the potential to extend Google's focus on digital innovation to the physical world, writing its own rules along the way to entrench its power. Minimal, Disney-style regulation could result in lower operating costs, which meant greater returns: if such cities were replicated the world over, the profits would be unfathomably massive.

Sidewalk's early ideas for the future of cities were gradually unveiling themselves to be as heavy-handed as the multinationals whose failures Townsend had studied for years. As the technologists, urbanists and dealmakers researched together, they were steered toward a massive land-development play filled with sci-fi technologies and control over traditional government regulations, all coated with the glossy sloganeering of collective progress.

"It was proposed as an urbanist's wet dream—all these great people coming together sharing super-innovative ideas," says another person who was around for Sidewalk's early days. "It was really just a real estate project with a tech veneer and Google money behind it."

By the fall of 2015, as the growing ranks of staff, contractors and consultants began to put together draft plans and renderings for Sidewalk, some were surprised by how hard it could be to get their ideas across, as if the company leadership was so obsessed with big ideas that they shrugged off the subtleties. As ambitious as Sidewalk's ideas were, getting even the simplest things done—from hiring staff to getting a response to an e-mail—could feel exhausting.

There was one set of people that Sidewalk consulted, however, that got VIP treatment. Over the course of 2015, Sidewalk set up a half-dozen "advisory groups" to share ideas and give feedback on the company's work. Former New York transportation commissioner Janette Sadik-Khan co-chaired the mobility advisory group. Ken Greenberg, a Toronto-based acolyte of renowned urbanist Jane Jacobs, co-led the group on quality of place. Sirefman and Seth Pinsky, a senior New York economic development bureaucrat from the Bloomberg administration, headed up the group studying economic development, jobs and the workplace. That team also included Richard Florida—the Carnegie Mellon turned University of Toronto professor whose 2002 book *The Rise of the Creative Class* convinced countless cities to reinvent their cores, inadvertently encouraging a massive wave of gentrification along the way. Sidewalk kept in regular touch with these panellists leading up to a massive meeting in November 2015 where many of the nearly sixty advisers gathered. They talked about Sidewalk as a potential city operator (which sent at least one conversation off the rails) and Sidewalk as an urban idea incubator (which generated much more positive feedback). They talked about design, affordability, inclusivity and ambition, too. The panellists would later get credit in official Sidewalk documents for their help in developing the company's vision. This was surprising to some staff, who noted that many of the panellists' suggestions were cast aside.

This was the first, but hardly the last, time that Sidewalk would be accused of bringing in big minds as little more than window dressing. And it wasn't the only way these experts were quickly forgotten, either. They were supposed to get high-end American Express gift cards as thanks for their contributions. But more than $100,000 worth of these cards were forgotten in a drawer for months as Sidewalk staff scrambled to come up with an actual plan for a city.

Anthony Townsend walked into the LMHQ co-working space downtown near Zuccotti Park thinking about ways Sidewalk could beat the elements. It was a cold day in early February 2016. He was more than half a year into his contract with the company, his mind still racing through all the high-level concepts they'd eventually need to get specific with. Many of the company's staff were expected at the day's meeting to brainstorm ideas for what a Sidewalk city might look like.

After warming up in the lobby, Townsend took the elevator to the twentieth floor. As infeasibly *Simpsons*-esque as a dome over a city might have seemed, Javelin had spent a lot of time and money studying the idea, and Sidewalk's management was keen to incorporate their predecessor's concept into their plans. Townsend had grown up in a town full of nineteenth-century homes with sleeping porches, and he realized that a dome wouldn't just make it easier to build buildings, it could allow for more of life to be lived outdoors—or at least as outdoors as one could be in an airlocked space underneath a hemisphere of air-inflated cushions. At a Sidewalk development, there should be nothing stopping residents from sleeping under the stars. Sidewalk could find ways to encourage urban camping. To Townsend, this felt like the kind of bold idea Larry Page might love.

Dan Doctoroff wanted to impress Page, too. But countless people who interacted with Doctoroff noticed that the perfectionist within him didn't like to settle for anything less than what *he* considered the best idea. He'd spar with enemies and friends alike, screaming until he was shaking, even if it left friends as high-placed as Senator Chuck Schumer calling him out for his arrogance. (That incident was a dispute over how many subway stops to build near Hudson Yards.) As deputy mayor, Doctoroff regularly shouted at colleagues, and he once joked that his staff got to bond "over the shared adversity of having to deal with me." Doctoroff's assistant often needed to take a "mood reading" each morning to pass on to colleagues. Outbursts at City Hall were so normal that Michael Bloomberg once sat back and started chomping on popcorn to watch one unfold.

Doctoroff would be the first to admit to being a bad listener with limited patience—and that he would wield his anger and arrogance as a weapon in professional settings. Even when asked in 2021 if he believed such behaviour is acceptable in an era when people are fighting back against toxic workplaces, he called it "a natural part of my management style" and "a strategic tool to move people forward," though it's one, he says, that he uses less than in the past. Yet he has a history of bragging about this trait. As deputy mayor, he revealed in his memoir, he once gave a wink to Sirefman as they headed into a meeting with Port Authority staff about World Trade Center delays. "Want to see me get angry?" he asked. As discussions in the meeting stalled, Doctoroff slammed his fist on the table in fury to show his dissatisfaction. According to Doctoroff, this got talks moving again.

This so-called strategy was less effective at the February 2016 brainstorming meeting at LMHQ, as a deadline loomed for Sidewalk leadership to share their vision with Page. At one point, as they traded ideas, Doctoroff paused to accuse his staff of being lazy. Yet when Anthony Townsend offered up a specific, well-thought-out idea—that they could market the dome as a place where people could sleep outdoors in the middle of the city—Doctoroff started to get worked up. Townsend tried to remain composed and self-assured. Young creatives would embrace the idea, he told the CEO. So would the bold, progressive types they'd want to lure to a Sidewalk city, he added, using the example of Occupy Wall Street encampments at neighbouring Zuccotti Park a few years earlier.

Doctoroff didn't care for the idea, and told Townsend it was crazy. Raising his voice, he began to belittle Townsend in front of his co-workers, swearing and slamming things on the table. "You're all idiots," he barked.

At least one person walked out.

Townsend was humiliated. *What was that?* he thought. *Why does Dan think that would motivate people?* "Things started to suddenly feel doomed," he recalls. "It was the first feeling I got that the DNA of the company was cooked badly."

Townsend wasn't the only person fed up with Sidewalk around this time. One of the outside consultants the company had brought in for help was a multinational engineering firm called Arup. It had been trying to elbow its way to the front of the smart city movement for years, particularly in Europe, where it worked on projects in cities from London to Bursa. The head of Arup's digital studio, a bearded Brit named Dan Hill, began flying to New York regularly to help guide Sidewalk's early ideas. His role shifted between design lead and dismissed outsider as he tried—sometimes in vain—to point out that some of the big ideas Sidewalk thought would revolutionize cities, like making some streets car-free and centralizing energy-efficient neighbourhood heating, already existed. Many of them were already in effect in Europe, where the governments of cities like Helsinki often had more control of their infrastructure than Americans were used to—which itself was a much different approach than Sidewalk's Disney-style dreams.

Hill had arrived with a hope that Sidewalk would build relationships with governments to reinforce democracy, but people noticed he seemed to be losing confidence that democracy had much to do with it. Doctoroff's brashness didn't appear to help. Soon after the LMHQ dressing-down, Sidewalk's chief source of sober second thought stopped coming into the office altogether.

—

Sidewalk Labs announced in March 2016 that it had partnered with the U.S. Department of Transportation to offer new software called Flow to the winner of the government's nationwide Smart City Challenge, which tasked cities with creating data-driven ideas to move people and things using less time and money. Flow sought to centralize as much data on traffic as possible, from sources like Google Maps and municipalities' own data sets, in order to better understand congestion. The city of Columbus, Ohio, won the contest, proposing a way for low-income residents to obtain better access to medical services through things like a trip-planning app and a Wi-Fi-laced bus rapid transit system—both of which assumed low-income residents would have easy access to smartphones.

Any criticism of this assumption, however, was overshadowed in June of that year, when the *Guardian* dug up public records that showed how Sidewalk was pitching Flow to cities like Columbus, offering them "new superpowers to extend access and mobility." Drivers could use the software to find available parking spots, reducing the need to drive around looking for one. This could reduce traffic by as much as 30 percent if perfectly executed, while helping Sidewalk observe how drivers behave. As a bonus, Alphabet would receive more information that it could pump into analytics tools to find patterns that could help it solve future traffic problems, and make driving and parking more efficient before roads became filled with self-driving vehicles. The company also wanted to extend its software to mass transit, proposing a version of Google Maps that would work out the costs of taking transit, a rideshare, a taxi or a shared bike to a destination, then allow people to pick their mode of transportation and pay for the ride through the app.

The mobility tech expert Alexei Pozdnoukhov—who would later work for Sidewalk Labs—warned in the *Guardian* that Flow could redirect money from public transit into ridesharing at a time when Google owned about a 5 percent stake in Uber. The technology would also let cities more efficiently punish residents for misbehaving.

Through Flow, Sidewalk promised that a city's parking enforcement could be "optimised" with artificial-intelligence algorithms to better find parking violators, bringing in as much as $4 million in extra revenue a year.

As part of the free three-year Flow demo it offered to Columbus, Sidewalk said it could install one hundred streetside Wi-Fi kiosks for residents to use. Like Sidewalk itself, the kiosks had origins in New York City's Bloomberg administration. In 2012, the city began transmitting Wi-Fi signals from pay phones and started looking for new uses and designs to make them more useful for the smartphone era. Months after Bill de Blasio replaced Michael Bloomberg as mayor, de Blasio announced that a consortium called CityBridge would replace the city's pay phone network with free-to-use Wi-Fi kiosks. Branded LinkNYC, they would span the five boroughs and offer a range of information about city services. They also came with a promise to deliver half a billion dollars to taxpayers over twelve years from ads on each kiosk's 55-inch display.

The CityBridge consortium included hardware makers Qualcomm and Comark, plus Titan, a massive advertising company that focused on outdoor, transit and pay phone ads, and Control Group, which designed technology for urban spaces. The group began installing LinkNYC kiosks around the city in late 2015, and they quickly caused controversy. People were using them to watch porn, forcing CityBridge to limit access to web browsers on the kiosks. But the worst accusations faced by CityBridge and the city itself were much farther-reaching. The digital rights advocates at Electronic Frontier Foundation called out CityBridge for its "nearly limitless retention" of data from users; even when CityBridge updated its policy in 2017, the kiosks still held on to a range of data, including device identifiers, for up to sixty days, while cameras installed on the kiosks captured footage they retained for up to a week. Privacy did not appear to be a paramount concern. The foundation warned that the city was treating citizens as sources of data for the private sector, and LinkNYC kiosks became central to the debate over human rights in the digital age. The New York Civil Liberties Union had already warned that the kiosks created a significant trade-off in which connecting to the internet—"one of the basic necessities of modern life in New York City," the organization argued—could expose users to "unwarranted government surveillance."

In Dan Doctoroff's long-time neighbourhood, the Upper West Side, people began putting sticky notes on the kiosks to block the cameras. Meanwhile, in some of the city's less-affluent areas there were no kiosks to use at all. As of March 2020, more than four years after the LinkNYC rollout began, hundreds of kiosks intended largely for the outer boroughs had yet to be installed. And the city's information technology commissioner told a council committee that month that CityBridge owed the city nearly $44 million dollars that fiscal year, despite having made $105 million in ad sales over the previous eighteen months.

Soon after Doctoroff helped launch Sidewalk, the company proposed investing in two key players in the kiosk consortium, Titan and Control Group, and merging them. The combination of urban technology expertise and advertising was an enticing investment for a business that wanted to profit from the future of cities. Doctoroff imagined the company that would result from this merger developing a platform that could connect people to their cities in new ways over time as both cities and technology evolved.

But Doctoroff's ambitions sometimes spread him thin. At the same time, he was working on his memoir and chairing the organization tasked with building the Shed—an arts centre at Hudson Yards that branded itself as a destination for all New Yorkers, but would later be used in the HBO show *Succession* as the scene of a comically opulent party for the ultra-rich. To Sidewalkers, Doctoroff could seem deeply distracted, and for long stretches of time was rarely seen in the company's offices. Sometimes that meant staff had to work frustrating hours to accommodate rare gaps in his schedule to give him updates on the company. And when the Titan–Control Group merger closed in late 2015, Doctoroff wasn't done with his extracurricular interest.

The new business, Intersection, was officially "a Sidewalk Labs company," and he announced that he would chair it, too.

These distractions were not ideal for his Sidewalk employees. They needed to show Larry Page what his hundreds of millions of dollars could buy.

CHAPTER 4

Is Anybody Home?

ANTHONY TOWNSEND WAS staring at Google Earth and thinking about churches. It was the fall of 2015, and the smart-city consultant was sitting in a northeast-facing conference room at Intersection's offices in the Woolworth Building, overlooking City Hall Park. It was a sunny day, and he was clicking around satellite images from a different sunny day in Detroit. Townsend was trying to figure out what a hollowed chunk of the city might look like if Sidewalk had its way with it, building a neighbourhood out of what the company called a "street mesh"—city blocks divided by a mix of avenues, laneways and promenades, with various levels of access for people and self-driving cars. At 1,000 feet long by 1,000 feet wide, each city block could ostensibly house about 3,500 people amid a mix of park space, apartments and offices.

With help from outside consultants, Sidewalk staff were working on feasibility studies to show what a "Project Sidewalk" community might look like in a real city. They were studying a few sites, including one on the outskirts of Denver and the site of a former naval air station in Alameda, California, across the bay from San Francisco. If the company put in a $2 billion investment and gradually reinvested some of its proceeds, the team calculated that Denver could give Alphabet a return on its investment of 11.9 percent over thirty years from real estate alone. Thanks in part to the Bay Area's soaring land values, Alameda could deliver 28.4 percent.

Detroit had the least potential estimated return for the company, at 10.5 percent, and staff freely admitted that it would take longer there to achieve their goals. But they also saw it as a chance to shake the rust off the Rust Belt and revive one of America's lost cities of industry. Detroit was also home to huge swaths of underused land close to its core and remained the centre of gravity of America's legacy automakers. Given Alphabet's massive cash reserves and valuation—at nearly half a trillion dollars in February 2016, it was worth more than ten times what Ford or General Motors were—some Sidewalk staff thought the company could buy one of those manufacturers to help accelerate Alphabet's self-driving dreams.

After studying the whole city, Sidewalk staff had found what they were looking for: a 1,307-acre site they believed to be nearly 85 percent vacant, more than half of which was owned by the city itself, wedged between downtown Detroit and GM's Detroit-Hamtramck Assembly. Much of it was in a neighbourhood called Poletown East. At the corner of the site sat the Packard Automotive Plant, considered the largest abandoned factory in the world, its forty thousand employees long gone. The company mulled buying the plant and repurposing some of the structure into Project Sidewalk, which staff wrote in an internal document would "provide an eloquent opportunity to bridge Detroit's past and future." They had run the numbers, too: among the 7,132 parcels of land on the site, only 1,132 were believed to be occupied. That wasn't even 16 percent of them. It couldn't be that hard to move people out, raze the neighbourhood and build out a street mesh.

But as Townsend clicked around Poletown East on Google Earth, he couldn't stop dwelling on something it seemed no one else had bothered to consider: the churches. Townsend counted at least a dozen of them. They all had names: Faithway United Ministries, Harper Avenue Church of God in Christ, Sweet Kingdom Missionary Baptist Church. A few were historically Polish churches. Many had

Black congregations, and were still active. These were people's spiritual sanctuaries. Even though thousands had left the community, many still returned weekly to the place they once called home.

Townsend was stunned—not because of the controversy that would clearly emerge if Sidewalk tried to displace the churches, but because no one at Sidewalk had raised this as an issue. They hadn't considered how real people might be hurt by the utopian city they were planning to satisfy their billionaire patron. The many New Yorkers and Silicon Valley residents who shaped Sidewalk were so hell-bent on making ordinary cities better that they'd overlooked what made people come together in cities in the first place.

Despite caution from the urbanists on the team about the necessarily slow nature of city-building, Sidewalk was under pressure to deliver a vision to Larry Page. He rarely interacted with the rank-and-file staff, but his presence was constantly felt—especially when he appeared on a massive screen in glass-walled rooms for virtual meetings with Dan Doctoroff, sometimes while walking on a treadmill. By the start of 2016, he had made it clear that he was impressed with the team's ideas but wanted his city-building spinoff to work faster. He seemed preoccupied with starting in the wildly expensive Bay Area, as though an exclusive Sidewalk community there could build high-end demand. As Elon Musk had done with Tesla, this could help the company gradually gain enough experience and profit with a VIP product to finance and develop a mass-market product for the everyman. But Page also wanted Sidewalk to step back and think about the pieces of a city in wholly new terms. Every piece, every unit, could be movable or rearrangeable. They wouldn't be restrained by governments, real estate prices or people—they would abide only by the constraints of physics.

Dan Doctoroff also wanted to make an ambitious VIP product. A city, he once said, was "like any other product. It had customers. It had competitors. It had to be marketed." While he was working on New York's Olympic bid in the nineties, he had become determined to find the detailed bid books for the 1992 Games in Barcelona and Atlanta's 1996 Olympics. In the run-up to the Atlanta Games, a member of his team pressed used-book dealers to pin down a copy of that city's bid, eventually spending $800 to get one.

Storytelling meant everything to Doctoroff. Changing the world meant everything to Page. So Sidewalkers set out to build a bid book that suited them both.

Some of the staff and consultants started with feasibility studies. Others began diving into data collection. Water recycling. Those street mesh road grids. Utility tunnels. In particular, the dome was a crucial component of the neighbourhood they would design. It became a running joke inside the company after a *Saturday Night Live* sketch about city-dwelling progressive elites living inside an actual bubble, one that would become "a fully functioning city-state" after Donald Trump's presidential election. "It's Brooklyn," the faux ad sloganeered, "with a bubble on it."

It took more than a year to build out the vision. Dozens of staff, contractors and consultants worked late nights and weekends to synthesize the details they'd gathered and brainstormed, preparing a 437-page document that would become known as the Yellow Book. Peppered with references to Disney and the dome-obsessed futurist Buckminster Fuller, the Yellow Book declared that Sidewalk would help cities "overcome cynicism about the future" and build "a city from the internet up." The plans proposed a massive community that could house 100,000 people across 1,000 acres. It viewed people through the lenses of efficiency and profit, even as it promised residents better lives through novel technologies. And the document was sheathed in whimsy, in a bright-lemon jacket adorned with a tribute to a 1935 Frank Lloyd Wright exhibition called Broadacre City, in which he suggested dispersing the entire population of the United States into decentralized homesteads. The Broadacre exhibition promised "No realtors except the state," "No private ownership of public needs" and "No traffic problem." The Yellow Book's back flap imagined a world with "No glass ceiling. No fossil fuels. No taking out the trash. No forgotten passwords." And even "No bowling alone," paying tribute to Robert Putnam's landmark research on the decline of social engagement in America.

Much of the Yellow Book reimagined society in progressive ways, proposing schools that would account for different learning styles and social backgrounds, and outlining a justice system that could divert people from prisons and into programs to address substance abuse or mental illness. But it also described a dome-covered dream world built on principles that were deeply skeptical of government. "Fearing risk, bureaucracies stifle innovation," the authors wrote, several pages after Dan Doctoroff's introductory letter, in which he lamented "how difficult it is to get anything done in a city." Though Sidewalk promised to build an inclusive world, it proposed handing enormous power to the private sector. It would resemble a kind of fiefdom—one where people would be monitored from the moment they looked in the mirror in the morning.

Literally. The mirror would collect data on residents using a hidden sensor to monitor for stress marks on a person's face or changes in skin colour that could indicate fluctuations in blood oxygen levels. Like many other technologies proposed in the Yellow Book, this "smart mirror" didn't exist, but many of the components to make one did. The mirror was framed in the language of kindness and care: it would give doctors a clearer picture of a person's health. But the city the Yellow Book envisioned would also be a for-profit enterprise. Users would be encouraged to share as much data about themselves as possible, collected by everything from those smart mirrors to their smartphones. Project Sidewalk would create "a new market for data" to learn more about city living, and Sidewalk and other companies would reap economic rewards from learning about people's day-to-day lives. Insurance company executives, for example, would trip over themselves for the type of smart-mirror data Sidewalk wanted to collect.

The Yellow Book took Silicon Valley's typical disdain for government bureaucracy a step further, weaving data collection into many of its plans. It wanted power on par with government—and in some cases, even more power than that. It wanted to levy its own taxes, track and predict people's movements, and control public services, including law enforcement. The ideas were dressed as progressive but gave unprecedented control to Alphabet and its partners. For instance, a police accountability system that tracked cops' movements and included a rating program like a "Yelp for police officers" might help officers build trust in their community. But the system would only work because the cops would be working under constant surveillance.

Sidewalk's insatiable hunger for data as detailed in the Yellow Book contrasted with how Alphabet executives sometimes discussed data collection in public. Eric Schmidt, for one, liked to point out the risks of mass collection. In his 2013 book with Jared Cohen, *The New Digital Age*, he warned of the "cat and mouse game" between states and bad actors that could threaten user data "as long as the Internet exists."

It's not that privacy was entirely an afterthought in the early days of Sidewalk. The company promised to follow Privacy by Design, a set of principles meant to ensure that privacy protection is built into systems and technologies from the jump. Some of the Yellow Book's data-sharing ideas had the potential to be very helpful—like a smoke alarm that, when alerting the fire department to a fire, would give the resident the option to automatically let first responders know other crucial details, such as if anyone in the household had asthma. Yet despite a promise to let residents opt into data sharing, people living in a Project Sidewalk community would also have to endure tiered access to their own neighbourhood based in large part on how much data about themselves they were willing to share. In this regard, data would be a kind of currency: people who chose not to share anything about themselves when they visited a friend's apartment in Project Sidewalk might not get access to its self-driving "taxibots" or be able to buy items from certain stores.

Though trading data for access to services is already common in the digital world, Sidewalk envisioned the exchange in much broader swaths of life. Data could ultimately be used to reward "good behaviour," too. Business licences could be more easily renewed for those offering good customer service. Interest rates on loans could be determined by "digital reputation ratings." A nightclub could be instantaneously fined if sensors found it was too noisy, and people could be automatically blocked from accessing certain kinds of housing if someone filed a complaint about them. With everyone and everything logged in a central registry, someone wearing an enhanced version of Google Glass might be able to see the entire community annotated by their relationships or shared interests. They could also "replay past events"—presumably by video—and maybe even view a simulation of what a neighbour was about to do.

The Yellow Book dream could have pulled in as much as $83 billion in revenue from building leases and eventual land sales, Sidewalk staff believed. But to *really* sell Larry Page on the community they could build, Sidewalk staff dreamed up a dream day in their dream community, giving it its own chapter in the Yellow Book.

It would begin with the rumbling of automated taxibots coming to life at 6 a.m., preparing to shuttle people to work. Sidewalk would own the platform that dispatched the vehicles, but investors could place fleets in the city, adding to the morning din. Many other pieces of life were automated or reimagined. Garbage bag collection was too "costly, wasteful and unsanitary" a technology to fit into Project Sidewalk, the authors wrote, and would be replaced by a disposal system where inorganic material would be thrown into chutes and Insinkerator-style technology would grind organic waste to be pumped and shipped out of buildings. "Biological engines" would then "cook a concoction" from it to make electricity and methane, while basement-dwelling, garbage-moving robots hauled the rest from standardized garbage chutes at low cost, since the city wouldn't need to pay night-shift differentials to the robots. Also automated would be the city's centralized computing core, "constantly thinking three steps ahead about what could go wrong and nudging us collectively down a more optimal path."

By 8 a.m., the city's digital brain would be commanding even more taxibots to zip through Project Sidewalk, flowing "like liquid in packs around the clusters of human activity." Robotic cranes would lift prefabricated building pieces and stack them onto building spines with connections for power, water and waste, docking them into place like the Legos that Page once used to build a functional printer.

When Project Sidewalk's residents took a lunch break, restaurants would unfold onto malleable streetscapes. People working and living under the community's "canopy" wouldn't even need to worry about the elements; they would be sheltered by the airlocked dome built from carbon bands and transparent, air-filled cushions. Sun-shielding screens would pop out from buildings as people went back to work—with many ditching their offices for pre-booked outdoor workspaces. Some would return to work-live co-operatives that "feel more like Burning Man than Bushwick."

As the work day wound to a close, teenagers would roll out of bed for night school, learning at the time that works best for them, with local professionals who would teach them on-the-job expertise. Members of the "Guardian Police Force," meanwhile, would mingle with residents, hoping to boost their performance stats as they tried to "gain cred with the kids" in the neighbourhood in case they needed to address any behavioural problems someday.

By nightfall, adaptive spaces would transform themselves: a ground-floor apartment would become a tapas bar as "chic speakeasies" emerged elsewhere. But Project Sidewalk would be a model of productivity, shifting streetlights to a more orangey colour, like smartphone-dimming software, to lull residents to sleep. Then the robots would return to clear the streets of trash and deliver freight, preparing Project Sidewalk for a whole new day.

A dream day like that depended on everyone in the city buying into Sidewalk's vision, and on every aspect of daily life—from the technological to the social—going according to plan. In real life, few things go according to plan. The Yellow Book presented Project Sidewalk as a utopia, but the community would always have been a handful of malfunctions and unintended consequences away from a nightmare.

Let's imagine a day in Project Sidewalk from a different perspective: that of a guy named José who owned a bar on one of the community's laneways, a few blocks outside the dome that covered much of the neighbourhood. On this day, he'd woken up at noon in his "pod" apartment, not because he worked nights, but because he had no job to wake up for anymore. Project Sidewalk had shut down his bar after noise sensors caught his DJs breaking the decibel limit too many times in one week, automatically handing him bigger and bigger fines. José didn't pay the fines—not because he didn't want to, but because he didn't get the usual notifications due to recurring network errors that month. That malfunctions didn't matter to the city's central registry, apparently. Once the network was fixed, José lost his business licence.

When José got out of bed, he tripped over a bunch of non-recyclable plastic containers that he couldn't afford to throw out, since he was running out of money and the neighbourhood operated under a pay-what-you-throw system. He grabbed a beer; it was a little early for one, but Project Sidewalk charged more for potable water than the water he showered with, and he'd already paid for the beer. He logged on to the City Assistant app on his phone to apply for a new business licence. For the sixth day in a row, he was denied because his noise violations had ruined the bar's reputation rating in the city registry. The city's anti-bureaucracy philosophy meant he couldn't get in touch with a real person to plead his case.

So José called a friend who lived nearby, Edwin, who had always wanted to go into business with him. Maybe they could register a new business licence under Edwin's name and open something new, José thought. But Edwin couldn't get a loan from any of the banks in Project Sidewalk because he'd maxed out a credit card in his twenties. He hadn't been living in Project Sidewalk at the time, but the data followed him when he signed up for the city registry. It turned out his history didn't align with the registry's preprogrammed parameters of "good behaviour." The algorithms wouldn't budge in their decision.

Without jobs to go to, José and Edwin decided to grab lunch. They arranged to meet up with their friend Troy at a tapas joint under the dome. José was glad to get out of his place—he was tired of the garbage he couldn't afford to throw out, and the building structure his tiny pod apartment was attached to was rusting. Compared to the palatial apartments across the street where local executives made their homes, the stack of pods he lived in reminded him more of a trailer park than a gleaming condo tower.

Troy had come to visit from nearby Sunnyvale. This was his first time in Project Sidewalk, and something seemed strange as he parked his car at the edge of the neighbourhood. He had skipped registering for an ID card—even though it was supposed to keep his identity anonymous, he didn't like the idea of being tracked. A few blocks into his walk, he stepped onto the road and was nearly hit by a Cadillac Escalade hauling an older couple being chauffeured by a human driver. "Huh," Troy mouthed. He'd thought everyone was supposed to be carted around by taxibots; no one had told him that was more marketing than reality. The future wasn't as robotic as he'd believed.

The robotic cars that *were* zipping around, though, ignored Troy. Because he hadn't gotten a card and wasn't sharing data with Project Sidewalk, he couldn't use a bunch of services, including taxibots. So he trudged a dozen blocks and showed up to lunch late. He was sopping wet, too. The canopy was leaking, and it was impossibly humid inside the dome—something had screwed up with the air-exchange technology.

The three friends sat down to eat. They talked about José's bar and what they'd do for work. Troy wanted to spend the rest of the day checking out the neighbourhood, but he got an e-mail at lunch asking him to finish a project by 5 p.m. Paying the bill was awkward—Troy wanted to cover it for his cash-strapped friends, but the restaurant wouldn't let him without a city registry account. José paid for the meal without telling his pals it would mean he'd have to go another week before he could throw out his trash.

Troy stepped into a park and tried to get to work, but city staff told him he couldn't use any of the laptop stations because they'd been pre-booked by a local start-up. He gave up and began the long, sweaty trudge to his car.

When José got back to his pod outside the canopy, he tapped his phone at the entrance, but the door was locked tight. He knocked furiously, but there was no human inside to unlock the door. So he walked across the street and tapped his phone at the City Assistant kiosk. It didn't recognize him, but there was a red news alert warning on the touch screen. José gasped when he clicked through: the city registry had been breached by unknown hackers, who'd stolen years of personal information and login details from Project Sidewalk's thousands of residents. Sidewalk Labs had claimed its city would open up "a new market for data." That data was now on the black market and would be sold to the highest bidder.

At least he wasn't worried about losing money. Project Sidewalk had taken away his ability to make any. Mostly he was worried about getting inside to rest. As neighbours gathered around the kiosk, José turned around and looked across the street, up through the window of his unit. The pictures of his family that lit up the digital photo frames built into the walls flickered to black.

The Yellow Book vision for Project Sidewalk would have become, like Hudson Yards, one of the largest real estate developments in U.S. history. Sidewalk planned to invite cities to bid on hosting its community through a public competition—much like the International Olympic Committee did, and as Amazon would later do in its quest for tax breaks for a second headquarters. The winning local government would need to give Sidewalk control over road infrastructure and public transit, as well as water, electrical and other utilities. Sidewalk also wanted the authority to tax, issue bonds and work with its own building, health and environmental safety codes. Meanwhile, its expert advisory groups, like the ones staff tapped into while writing the Yellow Book, would "provide credibility" for Sidewalk's vision and "engender trust" with locals.

But first Sidewalkers had to sell their idea to Larry Page.

The company set a meeting for May 2016 to show the plan to the Alphabet CEO and a handful of key executives. With the Yellow Book as its guide, Sidewalk had put in almost as much effort preparing for the meeting as Doctoroff had for his Olympic bid presentation eleven years earlier. Between the advisers Sidewalk solicited, the graphic design, the printing costs, and the fictional essays about characters who might live in Project Sidewalk commissioned from future-focused novelists like Madeline Ashby and Robin Sloan, preparations had cost hundreds of thousands of dollars, likely millions, to complete. The project had consumed many Sidewalkers' lives, keeping them away from families on evenings and weekends for more than a year. Thanks to the company's place within Alphabet, there was additional pressure: Google's other side projects that pushed into the physical world, like Waymo's self-driving cars, had been getting little traction in reality—the side projects needed a win. Combined with the intense, no-room-for-error environment Doctoroff had fostered, Sidewalk's offices were filled with nervous energy leading up to the May meeting. After they'd put so much of their lives into the project, all the Sidewalkers could do was hope for Larry Page's stamp of approval to go scout for a city that would let them build a neighbourhood.

It was a big ask. To build out everything they'd planned in the Yellow Book, they'd need as much as $7.4 billion in commitments.

In Mountain View, Doctoroff and a handful of deputies, including Anand Babu and Josh Sirefman, gave a well-choreographed keynote presentation. Doctoroff remembers the proposals were met with "unanimous excitement" from the Alphabet brass, with an agreement to move forward and build something. His recollection is not universal. Others with direct knowledge of what happened in the meeting and in the days that followed say the pitch appeared to neither capture

Page's heart nor sway the bean-counters in the room. The Alphabet executives' response struck those people as, at best, tepid.

At the meeting, Sidewalkers back in New York were told, Page spent much of the time pushing different pet ideas for the future of cities, with innovations like buildings on wheels. This didn't bode well. Staff already felt that Page had struggled with some of the basic ideas they presented him, like a vibrant city's need for density. And he wanted to build in the Bay Area, close to home, rather than on cheap land near Denver or in Detroit.

They wouldn't be launching a contest for a host city anytime soon, after all.

After a year and a half of research—of scouring cities for usable land and calculating their probable rates of return, of ignoring churches on that land, of dreaming up whole new ways to run law enforcement and recycle water and manage society, of convincing dozens of global experts to share their wisdom and forgetting to compensate them—the next steps for Sidewalk Labs were unclear. For some Sidewalkers, Page's response was a breaking point. The ensuing weeks were like walking through a fog or battling a hangover. The sudden slowdown felt like an existential crisis. After so many hurried, harried months working on the Yellow Book, numerous workers began to feel like they'd been set up for failure, as though they'd wasted more than a year on a billionaire's half-baked fantasy only to find out their ideas weren't fantastical enough.

Anthony Townsend was burned out. He couldn't see a future for himself at a company that operated this way. He'd been turning down public-speaking requests for his book, *Smart Cities*, for months, but finally he started to say yes. He spent the summer flying to Argentina, to Russia, to Quebec, to Sweden, to England, to Spain. By the time he'd wrapped up this one-man tour in September, his contract with Sidewalk had expired. More staffers left in the months following. For all the time, money and optimism they'd been handed to connect the disparate worlds of technology and city-building, they hadn't come close. "We had not bridged the most fundamental urbanist-technologist gap we faced, which was with our own benefactors," Townsend says.

But a few weeks after the Yellow Book meeting, an old colleague from Doctoroff's Olympic bid sent him an e-mail. He was working with a public agency in charge of developing a city's underused waterfront up in Canada. "My new CEO and I are very interested in what you are doing at Google," he wrote, "and would like to talk to you about a potential pilot in Toronto."

CHAPTER 5

Shoreline

TORONTO IS A CITY THAT loves to tell stories about itself, regardless of how true they are. Thanks to an error by an influential local historian, for decades many believed the city's name was rooted in the Huron word *toronton*, which the historian translated to something like "meeting place." That translation squared nicely with how the city preferred to be seen—as a place that valued diversity, community and cosmopolitanism. But more recent historians have come to insist that Toronto's name originated from the Mohawk word *tkaronto*, which means something far more prosaic: "where there are trees in the water."

Even the story of how the British "bought" the initial land along what's now called Lake Ontario from Indigenous people at the turn of the nineteenth century is laced with boosterish self-deception. To this day, the city's official version of history insists that members of the Mississaugas of the Credit got £1,700 in cash and goods for the land that colonialist settlers would eventually fashion into Canada's biggest city. But anyone who listens to the Mississaugas themselves—and it turns out to be very few people—learns that the cash and goods were an unrelated gift. It wasn't a sale at all. And the supposed deed for the land, found years later, was blank. It took until 2010 for the Canadian government to truly acknowledge that its predecessors had cheated the Mississaugas and give them a C$145 million settlement for two tracts of land that included the "Toronto purchase."

The piece of land later called Toronto spent a while as the capital of the British colony that became Canada, and in the century and a half since it lost that title, Toronto has tried endlessly to be the best at different things in different ways. In the nineteenth century, it was "Toronto the Good," where Victorian values were so entrenched that the city's most prominent department store drew its curtains on Sundays to discourage window shopping on the day of worship. Even as the twentieth century progressed, it remained so good at being "good" that people who didn't fall in line were mocked, beaten or murdered—a trend that especially afflicted LGBTQ+ and immigrant communities. Things have evened out a bit in the last few decades. Mass media invaded, punks became politicians, and out, proud LGBTQ+ communities flooded into a central neighbourhood that become so popular, its LGBTQ+ residents are now getting gentrified out.

Toronto's cultural cachet has never been greater, thanks in large part to Black talent coming to the fore after years of white gatekeeping. The NBA champion Raptors eclipsed the NHL's woebegone Maple Leafs as the kingpins of the city's biggest arena; Drake and the Weeknd have reshaped pop music several times over in just a decade and a half. Toronto is a magnet, and the nearly three million people living there today could be joined by another half million by 2030. In 2013, the city celebrated its population surpassing Chicago's—an oversimplification that ignored the economic and demographic might of Chicago's massive suburbs and exurbs, but which, of course, Torontonians still bragged about. That isn't even Toronto's most blatant attempt to co-opt the reputation of a midwestern American city. In the mid-2010s, it was impossible to walk down the street without seeing someone wearing a shirt emblazoned with *Toronto vs Everybody*, a slogan lifted from Detroit.

Two of Toronto's defining, competing attributes—ambition and insecurity—have become clearer with each passing year. It wants the kind of global renown that benefits New York and London, and the city increasingly gets that renown—or at least a reputation that's more Melbourne than Cleveland. But Torontonians won't stop talking about it, undermining a confidence that might speak for itself elsewhere. Structural problems hold back Toronto's potential with equal heft: decades of short-sighted city planning left behind a city of single-family homes, as present-day developers rake in millions filling in any gaps they find with towers of paint-by-numbers single-bedroom condos that are increasingly unaffordable to the average person. Many of the cities Toronto gets compared to had more foresight. It lacks the sheer density of New York, the triplexes of Montreal, the mid-rise apartment blocks of western European capitals, the freeways of Los Angeles, or even a subway system with more than one east-west line through the city centre.

There are consequences to the city's swelling. Toronto's core is beginning to fill with tech start-ups and Silicon Valley branch-plant offices, strengthening the flood of thousands of graduates and immigrants who move to the city each year. In the 2010s, this influx ensured Toronto joined the ranks of New York and San Francisco in at least one metric: real estate prices. Income inequality has soared, and many people who arrived seeking better lives have been pushed to the literal and figurative margins of the city as others with greater financial means race to outbid each other for century-old semi-detached houses. The city wasn't built for the volume of people it attracts or the ambition it has today.

Toronto is also hemmed in by Lake Ontario to the south, limiting how the city can grow. A lot of people spent a lot of the twentieth century trying to figure out what to do with the shoreline. With shipping activity on the decline since 1969, by 2001, the port was getting less traffic than Vancouver, Montreal or even neighbouring Hamilton, home to a population just a fifth the size of Toronto's. Enough of the lakeshore's industrial inhabitants had moved to cheap suburban land—or out of North America entirely—that it became a key site for the city's 1996 Olympic bid, with an eye toward the kind of redevelopment Barcelona had done before the 1992 Games (and that Dan Doctoroff had set in motion in New York ahead of his own bid). Toronto's bid fizzled, but it amplified concerns that would only become more prescient with time: that development along the waterfront might benefit only a privileged few.

Governments zeroed in on the lakeshore again in 1999, this time to bid for the 2008 Summer Games. They formed a task force to do the heavy lifting—the latest in a long history of groups and commissions that had tried to solve the waterfront riddle. The task force decided that redeveloping that slice of Lake Ontario shoreline was necessary for Toronto to be competitive on a global scale. This approach hedged against an Olympic-bid loss: even if the effort to attract the public spectacle failed, the business community would still have a reason to invest, and the rest of Toronto would finally get access to a reimagined waterfront.

The hedge worked out. Toronto *did* fail in its bid for the 2008 Olympics—losing to Beijing—just after the Canadian, Ontario and Toronto governments pledged C$1.5 billion to the cause. The ordeal had drawn attention to all the potential on the waterfront, and governments approved the creation of what they called the Toronto Waterfront Revitalization Corporation. "We didn't cry, because we already had Plan B so tightly interwoven into Plan A," says Robert Fung, a long-time financier who ran the task force and later chaired the corporation. "The consolation prize was pretty big."

Within a couple of years, the corporation got access to that C$1.5 billion, received some guiding provincial legislation and was granted stewardship to government land along the waterfront—so long as it walked a fine line between economic growth and social benefits in how it applied those resources. Some of the task force's recommendations never materialized, like tearing down the expressway that blocked much of Toronto from its own waterfront and building an innovation-focused district on underused industrial lands just east of the core—which Fung hoped would bring like-minded people and companies together the way Silicon Valley did. A few of Fung's roadblocks were political, but others were structural. The agency had been modelled in part on similar initiatives in cities like New York, Barcelona and Sydney, designed to balance the authority of government and the operational style of private business. But it was given less power than many of those similar agencies, sometimes forcing it to fight with the governments that were supposed to give it support.

At least the agency eventually got a catchier name: Waterfront Toronto. It was ostensibly a tool to get development done faster when forced to deal with a bunch of levels of government. Steered by a board of mostly private-sector professionals appointed by those governments, Waterfront took on dozens of projects in its first decade and a half across the 2,000 acres over which it had power. Its strategy was to build out public spaces and infrastructure—not just physical infrastructure like water and sewage pipes, but broadband cables, too, to make internet access widely available to everyone there, including people in affordable housing. In laying this groundwork, Waterfront raised the value of the land it controlled while making it more enticing to real estate developers.

As Waterfront gradually spent its C$1.5 billion on these kinds of projects, its staff hoped that the income from land sales would gradually replace government funding. In turn, the agency would keep making more money to spend on building out *more* infrastructure and public space. "Great cities aren't necessarily created by buildings—they're created by the public areas," says John Campbell, Waterfront's long-serving early CEO.

In its first dozen or so years, Waterfront cleaned up a prominent beach in the Port Lands and turned a parking lot into another beach. It helped get parks built and worked with other organizations to build a trail underneath the city's aging downtown expressway. It made the city's lakefront street a lot more friendly to cyclists and pedestrians. It guaranteed a bunch of affordable housing near the lake. It found a home for the athletes' village for the 2015 Pan American and Parapan American Games—a sporting event less ambitious than the Olympics, but still big enough that Kanye West played the closing ceremonies—then used the village to accelerate the development of a mixed-use neighbourhood. And Waterfront ensured that new developments, like a media conglomerate headquarters and a college campus, were energy-efficient and sustainable.

The projects were a mix of the necessary and the ambitious—often overdue city-building exercises that needed a catalyst. Waterfront Toronto was happy to be that catalyst. Then, in 2015, Campbell decided to retire as CEO. He'd been beloved by staff and many others from across the city. The *Toronto Star*'s architecture critic declared that he'd mastered his way through political infighting to dream up a lakeshore that was "vastly superior" to that of the past. "Nice guys *do* finish first," the critic wrote. Adored as Campbell was, his exit represented a significant opportunity for the agency. Its original funding was dwindling, and Waterfront's board wanted a successor who had specialized in other kinds of funding models—and who could use the lakeshore to help Toronto become the global city that so many people wanted it to be. The directors wanted someone who could handle complex projects and real estate deals. And in the job description the board shared with a headhunting firm, the directors said they wanted someone who would bring "vision and innovation" to the role.

That word—*innovation*—would haunt Waterfront Toronto for years.

Will Fleissig first visited Toronto in 2011. As his car slid between the towers that line the city centre, he noticed he could catch only glimpses of Lake Ontario and Toronto Island. The experience, he felt, was almost like a haiku—just enough details teased out to make sense, but not enough to understand the city. He'd spent his whole career studying cities. Places like Boston, New York and his most recent home of San Francisco had well-used shorelines. As he learned more about Toronto's, it felt fragmented. But something else struck him about the city: he admired how much Toronto embraced its diversity and wasn't ravaged by income inequality. (Much. Yet.)

A Harvard alumnus with a broad smile, Fleissig had spent his working life trying to coax a better future from the stubborn present.

His career took off in Colorado, and he became Boulder's director for planning and development in 1994. Boulder was a city so concerned about suburban sprawl that residents had voted *for* tax hikes in order to buy land at its outskirts. Real estate prices soared, and residents sometimes wondered if the economy might be too strong to keep the city as dense and close to nature as they liked. Rather than get stuck in the middle of that debate, Fleissig tried to implore the people around him to think about sustainability and inclusion. Tom Clark, who became president of the Boulder Chamber of Commerce that same year, first thought Fleissig might be anti-business and was wary of him, but later came to see him as forward-thinking. "He was way ahead of everybody in terms of what the future would be," Clark says.

Fleissig then entered the private sector, working on development and transit plans for a handful of cities. He often brought the private and public sectors together. And in 2015, that's just what Waterfront Toronto was looking for: a new champion to tie them together in Toronto, too.

Nearly two decades after he'd left Boulder, society's divisions were still on Fleissig's mind. The San Francisco Bay Area's tech boom had only served to underline those divisions, as the nouveau riche pushed lower-income Californians to the fringes. His daughters were going to private school with what he called the "one percent of the one percent," and he was getting uncomfortable. When Waterfront Toronto came knocking, he remembered his first impression of the city in 2011. He saw an opportunity to be a city-builder in a city that actually seemed to work for everyone.

One piece of the CEO job description etched itself into Fleissig's memory. The agency said it wanted to change how Toronto, the rest of Canada and the world saw the city. "The hair on the back of my neck stood up," Fleissig later said. "I've never read that before anywhere, and I've done work all over the U.S. and the world. That was unique. This was the Magna Carta for the waterfront."

More than three hundred candidates, including at least one internal Waterfront executive, applied for the job. Fleissig got it. Some directors were surprised, even nervous, that the job had gone to a guy they felt had built a few small, interesting projects, as opposed to a Canadian applicant who knew how to work with local governments and developers. But the search committee was thrilled. "He was head and shoulders above all other candidates," says Mark Wilson, the board chair, who led the committee.

Waterfront announced Fleissig as its new CEO in December 2015 and gave him a starting salary of C$350,000. Until then, the agency was still known mostly for its work building out parks and infrastructure to entice developers to the edge of Lake Ontario. For a long time, it had been a process that worked for an agency whose three shareholders were governments with massive bureaucracies and competing interests. But just as Fleissig arrived, Waterfront had a chance to think bigger.

Cutting through the city's east end, the Don Valley is one of Toronto's most striking natural features. It draws two different kinds of awe: one from the rolling green slopes that surround it, the other from the thick streams of traffic on the highway that runs through it. The river that shaped the valley has long suffered from neglect. Over the course of a couple hundred years, early Torontonians blocked the paths of migrating salmon with mills, destroyed the river's ecosystems with sewage and industrial waste, and rerouted its natural mouth from a marsh, steering its waters westward with a concrete channel. The city filled in the marsh to make room for a new industrial neighbourhood.

By the 2010s, the rapidly growing city needed to use any land it could to support all the people and businesses that were showing up. As industry disappeared from the city's core in the twentieth century, the land around the Don's artificial mouth—usually called the Port Lands—was a natural choice to be redeveloped. But humans had refashioned the river to spill into Lake Ontario at a ninety-degree turn, leaving the land around it very prone to flooding.

By 2016, Waterfront's original C$1.5 billion was running out, the land available for it to develop was dwindling, and it was nearly three-quarters into its twenty-year mandate. One of the reasons John Campbell retired when he did was to give a new CEO enough time to extend that mandate—possibly by attracting fresh funding to reduce flooding from the Don River, allowing Waterfront to take on more projects in the Port Lands. Among Will Fleissig's first tasks, then, was to finish the job of wrangling hundreds of millions of dollars and a longer life for the agency.

But Fleissig's vision for Waterfront was grander than that. He was still obsessed with sustainability, had been asked to think about innovation, and was hired for his willingness to find new forms of funding. And just next to the Port Lands, at the foot of Parliament Street, Waterfront actually owned a couple of pieces of property. The agency had bought the plots a decade earlier for C$68 million and had been waiting for the right opportunity to use them. In the mid-2010s, Waterfront had begun hatching plans for a 4.5-acre patch of parking lots and low-rise commercial buildings across the street from the lakeshore. Then-CEO Campbell expected it would become a pretty traditional development: sustainably designed towers with retail at the bottom and apartment units above, with a bunch of affordable housing. It lay next to two pieces of land the agency called Bayside and Parkside; Waterfront named this little parcel Quayside (pronounced *key*-side). Across the street, on the other side of a little inlet called Parliament Slip, sat another 6 acres owned by Waterfront: a muddy parking lot wedged between some old soybean processing silos and the loading area for a party boat. In between the two plots sat a little 1.5-acre triangle with a bunch of different owners, including the city. The three pieces combined to form an L shape, like a Tetris block that had landed at the edge of the Great Lake. Tasked with finding ways to extend Waterfront's life, and imbued with the belief that visionary city-building could make society more fair, Will Fleissig looked at the L shape and saw potential that no one had seen before.

Fleissig had already spent years thinking about sustainability, but his arrival in Toronto coincided with a high-profile, 196-nation treaty ratified at the United Nations Climate Change Conference in Paris in 2015 to curb greenhouse-gas emissions. One of the first projects he undertook was to work with the City of Toronto to develop a plan for a new island called Villiers that would be carved out by the rerouted Don River. He worked with the city to develop a climate-positive neighbourhood there, one that would avoid emitting greenhouse gas and actively offset others' emissions by exporting clean energy. But despite the help Waterfront gave in planning Villiers Island, its powers to actually do anything there were limited, and it would take years to reroute the river and open it up. So Fleissig began to wonder: What type of project could Waterfront entice to Quayside that would set the same kind of climate-positive example for future generations of Torontonians?

As Fleissig brainstormed, Toronto was watching homegrown start-ups pop up at a pace not seen since the dot-com boom (and bust) a decade and a half earlier. Many faced a significant problem: it was easier to find customers and venture capital in the United States. Companies often needed a first major customer to prove that their business model would work, in order to grow their business and find more customers—that is, in tech parlance, to scale. If Waterfront found a way to let local start-ups try out tech at Quayside, it could bring the city an economic boost, too. In 2016, Fleissig and his staff began to research how to bring it all together—cities, technology, sustainability—for a public-private partnership. They spoke with dozens of companies, including some from the infrastructure and real estate sectors. Most of those firms specialized in one or two, but not all, of the themes Waterfront hoped to focus on. There was one young start-up in New York, though, that seemed to match Waterfront's progressive, neighbourhood-scale ambitions on all fronts. It was called Sidewalk Labs.

Fleissig tried to contact its CEO, Dan Doctoroff, but had no luck. Waterfront's head of innovation in urban development, Kristina Verner, tried reaching out, too. No response. So in June, they took a more personal approach. The agency's chief planning officer, Christopher Glaisek, had spent a decade and a half as a planner in New York, much of it in Doctoroff's orbit. He'd put together designs and plans for New York's Olympic bid, later running the process that selected Daniel Libeskind's studio to oversee the master plan for the new World Trade Center. He sent Doctoroff an e-mail on June 27, inviting him to discuss the possibility of a pilot in Toronto.

It was just that—a possibility—but it came at an extremely opportune moment. Sidewalk had just lost the momentum and excitement that had defined its first year and a half. Staff were no longer racing to write a pitch book to convince Larry Page to let them build a dream city; they were taking long-overdue vacations and writing materials to explain to Page how cities worked in the first place.

Sidewalk executives still held out some hope of running an Olympic-style public competition to see which cities would offer them the best incentives and the most control. But doing so might be seen as imposing on cities rather than partnering with them. Optics were important; the company wanted a partner to prove itself with. Doctoroff's team soon responded to the e-mail with interest. It turned out that Sidewalk Labs and Waterfront Toronto each had something the other wanted: Waterfront's city-building work was consequential but not particularly exciting; Sidewalk was drumming up ideas that sounded exciting, but it needed a city to help it turn them into something of consequence. Neither knew if they were ready for commitment; both kept their eyes on other suitors. Still, they'd intrigued each other. And so the boring Canadian government agency and the Silicon Valley billionaire's pet project began a months-long fling to see if, maybe, commitment was worth it.

Navigating this tentative relationship alongside Will Fleissig were Kristina Verner and Waterfront's chief development officer, Meg Davis.

Verner was a tall, wide-eyed workaholic. She'd spent more than a dozen years figuring out ways to use technology for the public good in southwestern Ontario, with the University of Windsor, before John Campbell convinced her to bring her work ethic to the big city. She rarely held fewer than two jobs and continued teaching at the university even after she'd come to Toronto; in 2017, Waterfront would name her its vice-president in charge of sustainability, innovation and prosperity.

Davis had an inherited urbanist streak. Her father was Bill Davis, a much-admired Ontario premier who'd stopped an expressway from cleaving Toronto in half more than fifty years ago. In a *Toronto Telegram* front-page photo from 1971, seven-year-old Meg could be seen thrusting a rose under her father's nose at the opening of Ontario Place, an Expo-style public space on the west side of Toronto's lakeshore. Armed with both a planning degree and an MBA, she spent her career jumping between planning, development and procurement before Campbell offered her the chance to blend all of that work as a vice-president with Waterfront.

They were a good pair for the task of negotiating with Sidewalk: Davis brought planning and development expertise, and Verner knew how to talk tech with both developers and bureaucrats. Their early conversations with Sidewalk were mostly about big-picture details like bridging the worlds of technology and urbanism—the kind of optimistic stuff upon which Sidewalk had long tried to build its brand. But they soon started looking at public remarks that Dan Doctoroff had made about using a city as a lab for solving issues related to privacy and cybersecurity. That goal was ambiguous enough to raise concerns for Davis and Verner about that kind of community's effects on real people, and it suggested that Sidewalk might want more control over a neighbourhood than Waterfront was willing to give. At the time, the New York Civil Liberties Union was calling out LinkNYC for collecting too much data on its users, and the *Guardian* had just unearthed documents that Sidewalk had sent Columbus, Ohio, about its traffic-management app Flow. Smart-city experts warned in the newspaper's story that the app could force cities to make costly tech upgrades while undermining transit systems and potentially giving Sidewalk and Google an advantage in the burgeoning urban-tech marketplace.

Waterfront staff noticed that Sidewalk Labs talked a big game: *technology and urbanism, together at last!* Some of them even got to see the Yellow Book, since sharing material was normal for Waterfront in the early, informal stages of dreaming up a project. But when they drilled down into the 437-page book, Sidewalk's ideas raised their eyebrows. Verner thought it was a bit "fantastical," a blue-sky exercise that felt far detached from how cities actually function. She bristled at some of its sloganeering, like "bureaucracy stifles innovation," and worried about how plainly Sidewalk stated that it wanted some governance and regulatory powers to be handed to the company. "Some of what they were thinking was very out there," says Davis, for whom the proposed dome brought to mind the plot of *The Simpsons Movie*, and to whom autonomous vehicles seemed worlds away.

At points, Waterfront's CEO grew concerned. "Google has purportedly told other candidate communities that they want to control ALL data in this demonstration project area," Fleissig wrote in an internal e-mail. "Could present privacy issues and control issues." But he was clearly excited about the prospect of working with Sidewalk. Verner says she had to have a few conversations with Fleissig about recognizing how unrealistic it would be to implement the dome-covered utopian ideas the company was workshopping in 2016. She noted that those ideas, already aging poorly, made it seem like Sidewalk wanted to build a neighbourhood in a vacuum, not one attached to a very real city.

As the summer came to a close, Sidewalk executives came to Toronto for a meeting that was less formal than a pitch but more structured than a Q&A. Dan Doctoroff and Josh Sirefman reflected on the lessons they'd learned at New York's City Hall, presenting a slick slide deck and videos, just as they had in the Yellow Book meeting months earlier. In spite of Waterfront's concerns about data and regulation, it was clear that Sidewalk was thinking holistically about cities. They discussed using technology as a means, but not an end, for improving city life—although Verner noticed that the Sidewalk execs didn't dwell too much on individual technologies. It felt more like marketing than substance to Waterfront, but their optimism was aligned.

Doctoroff had questions of his own for Waterfront Toronto, many of which were tied to their common origin story: city-building resolve forged from a failed Olympic bid. His company was still in a post–Yellow Book limbo, but he was intrigued by what Will Fleissig wanted to do with Quayside. Sidewalk staff didn't shy away from acknowledging to Waterfront that they still wanted to run a competition to find a partner city. Nevertheless, the New Yorkers began to do some due diligence on how feasible it would be to work in Toronto. The extensive case studies Doctoroff's staff had done were limited to the United States, but Toronto was just across the border, and was bigger than most American cities.

Sidewalk's staff began talking to locals. One of the people they consulted was Ken Greenberg, who had once been the City of Toronto's head of urban design. A calm, bespectacled member of the Order of Canada with an honorary doctorate from the University of Toronto, Greenberg had spent four decades building projects across Europe and North America. It was through this work that he'd first met Dan Doctoroff about a decade and a half earlier, when his firm helped study how New York could better use Lower Manhattan's shore. Greenberg admired Doctoroff's ambition as deputy mayor, recognizing that his ability to marshal funding helped him accomplish projects that the public sector might otherwise have been too slow to pull together. They'd occasionally kept in touch, and Greenberg was one of the sixty-odd people consulted as part of the Yellow Book advisory groups in 2015. Greenberg saw his life's work as mitigating "the abuse of a technology" in cities—the car—and although he was skeptical that tech could be a panacea for future urban struggles, he was fascinated by Sidewalk's desire to bring cities and technology together with a single vision.

Stretching back to the Javelin days, Sidewalk and Google staff had long debated whether to pitch a project on empty land or build in an existing city. Greenberg inserted himself into this self-interrogation. "It had to be in a real city that had real politics, and already had momentum," he says. Otherwise, he warned, Sidewalk could fall into the same pattern of failure as the utopian city-builders of the past. When Greenberg learned that Sidewalk was talking with Waterfront Toronto, he realized the "real city" he'd been pushing for might turn out to be the one in which he lived. As he and Doctoroff discussed the potential of Sidewalk setting up in Toronto, Doctoroff made it clear that he was fascinated by Toronto's diversity and its growing tech sector.

Greenberg had worked with Waterfront numerous times, including in its plan to reroute the mouth of the Don River, and thought the agency could be the type of partner Sidewalk needed. Sidewalk's leadership thought so, too: because of Waterfront's strange tripartite structure, they believed it could be a one-stop shop to help them work through ideas touching on all aspects of city life. But the federal, provincial and city governments each often wanted their own say. In time, Sidewalk executives would discover that Waterfront was much less powerful than they'd first understood.

Before Sidewalk could bid to build in Toronto, it needed a blessing from Larry Page. Greenberg helped with that. Page agreed to come to Toronto to see Quayside in December 2016, shortly after meeting with American president-elect Donald Trump alongside other tech executives in New York. The Toronto plans were wrapped in secrecy: the Google co-founder didn't like attention, was awkward around most people and, in the words of one person who has worked closely with him, "doesn't go outside." Will Fleissig was thrilled that Page would visit and was cautious not to rub him the wrong way. He barred Waterfront staff from joining the tour or advertising it in any way. They knew Page was coming, but heard very little about it before or after.

The Waterfront CEO made plans to meet Page at a law firm to talk about the agency and its history, but he also wanted to show Page what he might have the chance to play with. It was freezing that day, and the ground was covered in snow and ice. So Fleissig, Doctoroff and Sirefman donned toques and scarves to take Page for a frigid walkabout. Page didn't appear to have the right gear for the weather, so Fleissig gave him a hat bearing the Toronto Raptors slogan "We the North" to help him feel warm—and, maybe, at home.

Greenberg trotted alongside as a kind of Toronto concierge, guiding Page through the waterfront's history and answering his many questions about a city he was interested in but didn't know very well. To Greenberg, Page seemed like a quick study. He was eager to grasp Waterfront's role in Toronto, how the lakeshore was changing, how the city was growing.

They wandered around the frozen mud of Quayside, marvelling at the unused waterfront land so close to downtown. They headed down through the Port Lands, too, to Rebel, a waterside music and entertainment complex that would certainly benefit from the density and transit access Sidewalk wanted to build. Everything around them in this half-developed, post-industrial prime real estate in one of North America's biggest cities could benefit from building the kind of neighbourhood Page had spent years dreaming of.

He looked northwest back to Quayside, and to all that land in between. He smiled.

CHAPTER 6

Rose-Coloured Glasses

WATERFRONT TORONTO HAD caught John Ruffolo by surprise. The venture capital executive showed up at the agency's lake-facing boardroom to pitch its executives, but he was thinking more about *their* pitch.

It was 2016, and he and his colleagues were scouting around the city to get buy-in for a hub to house tech companies as part of a complex that would stretch westward across downtown from the city's central train station. Ruffolo had become one of the country's most influential venture capitalists over the past half-decade, giving ordinary Canadians a share of the growing but risky tech ecosystem through his own wing of the gargantuan Ontario Municipal Employees' Retirement System (OMERS) pension fund. The pitch Ruffolo came in with was complex, but it had some precedent—thanks, in part, to Dan Doctoroff.

Hudson Yards, the Manhattan megaproject that Doctoroff had helped slowly push through on Manhattan's Far West Side, had by then earned global renown. It was being built on top of a rail yard, reclaiming prime real estate for more than just industrial use. As it happened, there was a web of connections between Ruffolo, OMERS and Hudson Yards. The pension fund's real estate arm, Oxford Properties, had co-developed the New York site. In Toronto, Oxford wanted to redevelop the long stretch from the train station to the

Rogers Centre, the domed stadium of Major League Baseball's Blue Jays. Ruffolo was about to move a renowned start-up centre he'd helped found into one of Oxford's existing buildings on this strip. But there was more potential: he envisioned a whole "innovation corridor" fitting into Oxford's plans for that stretch of downtown. And separating those buildings from the Rogers Centre, Lake Ontario and more underdeveloped land were rail lines.

Barrel-chested and nasal-voiced, Ruffolo walked into the meeting with Waterfront CEO Will Fleissig and his fellow executives looking to see what they thought of the idea of building something that might bridge those tracks. Such a project could have brought the lakeshore closer to Toronto's western core while letting OMERS and Oxford build their own technology-filled version of Hudson Yards in the Canadian city.

But Waterfront had ambitions to share, too. From the agency's boardroom just south of the station where the rail lines met, Fleissig gestured eastward, instead, toward the Port Lands. Governments were close to finally flood proofing the area, and the agency wanted to turn that slice of the lakeshore into a natural extension of downtown. Waterfront was thinking of bringing innovation to the fore, Fleissig explained, starting with an L-shaped block that the agency called Quayside.

It turned out the CEO and the venture capitalist shared the dream of building something that could showcase Toronto-made technology. Ruffolo's mind began turning: *Wow, this is interesting.* Maybe partnering with Waterfront would make a tech hub easier to build. Fleissig's team told him they wanted to invite companies to bid on the chance to get involved—it would be a while before the process would start, but it was worth keeping a conversation going.

Ruffolo wandered back to his office a few blocks away. Maybe, he thought, Quayside could become his tech hub. In the months that followed, he called some friends in the real estate business about the idea. He found that Oxford didn't seem to be interested in building something eastward, at Quayside—they had grand plans on the west side of downtown. But he also heard that a company called Sidewalk Labs, a Hudson Yards tenant whose CEO had made the Manhattan development possible, had some sort of connection to Waterfront's plans for Quayside. And Sidewalk, he learned, was somehow related to Google.

Google was the kind of company that consumed a lot of oxygen—the kind that could devote endless resources to any project it wanted, outsmarting and overtaking smaller companies that got in its way. The more people Ruffolo talked to, the more he felt that Sidewalk could win Quayside once Waterfront opened the site to bidders. He thought this was egregious. He'd spent years trying to boost the fortunes of Canadian start-ups. Why should such a massive opportunity to build a tech-first neighbourhood in Canada's biggest city go to an American company?

So Ruffolo made some more calls. He decided to bid on Quayside himself.

Waterfront Toronto's executives were less confident than John Ruffolo that Sidewalk wanted a shot at Quayside. They'd talked to fifty-two organizations, Sidewalk included, about what they could do with their 12-acre plot on Lake Ontario. When Waterfront hosted a meeting for all of them in November 2016, developers, builders and legacy smart-city companies like Cisco, IBM and Siemens showed up, and so did Oxford. Sidewalk was a no-show.

After getting input from those dozens of companies, Waterfront staff began writing out the request for proposals (RFP) for Quayside, and it became clear that their ideal fit would be a company trying to find ways to blend tech, sustainability and urbanism. The agency wanted an "innovation and funding partner"—something it had never before asked for—"with invention ingrained in its culture." It wasn't looking for a traditional developer. Not yet. It wanted ideas and financial backing.

This ideal partner would help Waterfront plot out a mixed-use, mixed-income neighbourhood that would showcase climate-positive building methods and materials to the world, making Toronto a beacon of sustainability. The agency wanted to try new business models for community-building and fund as many as eight hundred units of affordable housing. But it also hoped to build out an "urban innovation cluster" for businesses and people who wanted to test or exhibit emerging technologies. Thanks to the broadband infrastructure that Waterfront had mandated along Lake Ontario, there would be plenty of opportunities to install sensors—like, say, one that would notice motion near a streetlight and keep the light dim to reduce energy if no one was around. The age of the "industrial internet of things" was beginning, and all kinds of sensors could be deployed across the neighbourhood to enable the data-driven decision-making that was the backbone of many new start-ups and technologies. Waterfront also wanted the partner to help come up with ideas to fund long-awaited new transit to the eastern lakeshore.

"We were trying to see what we could do differently at Quayside that would actually make a difference," says Kristina Verner, Waterfront's VP in charge of sustainability and innovation, who helped write the RFP alongside chief development officer Meg Davis and a handful of others. But Verner says they were also trying to be inclusive: "We didn't want to preclude companies who had great ideas about, maybe, energy systems or building materials—or it could have been tech—from being able to actually respond."

Waterfront's ambitions for the RFP were so broad, however, that they lacked clarity. The agency usually sought partners for specific projects with obvious mandates, like developers to build a tower or designers for public art. The ambiguities in this RFP were rife for misalignment or misinterpretation. In all its excitement about hosting world-leading technology, Waterfront left significant details up for future negotiation. These included terms regarding data sharing and privacy—issues that are subject to complicated government regulations that become even more complicated when crossing international borders—and the sharing of intellectual property such as patents, which were becoming a critical source of income in a technology-driven global economy. "Critics have taken that as an invitation for the innovation and funding partner to create policy. That was not the spirit," Verner says. For the vast majority of companies who they'd talked to, policy "wouldn't have been an area of focus."

Quayside was—maybe—only the beginning. Given the relationships Fleissig had been developing with City Hall and other governments as they worked together to plan Villiers Island and protect the Port Lands from flooding, there was reason to believe those governments would be interested in whatever innovations came to life at Quayside. So the RFP team wrote that Waterfront was willing to consider extending those technologies and ideas across hundreds more acres through the Port Lands, so long as the ideas were successful at Quayside, and assuming the city eventually gave Waterfront powers over that additional land. Toronto's history of slow, careful procurement was not mentioned.

The opportunity was enormous. It was also confusing and opaque.

Waterfront released the RFP on March 17, 2017—St. Patrick's Day. Fleissig and Waterfront's chair, Mark Wilson, went on a media blitz, calling Quayside the chance to build a "21st-century city." They held an information session for potential bidders on March 30. IBM showed up. So did engineering, architecture and construction firms. Even the Swedish consulate sent an envoy.

If Sidewalk sent someone, Waterfront executives didn't recognize them.

Waterfront Toronto didn't know it, but the twenty-sixth floor of 10 Hudson Yards was humming. After his secretive visit to Toronto in December, Larry Page had given his blessing to bid for Quayside, putting aside, at least temporarily, his dream of building something in California. Soon thereafter, Dan Doctoroff gathered his staff at Sidewalk's new offices in the New York neighbourhood he'd helped build from nothing and gave them an edict: they weren't just going to bid for Quayside, they were going to win.

Early 2017 was a strange time to work in Sidewalk's glass-walled perch in the brand-new tower, which was shaped like a Cubist interpretation of an X-Acto knife rendered into reality. Many staff had left after the Yellow Book failure, and newer employees were a bit shocked: some had joined to work on the start-ups Sidewalk was incubating, thinking of the company like a venture fund with an in-house innovation accelerator. Instead, Doctoroff threw almost all of Sidewalk's resources toward bidding on Quayside. This meant working on a long, daunting project instead of at the move-fast-and-break-things pace of a young start-up.

But as the company staffed up in the shadow of the uncertain Yellow Book meeting, some of its most recent hires were perfect fits for a slow, calculated city-building project. A New Yorker named Micah Lasher was hired to oversee communications and policy. A bespectacled political operative with an eye for detail, he'd been a fixer for Michael Bloomberg in Albany and a fixture in Democratic election campaigns; just a few months earlier he'd narrowly lost a primary race to become a candidate for the New York State Senate. Alyssa Harvey Dawson, a lawyer who'd worked on intellectual property issues and licensing deals for companies like Netflix and eBay, was tapped to oversee legal, privacy and data matters. Diligent and compassionate, she was sometimes seen interjecting herself into discussions as the adult in the room when the people around her—usually men—dithered or argued. Jesse Shapins, a long-time designer with an enormous beard and an architecture Ph.D. from Harvard, and who'd overseen product development at BuzzFeed, was hired to look after Sidewalk's products and design. And a year earlier, the company had poached Eric Jaffe, the New York bureau chief of *The Atlantic*'s future-of-cities website CityLab, to give its marketing materials a progressive, professional polish.

When Waterfront Toronto published its RFP on St. Patrick's Day 2017, it was Jaffe who was tasked with writing Sidewalk's bid. It needed to be very different from the Yellow Book, which the company, when asked about it, had begun to spin as more of a brainstorming exercise than a serious plan. But this revisionist framing downplayed one of the most important efforts in Sidewalk's history: dozens of people had spent thousands of hours researching and writing hundreds of pages of detail that included robust financial analyses about deploying very specific ideas in very specific cities with very specific rates of return. It also ignored the Yellow Book's extensive plans to launch a contest for cities looking to host the company and "secure necessary regulatory and legislative relief." Whatever Sidewalk wrote in its bid for Waterfront would inevitably be consumed by the public. For that reason alone, the plan needed to be different in both detail and tone—it had to work within the constraints of democracy, reality and acceptability.

Jaffe and his team spent months researching and writing Sidewalk's bid for Quayside, doing fresh research for new ideas and plans with the help of firms like Deloitte, consulting engineers Buro Happold, and Beyer Blinder Belle Architects & Planners—all of whose work had also informed the Yellow Book. Some of the plan's ideas, in fact, were copied and pasted nearly word for word from the Yellow Book, including descriptions and diagrams of what Sidewalk called a "smart disposal chain" system, which promised to divert as much as 90 percent of solid waste from landfills through the industrial "digestion" of organic waste and imposing pay-as-you-throw fees on residents. (In forcing consumers to pay to get rid of inorganic garbage, Sidewalk used the progressive language of waste reduction to sell a conservative philosophy of individual responsibility.) There would still be "outcome-based" building codes that would use sensors to monitor for noise, emissions and temperature violations, punishing violators but otherwise allowing business and light industry to be located next to homes. The "canopy" dome was gone, replaced in the Quayside bid with actual retractable canopies to protect pedestrians and cyclists from bad weather. And the public realm would still be filled with sensors to help Sidewalk learn how people moved and lived in the community and develop new technologies that might improve it.

Jaffe's team also put great effort into showcasing the energy-related innovations it wanted to build in Toronto. The city was already using low-temperature water from Lake Ontario to help cool dozens of downtown buildings, and Sidewalk hoped to expand on this "district energy" endeavour with a grid that could centralize heating and cooling across Quayside, using thermal sources like the organic food digester and the community's own sewage. Prefabricated housing pieces, meanwhile, made their return from the Yellow Book. The Toronto bid also celebrated mid-rise, wood-structured buildings as a chance to reduce costs and environmental impact, and sought to make a regulatory case for building even taller buildings out of wood. Though modular homes and (somewhat) tall wooden buildings were already being built elsewhere, Sidewalk pitched them as part of a bespoke package—a series of innovations across technology, materials, construction and energy use—that only it could deliver.

The bid included a centralized traffic system that would minimize both car ownership and trip time thanks to ridesharing services and, eventually, those self-driving taxibots from the Yellow Book—plus, maybe, a system of suspended aerial pods to cart people around, not unlike Larry Page's monorail idea for Ann Arbor. And if Sidewalk's building, energy use and transit plans made Quayside *nearly* carbon-neutral, Jaffe's team determined that deploying the plan more ambitiously and broadly—across much of the Port Lands—would remove reduce greenhouse gases, just as Fleissig was hoping to accomplish at Villiers Island. The Sidewalkers believed the cost reductions from such wide-scale expansion could even bring the cost of living down 14 percent from the rest of the region.

This eye to expansion was central to Sidewalk's bid. When Larry Page had stood on the icy path near Toronto's least-conveniently-located major concert venue in December and looked back at Quayside, much of Toronto's underdeveloped Port Lands lay in front of him. Its hundreds of acres beckoned. Sidewalk Labs wanted to heed that call.

Even as it sought a broad vision from its suitors, Waterfront had only intended to make promises for planning rights for the 12 acres of Quayside. The agency had decision-making power over Quayside, which it owned most of. Granting power over anything beyond that would require permission from City Hall—and a lot of time to attain it. But this was not always the impression that Waterfront gave. Will Fleissig had had great success in working with governments to make big, progressive projects happen. He was confident in his ability to bring people together, as he had for Villiers Island and the Port Lands flood protection plan. In a presentation that Waterfront showed to potential partners in 2016, too, the agency demonstrated how it wanted to scale the tech, sustainability and housing ideas it would pilot at Quayside. One slide depicted an arrow charting a course from a small pilot site through the rest of Quayside, through the neighbouring 35 hectares of Villiers Island, and ending with the whole Port Lands. The RFP made no promises that Waterfront would hand over all that land—the agency simply couldn't—but the document was written vaguely enough so as not to dissuade a bidder as bold as Alphabet from thinking that Waterfront could hand over more than 12 acres. Even years later, in retrospect, Dan Doctoroff says that because the RFP wanted bidders to "consider how innovations could scale for greater impact," he didn't believe that "there was actually any misalignment whatsoever about the scale."

Yet many people close to the project, including Doctoroff's own employees, would come to acknowledge that this was not just a misalignment, but an enormous one. Some went so far as to call it the "original sin" of Sidewalk's relationship with Waterfront.

Quayside just wasn't big enough for many of Sidewalk's ideas to make financial sense. Self-driving taxibots might be decades away, but even then, the machines would need far more road to prove their effectiveness or pay back the investment needed to develop them than 12 acres could offer. Those sprawling Port Lands, removed from the harried and gridlocked downtown core, made far better sense as a testing ground for a new kind of autonomous vehicle.

So as Jaffe's team wrote out Sidewalk's Quayside bid, they repeatedly cautioned that Quayside itself was too small to deploy many of their bolder ideas. They would bid for 12 acres but plead for more, using the careful, affirmative language that companies and political candidates use to create the impression of inevitability. "What happens in Quayside will not stay in Quayside," they quipped, before getting serious. "The ideas first tested there will take on new life when deployed at scale across the Eastern Waterfront district." They made this clearer in a part of their response to Waterfront's RFP that would not be made public, in which the Canadian agency asked how dependent Sidewalk's vision was on scaling eastward across the Port Lands. The question was intended to gauge eagerness, but created an opening for Sidewalk to create the impression that expanding beyond Quayside was inevitable. The company responded that "The Eastern Waterfront is critically important to Sidewalk's vision and the shared objectives that Sidewalk and Waterfront Toronto have for Toronto."

The procurement rules warned bidders not to contact any government officials connected with Waterfront during the RFP process, to avoid prejudicing the result. Had Sidewalk staff been free to speak with local politicians and civil servants, they might have learned that the only way to get to the Port Lands was by going through hell.

John Ruffolo, meanwhile, was under the impression that he wasn't allowed to touch *anything* beyond the 12 acres in question. But he still thought he could fill Quayside itself with big dreams.

He relished every chance he had to be the first person backing bright ideas. He led the accounting firm Arthur Andersen's Canadian high-tech practice by age twenty-five. By the time the firm collapsed alongside Enron nine years later and Ruffolo's office merged with Deloitte, he'd become a sought-after adviser in the post-dot-com-bust tech sector. Less than a decade after that, the head of the OMERS pension fund, Michael Nobrega, reached out. They bonded over immigrant roots—Ruffolo was Italian, Nobrega was Portuguese—and the pension fund CEO told Ruffolo that he saw Canadian tech as an underfunded investment opportunity. Ruffolo, who'd been advising the sector through the Great Recession, saw the same untapped potential. So they went into business: Ruffolo would run the brand-new venture wing of one of Canada's most influential pension funds. Some people began to assume that stocky, gruff, attention-loving Ruffolo ran all of OMERS, a misunderstanding that Nobrega, a quiet executive who didn't care for the spotlight, was happy to leave uncorrected.

For the first three-quarters of the 2010s, if a Canadian start-up had a bright idea, John Ruffolo was probably studying its books to see if he should fund it. Canada didn't have as many splashy tech brands as the United States, but it was building software that other industries were becoming dependent on, like the online learning platform D2L, storytelling community Wattpad and social media dashboard Hootsuite. Ruffolo cut the biggest cheque of Shopify's $100 million financing round in 2013, a year and a half before the small-business retail platform went public and seven years before a pandemic prompted investors to temporarily make it Canada's most valuable public company.

He also wanted to build a place where all these companies could come together and share ideas. That's why he tried to pitch Waterfront Toronto to help with an innovation corridor on Oxford land in the western part of downtown—and why, once he saw Waterfront's RFP for Quayside, he began plotting a bid of his own for the eastern site.

Ruffolo first called up Peter Gilgan, a long-time cycling friend who ran Mattamy Group, a company that often boasted that it was North America's biggest privately owned homebuilder. Like Waterfront, Mattamy had come to embrace certifications from Leadership in Energy and Environmental Design (LEED) in its construction. Ruffolo also reached out to Conundrum Capital, a private-equity firm that focused on rental housing. The year before, Conundrum had struck up a partnership with the major Canadian insurer Manulife called Q Management, and it owned C$1.5 billion worth of apartment buildings in the Toronto and Ottawa areas. Ruffolo pitched a partnership to the other companies: his OMERS Ventures organization could handle tech at Quayside, Mattamy could be a builder, and Q could oversee rental units. Together, they could combine ideas around fast-moving technology, sustainability and the full spectrum of a mixed-income neighbourhood.

For nearly six weeks, Ruffolo and the partner companies' senior staff buzzed between each others' offices, rushing together a fifty-seven-page bid just before Waterfront's late-April 2017 deadline. "We barely had enough time to put résumés together," Ruffolo says. That first submission was, expectedly, light on details. It talked about harvesting rainwater, about "smart" energy grids, about building with LEED standards and showcasing the best of Toronto-made technologies. Like Sidewalk, they imagined a community with fewer regulations, though their imaginations had more restraint; for instance, they proposed a zone where building codes wouldn't apply, in order to try out "innovative building technology," but did not go so far as to detail the kinds of boundary-pushing wood-structured buildings of Sidewalk's bid. And to a greater extent than Sidewalk, they proposed making numerous major decisions later, in collaboration with Waterfront.

The members of the newly minted consortium felt like they brought the right experience to the bid, one that would use local talent to develop a globally significant project. They were so local, they joked, they could all see Quayside from their offices. Hudson Yards was hundreds of kilometres away.

The summer of 2017 was a big one for Will Fleissig. A year and a half into the job, he secured Waterfront Toronto's future. The federal, Ontario and Toronto governments chipped in nearly C$400 million apiece to reroute the Don River and flood proof the Port Lands. The press conference at Polson Pier, where Larry Page had stood months earlier, projected the image of Fleissig as a chief executive who could bring powerful people together to get impossible things done. In late June, with the city skyline behind him, Fleissig stepped off a commuter boat alongside Prime Minister Justin Trudeau, Premier Kathleen Wynne and Mayor John Tory to make the announcement. "The people we collectively serve are the happiest when they see their three governments agreeing on priorities," Tory said once he reached the stage.

The news cemented Fleissig's image as someone who could make big things happen. He'd navigated the personalities within three governments, and their three massive bureaucracies, to secure more than a billion dollars. It helped that Fleissig had joined Waterfront at a rare moment when the three governments were philosophically aligned, each relatively close to the centre of the political spectrum. Their money would open up hundreds of acres next to downtown for future redevelopment that Waterfront could lead, and effectively extended Waterfront's life beyond its twenty-year mandate, until at least 2028. For the optics alone, the deal was a coup.

The future of Quayside was coming together, too. Six bidders had come through before the deadline. One was disqualified for not following the process, and two didn't make Waterfront Toronto's short list a few weeks later. The three that remained came from very different worlds. One was the OMERS-Mattamy-Q consortium's made-in-Toronto bid. One was from the Canadian division of Siemens, a vanguard of the original smart city movement. The third was from Sidewalk Labs. Their submission surprised some Waterfront staff, who hadn't heard much from the company for months. But they'd soon spend much more time together. There was a rigid, summer-long schedule that each finalist had to follow as they put together more extensive bids. Starting in late May, there would be four rounds of meetings, each with precisely defined durations—seven hours, seven hours, sixteen hours over two days, seven hours—so that no one received preferential treatment. Former Ontario associate chief justice Coulter Osborne would oversee the process as a "fairness adviser." And when the process was over, Waterfront's government-appointed board would review the deal to give it one last layer of scrutiny.

At first, the Toronto consortium team was excited. There was easy money to be gained in the city from buying land and building generic condo towers, but the mission here seemed almost noble by comparison. Waterfront was pitching the opportunity at Quayside as potentially visionary: a dedicated blend of mixed-income housing and a chance to support the local tech scene. But at the mandated bidder meetings, some of the team felt there wasn't enough discussion of what Waterfront specifically wanted—there was plenty of talk about things like affordable housing, but not enough details about what specific kinds of affordability they wanted to see in bids.

By this point, someone had leaked details of Sidewalk's application to Bloomberg News, the company Dan Doctoroff had once overseen. The Toronto consortium members were annoyed that the leak might have breached the confidentiality of the bid process. Moreover, the story confirmed John Ruffolo's fears. There was little known publicly about Sidewalk in mid-2017, save for its connection to Google through Alphabet. Ruffolo, whose years of local tech boosterism had pushed his world view toward economic nationalism, worried the company might take whatever money was made at Quayside and carry it back to Mountain View, depriving Torontonians and their tax-collecting governments of wealth they had a hand in generating. With Sidewalk now confirmed to be in the running, Ruffolo's spirits were dampened. His instincts told him Sidewalk would win. So the highly competitive venture capitalist asked his team to do something he'd never requested before: "Do not embarrass us," he recalls telling them, "but do the minimum amount of effort to be number two." That way, he says, "if the Sidewalk shit didn't go through, we could pounce."

Another member of the consortium felt like the bidder meetings were filled with "empty discussion," like being the fourth person to interview for a job when it was clear the first candidate would get the gig. They also became concerned about the funding requirement of the "innovation and funding partner" Waterfront was looking for, which requested at least a seven-figure commitment to help with what seemed to be vaguely defined project costs. (Sidewalk had no problem with this, offering $50 million to cover the costs of drafting a plan for Quayside.)

By the time Waterfront began extensive two-day bidder meetings in July, the agency was losing confidence in the consortium's bid, too. These meetings included on-site visits meant to showcase the bidders' previous work. The consortium took Waterfront executives to a housing development near Toronto to demonstrate Mattamy's building expertise, and to Hudson Yards in New York, where they visited a multimedia-laced condo sales office for the OMERS-related development. This was not the only trip Waterfront took to Hudson Yards—Sidewalk brought them, too, for a much broader tour that included the Shed, the performance and art-installation space run by a non-profit that Dan Doctoroff chaired. "There was a slightly different level of effort," Kristina Verner says. Sidewalk even took Waterfront to Mountain View to see some of Alphabet's other moonshots.

As the selection team reviewed and scored the bids, "it was not even a discussion," Meg Davis says, as to which bidder was the best fit. Even when the third bidder, Siemens—a one-time leader of the smart city movement—invited Waterfront to see some of its projects in Munich and Vienna, Waterfront was underwhelmed. Siemens didn't seem to have enough focus on a holistic vision for the future of cities.

A second committee that oversaw the selection team and double-checked their process in great detail came to a similar conclusion to Davis's. Only one bid had a fully integrated, design-minded vision for the future of cities, says Val Fox, a local tech incubator co-founder who was on the oversight committee. It came from a company with a mountainous budget, an army of in-house experts and a monomaniacal dedication to its bid—a bid that also happened to make clear that Quayside wasn't enough for its dreams.

Waterfront let the bidders know in mid-September: the winner was Sidewalk Labs. All that was left was for Waterfront's board to review the deal.

Though the members of the made-in-Canada consortium were disappointed that Waterfront hadn't picked a local winner, John

Ruffolo had long ago given up hope. "It was a bit of a joke," he says. "Congratulations, we got where we fully expected, and we spent zero time on a horrible response." Instead, he started sharing his fears about Sidewalk with friends.

CHAPTER 7

Limelight

DAYS BEFORE WATERFRONT TORONTO's selection committee picked Sidewalk to be its Quayside partner in September 2017, the online retailer Amazon announced it would hunt for a city to host a second headquarters somewhere outside of Seattle. Though some close watchers of the contest saw it as a chance for the massive company to help a down-on-its-luck city like Detroit breathe new life into its local economy, it was viewed by more cynical—or more astute—observers as an extension of Amazon's years-long quest to minimize the taxes it had to pay: a hunt for the cheapest place to sustain its unprecedented growth.

The "HQ2" contest was not unlike the one that Sidewalk Labs had detailed in the Yellow Book. In one way, Sidewalk had aspired to go a step further than Amazon was proposing: it wanted power *over* taxation. "Project Sidewalk will require tax and financing authority to finance infrastructure and provide services," the Yellow Book explained, "including the ability to impose, capture, and reinvest property taxes and the local portion of sales taxes in the district." Once the public began calling out Amazon's intentions with the HQ2 contest, a sense of relief emerged among some Sidewalkers that they hadn't held a contest of their own. The way cities were tripping over themselves to offer deals to Amazon felt to them like a race to the bottom; Sidewalk preferred to cast itself as helping to build great cities, not taking advantage of weak ones.

When Dan Doctoroff spoke at a public event for urban planners in San Francisco that September, he assailed Amazon's brazen campaign for tax incentives, arguing that it would compromise the host city's budget. And at the same event, after years of dropping hints at events and in the press, Doctoroff added that Sidewalk was "quite far along" in finding a city that would let it build a neighbourhood—but, distancing the company's efforts from the HQ2 race, added that Sidewalk was "not looking for huge handouts."

The *Financial Times* wrote up the remarks, and they landed in front of Duane McMullen, the director general of Canada's Trade Commissioner Service, on September 21. He e-mailed colleagues "pfff" and shared a link to the story, adding: "amazon, is that all you're aiming for? Google wants to build a whole new city, and is 'quite far along' in its search."

"I welcome any intelligence on this," Canada's chief trade commissioner and future European Union ambassador, Ailish Campbell, wrote back. "Believe this cross-walks to discussions on Toronto waterfront."

Waterfront's board had yet to vote to confirm the winning bid, but with Bloomberg News having leaked word of Sidewalk's Quayside pitch months earlier, some of Canada's most influential bureaucrats were drawing a connection between the reports. They began a rushed thread of e-mails. Infrastructure Canada, the federal department that oversaw Waterfront Toronto, "would have the latest on this," wrote an official from the Privy Council Office, the central hub of the Canadian government's operations. "As far as I know, the Waterfront process is close to a decision," another person replied. "Can we find out more from the Ottawa reps on the board?"

Learning more proved difficult. Despite being appointed by the government, the Waterfront directors were holding to a mandatory blackout period before announcing their winner. So the officials set up a conference call with Waterfront's overseers in the infrastructure department. A Privy Council staffer filled in colleagues five days later. The selection process for Quayside "is still ongoing," the staffer said.

No candidates were named. "Announcement was initially anticipated by end of calendar year, but could now be advanced to mid-October."

The bureaucrats got the answer they were looking for a week later. The *Wall Street Journal* reported that Sidewalk was "nearing a deal" to build something in Toronto. Just in case the story's anonymous sources left any doubt, the *Globe and Mail* confirmed the scoop. Federal staff shot the stories back and forth on October 4. A senior Infrastructure Canada official told his colleagues that the news was "speculation" and that Waterfront was "increasingly concerned" about leaks.

"To my knowledge, no one in the federal family has had any confirmation regarding the identity of the proponent," the infrastructure department staffer wrote.

Discussions died down, but the newspaper leaks drove excitement through the highest ranks of the Canadian government. No one was more exuberant than prime minister Justin Trudeau. He'd been hoping Sidewalk might win—because Alphabet chairman Eric Schmidt had already secretly won Trudeau's support for the project.

It was 2017, two years into his first term, and the son of former Liberal prime minister Pierre Elliott Trudeau had already crisscrossed the globe seeking connections that would make him look more progressive and more interesting than his staid Conservative predecessor Stephen Harper. And in the mid-2010s, Big Tech could do no wrong: nothing looked more progressive or more interesting than a job at a place like Facebook or Google, who were connecting the world's people and information. So Trudeau would routinely try to get onto the schedules of global tech leaders like Amazon's Jeff Bezos and Apple's Tim Cook. Just months after he was elected, he even showed up to the grand opening of Google's new office in the tech-filled Waterloo region, an hour and a half west of Toronto. It helped create the image of a forward-thinking government focused on giving Canadians forward-thinking jobs. Big Tech was a big part of Trudeau's brand.

So there would be little surprise that, on January 16, 2017, the prime minister had a call with Eric Schmidt—except for the fact that they kept it hidden from the public. There was no record of it in either the federal lobbying database or in the many access-to-information requests I filed for years about the Quayside project. But one set of memos for Trudeau about Quayside that I received in late 2018 from the Privy Council Office, which manages access-to-information requests for the prime minister, included several curiously placed paragraphs that had been blacked out. On a hunch that these paragraphs would reveal something about Trudeau's interest in Sidewalk—some opponents were by then speculating that he may have interfered in the procurement process—I filed a complaint with Canada's information commissioner to make these paragraphs public.

It took three years, two re-elections for Trudeau and one very long, sometimes angry chain of correspondence with a federal investigator to get an answer. At points, it seemed like it would be impossible—as one deadline passed, the investigator told me "there is no mechanism to hold [the Privy Council Office] to account for their previous indication of a disclosure date." Then, one evening in January 2022, the government suddenly sent me a revised version of the memos. The paragraphs were no longer blacked out.

Someone from Alphabet had had a phone call with the prime minister two months before the Quayside procurement process even opened, to discuss the company's "interest in partnering with Waterfront Toronto in creating a smart city using only autonomous vehicles." When I asked for more records about the call, the Privy Council Office's access-to-information staff said they couldn't find any. Several weeks later, the Prime Minister's Office admitted to me that they *did* have records, but all they would share was that the call was with Schmidt. The call, meanwhile, had slipped through a gap in Canada's lobbying disclosure rules. Google had for years rigorously documented its staff's interactions with Trudeau, but after I found out about the January 2017 call through other means, the company insisted it was placed by the Prime Minister's Office, bypassing the need to put the meeting on public record.

The call didn't give credence to the rumours that Trudeau had somehow rigged the process for Sidewalk to win Quayside; it took place before the procurement process started, and numerous memos from his office pointed out the independence of Waterfront's process. Yet the consequences of the secret call were vast. This was a prime minister obsessed with aligning himself with tech companies, and one of the biggest tech companies in the world successfully planted a sense of excitement in his mind—one that helped ensure he and his office would give Sidewalk's ideas little scrutiny in the days to come. This, in turn, helped create a game of political football across the entire federal government in which no one would take responsibility for the controversies that would soon unfold at Quayside.

But before all that happened, there was an announcement to make. Soon after they found out that Sidewalk had won, Trudeau's schedulers asked a senior Liberal staffer from the infrastructure minister's office named John Brodhead, who was overseeing some of Waterfront's work in the Port Lands, to make an introduction to the agency. Brodhead connected them with Helen Burstyn, Waterfront's chair, to see if Waterfront's announcement could fit into the prime minister's schedule.

It seemed likely. Waterfront staff had been trying to pinpoint a date that would work for Dan Doctoroff to attend. Maybe even Eric Schmidt, too. It looked like Tuesday, October 17, was the earliest date that would work for the two Americans and the prime minister—and early suited Waterfront, since the agency wanted to avoid any more leaks.

But Waterfront's board of directors still needed to approve the selection of Sidewalk as the winning bidder, and its meeting to do so was set for October 20. So the board agreed to move the vote up by four days, to Monday, October 16. One of history's most consequential decisions about technology companies' role in the future of cities would be given less time for scrutiny because of an excited prime minister, Alphabet executives' work schedules and a fear of media getting a jump on the story. This created an uneasy feeling among some of the people involved. "It had to be approved on an accelerated basis," reflects Toronto mayor John Tory, who'd been invited to the rushed announcement. "That led to, perhaps, less precision in exactly what was being agreed to. That would later on cause trouble."

While American Thanksgiving marks the start of the holiday season, in Canada, Thanksgiving weekend tends to mark the final gasp of summer—one last chance to get away before the cold sets in, thanks to a statutory holiday on the second Monday of October. But when Julie Di Lorenzo took some guests to her cottage north of the city that weekend in 2017, she found herself far more frustrated than thankful.

She was one of a dozen directors appointed by the federal, provincial and city governments to oversee Waterfront. She'd just gotten the latest draft copy of the Sidewalk agreement, and spent the weekend fixated on it, hiding away from her guests. Though Di Lorenzo had a burgeoning fascination with data and privacy, it was the business terms in the two-and-a-half-dozen-page document that scared her. It referred to a handful of company names or descriptions mentioning Sidewalk, yet lacked an organizational chart to show how decisions would flow. She thought it was unacceptable that the title of "master developer" would be conferred to a limited partnership called Sidewalk Toronto, and not publicly accountable Waterfront Toronto, as the RFP had once stated. Though Waterfront would "share in the economics and governance of the master developer," the contract said, it also gave the master developer the opportunity to "divest, develop or otherwise monetize or develop project assets." And even the title "master developer" made little sense: this was a contract to devise a plan about affordability, sustainability and technology, and Waterfront had explicitly said it wasn't looking for a traditional developer. "This was not a simple transaction," Di Lorenzo says of the agreement she was reading. "There were implications to every sentence."

Her fellow board members came from many worlds—non-profit, post-secondary, infrastructure investment, consulting. One of Toronto's deputy mayors was a director. A handful of members had direct experience in real estate development.

Di Lorenzo was one of the latter group. She'd often tell the story about growing up in construction, learning the ropes from the cab of a Mack truck and launching her own company by the age of seventeen. By the mid-2000s, in her early forties, she had fashioned herself into a clever dealmaker—claiming she'd lost money on a project only once—and an advocate for quality architecture in a city whose defining physical characteristic was exhausting sameness, a sea of semi-detached brick homes pocked with green-glass condo towers. She was deeply analytical, the kind of corporate leader who actively sweated the details, preferring to spend as much time on construction sites as at the office.

As the president of the Diamante Development Corporation, she was one of the few women who'd reached the upper echelons of Canada's real estate sector. She considered herself lucky to have gotten where she did, and when Ontario's Liberal government invited her to join Waterfront's board in 2015, she saw it as a chance to give back to the public through an agency that she believed did great work. "Imagine if you could create a community with the distilled intelligence of the experiences of previous cities and the insights of visionary leaders of tomorrow," she told the committee of legislators who oversaw her appointment that year. "We have that template from which to work at Waterfront Toronto."

For her first couple years on Waterfront's board, Di Lorenzo was proud of the work she got to do. Thanks to her experience in development, she became chair of the committee that oversaw investments and real estate deals. The projects they reviewed were relatively routine: looking at plans for mixed-income housing and a waterfront college campus, and eventually looking at the potential of Quayside. By the summer of 2017, the committee was getting high-level briefings on the short list of bidders, and in September they learned that management had selected Sidewalk as its "preferred proponent."

When the board committee first heard about the Google sibling's interest in Quayside, they were thrilled. "It's nice to have a big international company be excited about what you're doing," Di Lorenzo says. Her committee met several times in September, including once at the office of Stephen Diamond, another developer on the board. Josh Sirefman and another staffer from Sidewalk briefed them on the project, and the committee members, including Di Lorenzo, were fairly satisfied with what they heard. Di Lorenzo asked if Sidewalk planned to bring in ideas from Europe, where, she understood, cities were a little more advanced at blending technology and sustainability. She was, as ever, looking for details. "I remember having really a strange feeling when the response was really ambivalent," she says. "That's how high-level that discussion was."

It was the board committee's job to review the agreement with Sidewalk—to scrutinize it, to suggest revisions, and to recommend that the board approve or reject it. As September turned to October, Diamond in particular pressed for some details to be included in the agreement that would protect Waterfront from unexpected surprises. He wanted language that confirmed Sidewalk could expect to deploy technology across only the 12 acres of Quayside, not the entire eastern waterfront. He also wanted to make sure Waterfront wouldn't accept the full $50 million until the two parties could supply more details about the partnership in a later agreement. They didn't want Waterfront to feel obligated to Sidewalk: "Nothing is free in life," Diamond says. Finally, he pushed for the unilateral right to walk away if something didn't feel right. With those pieces in place, the committee could be satisfied with the Sidewalk deal.

But the version of the draft agreement that Di Lorenzo saw on Thanksgiving weekend changed everything for her. It was not the kind of deal Waterfront usually got into, and she felt it needed far more scrutiny than it was getting. She began sending dozens of queries to Waterfront's management about the proposed relationship. Waterfront tried to keep up with her requests, and staff worked themselves to exhaustion writing up answers for her. Some thought she was dwelling on the details to a far greater extent than necessary, but that was how Di Lorenzo operated. She'd always done so.

The following week was one of the most hellish of Di Lorenzo's life. While obsessing over what could go wrong with Quayside, she was shuttling back and forth to a downtown hospital with a friend who had cancer and was about to undergo major surgery. All the while, Di Lorenzo was frantically contacting lawyers for advice about the board decision and going to Waterfront meetings, at one point dialling into a board briefing from a random table in a building near the hospital. Waterfront's management briefed the board twice that week. While some of the other directors shared Di Lorenzo's concerns, they largely saw the Sidewalk contract as an "agreement to agree": a deal that would let them start working on a draft plan for Quayside, figuring out the details along the way.

"If we do want to solve problems, we have to have the appetite for risk," says Mazyar Mortazavi, CEO of the design and development company TAS, who had been appointed to the board by Ottawa earlier that year. "The intention here was to drive innovation—you weren't going to have everything figured out. You can't set up one intent to explore and innovate, and then preclude that by being prescriptive and definitive."

Though most of the other directors were less concerned than Di Lorenzo about the contract, many felt that moving the Quayside vote four days earlier, to October 16, deprived them of time to do due diligence. Some felt that doing so in order to accommodate the schedule of the prime minister and others was egregious. "I thought that the optics of that would be unfortunate," says Ross McGregor, a lawyer and charity fundraiser who was on the board. It wasn't just the optics: an e-mail later found by Ontario's auditor general Bonnie Lysyk showed that the board had felt "urged—strongly" by both federal and provincial governments to get the framework agreement approved. "It seemed like the announcement was more important than the deal," says Denzil Minnan-Wong, Toronto's deputy mayor and the only politician who sat on Waterfront's board in 2017.

Four days before the vote, on Thursday, October 12, Di Lorenzo drafted a compromise that would task her committee to scrutinize the deal further before advancing beyond the current agreement with Sidewalk, given the limited time they'd had to review all the partnership materials. She sent it to the committee members and Waterfront executives early on Friday. The timing was awkward. They needed to meet and discuss it as soon as possible, but she also had an annual tradition to attend. Each year, she'd buy a few dozen seats for one of the Royal Conservatory of Music's biggest fundraising galas. So she had to step aside from entertaining her guests that Friday evening to call in to a teleconference.

The committee had already agreed to include language in the Sidewalk agreement addressing the three key issues that Diamond and others had been pushing for: that Waterfront had the right to walk away, that no land was available beyond Quayside, and that no money would be exchanged before the next agreement. With those safeguards in place, Di Lorenzo's compromise got no traction. Though she saw her proposal as an added layer of protection, others felt that her concerns had already been accommodated.

By the end of the teleconference, Di Lorenzo was refusing to endorse the Sidewalk deal, leaving the committee at a standstill. That meant the entire board would need to make the decision on Sidewalk together.

It was Friday the thirteenth. The events that unfolded that day weren't particularly eerie, but they *were* rather strange. Di Lorenzo returned to entertaining her guests, eventually taking her seat with them in the Royal Conservatory's flagship 1,135-seat performance hall with its wood-ribbon ceilings. While she watched the actor Bill Murray read Ernest Hemingway and Mark Twain and join the cellist Jan Vogler and a band for a few songs, Waterfront staff raced to compile documents for the rest of the board to review, including the latest draft agreement and letters from Waterfront's lawyers and its procurement and intellectual-property experts. Di Lorenzo spent the weekend in anguish, talking on the phone with a governance lawyer about whether this was a fair situation for an organization to put its board in. The rest of the directors rushed to brush up on the deal.

On Monday morning, Julie Di Lorenzo woke up with her mind racing, trying to pin down what she might say at the board meeting that day. Soon after 8 a.m., she hailed a cab to take her to Waterfront's office by the lake. She walked into its main boardroom, filled with echoing overhead microphones and a scale model of a past Waterfront development partly made of expired candy, and was shaking by the time the chair, the long-time public servant Helen Burstyn, called the meeting to order.

There was really only one matter to discuss. Burstyn began to go over the details of Waterfront's agreement with Sidewalk for what felt, to some, like the thousandth time that month. After reviewing advice from Waterfront's lawyers that explained some aspects of the deal, she ceded the floor to the board's real estate and investment committee—the committee that, after Di Lorenzo's refusal to recommend it, had sent the Sidewalk decision to the board for approval.

Di Lorenzo wanted to spell out her fears about the deal: namely, that Waterfront didn't really know what it was signing up for. But her mind was reeling after a tense week and intense conversations with her personal governance lawyer over how to handle the situation. She kept her comments short, saying there hadn't been enough time to review the details. Another committee member, an infrastructure project investor named Susie Henderson, admitted she, too, was concerned about how fast the board had to make its decision, but said she was satisfied with management's legal advice and would vote to approve the agreement. The third committee member, Stephen Diamond, wasn't at the meeting, but had let the chair know he was satisfied with the deal after its revision to include his three recommended amendments.

Around the table, a few people acknowledged Di Lorenzo's concerns. One director worried that the board wasn't experienced enough in the world of technology to make the right decision for Toronto. Another, Ross McGregor, said he was concerned about how data would be managed and how much land Sidewalk hoped it would get, but that he viewed the agreement to partner with Sidewalk as probationary. These were complicated issues, he thought, and Waterfront could work them out over time, and even walk away if it had to. Other directors felt that way, too. "It's why people date, engage, and *then* they marry," says director Mazyar Mortazavi. "The foundation of any relationship is the same."

As the meeting wore on, the discussion lacked the serious tone Di Lorenzo thought it should have. The board decided not to listen to management give another presentation on the project—they were tired of the constant flow of new information—and one director left the meeting early.

Finally, they agreed to take a vote. Of the nine directors present, eight voted to partner with Sidewalk.

As hands shot up around the board table, the conversation Di Lorenzo had had with her personal governance lawyer earlier that morning ran through her head. She couldn't understand why Waterfront Toronto was so eagerly signing on to plan a community with a company that had no track record in city-building. She was angry at being put in this situation. And she was afraid of what might go wrong if they partnered with this mysterious company that Waterfront hadn't properly studied.

But she pushed any thoughts of resigning out of her mind. As advised by her lawyer, she dissented, and insisted that her dissent be included in the minutes.

The meeting ended at 10:35 a.m. When the boardroom doors opened, Dan Doctoroff and his staff were already there, waiting on standby to answer questions—and, as it happened, to celebrate the news. That night, Julie Di Lorenzo stayed home while others from Waterfront met Sidewalkers at a downtown bar in the base of a condo tower built behind a historic Gothic-revival facade. It was a building that conveyed intrigue at first glance, but was unimpressive in its details. The Sidewalk partnership would soon face the same criticisms.

Twenty-five minutes after the board meeting ended, Justin Trudeau's policy staff were reviewing a draft of the remarks he'd make the next day. His involvement in the Sidewalk Labs announcement would be a coup for a prime minister who was carefully trying to convey a relatable, progressive image. In fact, this would turn out to be a week in which his eagerness to do so would shine. But it would also be a week in which that eagerness would repeatedly backfire.

That same morning, Trudeau stood in his shirt sleeves at a lectern in an Italian restaurant north of Toronto and did as much as he could to prevent journalists from talking to the man standing behind him. The man was his finance minister, Bill Morneau, a former HR company executive who, the *Globe and Mail* had just reported, hadn't put his vast holdings in a blind trust to keep them at arm's length—generally a requirement in Canada's federal cabinet, since ministers are in a position to devise laws that they could profit from. Two different reporters asked if they could talk to Morneau, and each time, Trudeau tried to dodge the question by playing up the coolness he'd been ascribed by fawning global press attention. The first time, people in the room laughed. The second time proved more awkward. "You have to ask a question of me first," he said to the reporter, "because you get a chance to talk to the prime minister." He was met by stunned silence, including from the reporter. (The reporter . . . was me.)

It was Trudeau's evasion that coverage focused on that day—not the charisma he'd been trying to show off.

Wednesday of that same week, Trudeau's eagerness was punctured once again, this time more subtly. Though the prime minister tried to cultivate a progressive image, his government's choices did not always reflect that; he once placed an ex-Greenpeace activist at the helm of his environment ministry, but he also bought an oil pipeline to nationalize it. Two days after the Italian restaurant press conference, the beloved Canadian radio-rock band the Tragically Hip announced that its frontman, Gord Downie, had died of terminal brain cancer. Under Downie's leadership, the Hip had become one of the rare bands to make a career in the country's small touring market by celebrating Canadian identity in song. Back in the halls of Parliament in Ottawa that day, Trudeau teared up over the news. "Gord was my friend," he said.

In his final years, Downie had used his platform to call out Canada's colonial history of systematically killing the Indigenous peoples on whose land the country was built. He saw Trudeau as responsible for reconciling Canada's relationship with First Nations. At the Hip's final concert in August 2016, Downie used some of his last-ever moments onstage to thank the prime minister for caring about reconciliation with Indigenous peoples, particularly those in northern communities, but warned him there was still work to be done. "What's going on up there ain't good," Downie told the crowd. "It may be worse than it's ever been."

The prime minister was happy to show off a photo of him and Downie embracing that night, yet subsequently did little to help many struggling First Nations access even the basic rights afforded to most Canadians, dragging his feet on a promise to ensure clean drinking water for dozens of struggling First Nations. In 2021, the Trudeau government faced legal action seeking compensation for Canada's well-established history of harming its Indigenous peoples—including through its notorious residential school system—and, in court, denied responsibility.

The image-oriented approach to governing extended to economic matters, too, dovetailing with Trudeau's enthusiasm about tech. In their first term, Trudeau's Liberals rarely scrutinized the business models of the American giants they aligned with, even as public opinion began to shift toward greater concern about the giants' data collection, market power and influence over society. Nine months after the prime minister's secret phone call with Eric Schmidt, and as Waterfront Toronto got ready for the Sidewalk Labs announcement on Tuesday of the same week, Trudeau's staff drafted him a memo that celebrated Sidewalk winning the Quayside project because it "complements federal efforts to grow Alphabet/Google partnerships and operational footprint in Canada."

And that's how, on the day after he'd awkwardly blocked his finance minister on national television, and the day before he tearfully gushed about a dead musician who may not have actually been his friend, the most powerful elected official of one of the world's biggest economies announced his support for a company that was created because one of the world's richest men thought governments got in his way.

The prime minister's tie was askew that Tuesday, but his smile shot straight ahead into the press conference audience. He was standing in the sun-filled, five-storey atrium of a media company's studio complex on the edge of Lake Ontario. The event was just down the block from the soggy plot of land that, a day earlier, Waterfront's board agreed would get a chance to become Sidewalk Labs' plaything.

Trudeau welcomed Sidewalk to Toronto, putting weight on the city's second *T*, which most Canadians avoid doing unless they're talking to Americans. In fact, the prime minister *was* talking to Americans—a pair of them in the front row—and as his glance turned to them, he began to deviate from the speech his team had prepared. He flashed another grin as he looked down at one of the Americans: Dan Doctoroff. "Dan, I've been a big fan for a long time, and of the things you've been doing in New York City," he said. Then he turned his gaze to his left, to Eric Schmidt. "Eric and I have been talking about collaborating on this for a few years now," Trudeau said. "Seeing it all come together today is extraordinarily exciting."

Standing in the crowd among local politicians, urban planners, students and media were Waterfront Toronto's staff, executives and directors. They'd just spent months picking out a winner for Quayside, in a tightly controlled process overseen by a retired senior judge, and Canada's most powerful politician had just implied he'd been planning the outcome with an Alphabet executive all along. They had no idea that Schmidt had personally wooed the prime minister months before they'd even opened a procurement process that they'd run independently with numerous layers of precautions. Board chair Helen Burstyn, who'd spent time as an interviewer on a local television show, remembers standing to the side of the stage, wishing she could ask the prime minister a few questions. In the audience, Kristina Verner turned to her husband. "That was an odd comment," she whispered.

"We looked all over the world for the perfect place to bring this vision to life, and we found it here in Toronto," Doctoroff said after stepping to the podium. His comment sidestepped the fact that it was actually Toronto that had picked Sidewalk. But the pistons of Sidewalk's mythmaking machine were firing, and Doctoroff filled his speech with praise, progressive aphorisms, a Jane Jacobs reference and language that hinted at a much grander vision than the 12 acres of Quayside would allow. "Our journey here will begin with close collaboration and strong community engagement and ends with us creating a place that creates a new model for urban life in the twenty-first century," Doctoroff said.

Then he introduced Schmidt. All Sidewalk wanted was a city, the Alphabet chairman said, and Toronto had offered the company what it needed. Rarely turning to notes, Schmidt gave the crowd a high-gloss version of Sidewalk's origin story. "Years ago, we were sitting there thinking, 'Wouldn't it be nice if we took the technical things we know and applied them to cities?'" he said. His purple tie was straighter than Trudeau's, his hair greyer, his hands conveying his excitement more than his face. "Our founders got really excited about this, and we started talking about all of these things that we could do if someone could just give us a city"—he raised his hands upward, then quickly threw them down—"and then put us in charge." He paused. "That's not how it works, guys," he said as he mimed shoving the idea away. "For all sorts of good reasons."

Though Schmidt spoke humbly of being welcomed into Toronto, the plans that brought him there were backed by the confidence that Google could once again change how the world works. "This is the culmination, on our side, of almost ten years of thinking about how technology could improve the quality of people's lives," Schmidt said, citing none of those thoughts or technologies, only vague things the company could improve—things like "inequality and access and opportunity and entrepreneurship."

Schmidt wound down his speech by admitting he hadn't been sure if Waterfront's board would vote Sidewalk in. "I was actually worried we weren't going to make it," he said. "Dan and I were thinking, 'What happens if they say no?'"

It was too late to say no, but a few minutes later, Waterfront CEO Will Fleissig tried to at least take back the narrative. He outlined the long, careful bidding process for Quayside, looking for "outcomes" around affordability and prosperity, then issued caution about the details Sidewalk was already planning to share with the public. "When you go to the web page, you will be seeing their proposal," Fleissig said. "It's not a plan—it's not something that we've been involved with. They were responding to the outcomes that I just identified. So now the question is, we think they have some pretty good ideas and they have some ways of how they want to approach it. Dan understands that it has to come out of the conversation—that the initial ideas that his team put together are just that."

The project had been public for an hour, and Waterfront's boss was already fighting to set the story straight.

CHAPTER 8

Northern Touch

THE QUAYSIDE ANNOUNCEMENT wasn't the first time Dan Doctoroff got excited about the crossroads of Canada and technology. He'd embedded one of Canada's most famous inventions into one of his favourite pieces of personal lore.

On the final day of 2005, on the border between Chile and Argentina, Doctoroff found himself cycling a 20-kilometre route through the Andes that rose 1,500 metres, the altitude thinning out the pines around him. He would eventually call this the most difficult physical test of his life; each 100-metre marker he passed was a blessing.

He was counting anything that might resemble a blessing. The humiliation of losing the Olympic bid for New York City—and the stadium at Hudson Yards—had followed him around for months. He desperately wanted a win. But he would later claim that as his mind raced during the exhausting ride, he came to realize that all the effort he'd put into the Olympics had paid off for New York. Projects that had been ignored or delayed for years were finally under way, from the High Line to the revitalization of the Queens and Brooklyn waterfronts. Affordable housing was getting built. So would Hudson Yards, even without a stadium. *Failure is only failure if you define it that way*, he thought. It didn't matter that his $35 million Olympic bid had gone nowhere or that the New York he was reshaping was increasingly unaffordable to the average person.

As he pedalled to the summit, he decided that he could rebrand failures as successes.

And when he crested the snow-capped mountain, he pulled out his BlackBerry and e-mailed his wife to mark his victory.

Doctoroff was addicted to his BlackBerry. Jetting around the globe to build support for his Olympic bid, he'd reach for the device as soon as he landed. Being out of range put him on edge, and he'd get cranky. His grounds for whether a country was "first world" was based on whether his device could get a signal.

The BlackBerry had shaped the life of another, equally calculating executive named Jim Balsillie. Balsillie may not have personally invented the BlackBerry, but he became a billionaire leading the company that sold it to the world. By the time he and Doctoroff found themselves at odds in Toronto in 2017, they were both established dealmakers accustomed to bending powerful people to their wills, but long past the highest-profile years of their careers. The BlackBerry was by then a relic of a bygone era, replaced by consumer smartphones with shiny touchscreens and utility broad enough to make them the Swiss Army knives of the digital era. The downfall of Doctoroff's favourite toy and its high-profile salesman would come to reshape Canada's relationship with tech companies—including Sidewalk.

Jim Balsillie both distrusted authority and craved attaining it. Growing up northeast of Toronto in Peterborough, Ontario, he hoovered up Peter C. Newman's 1975 book *The Canadian Establishment* to study the journeys of his country's executive class. He quickly wound up emulating those stories, getting degrees from the University of Toronto's Trinity College and Harvard before jumping into an executive position at an electronics equipment company in Ontario's Waterloo region. There, Balsillie became a tough negotiator, at one point selling the rights to a new gas compressor process for $2 million to a company owned by T. Boone Pickens. His negotiation skills came alongside a reputation of being difficult to work with, which preceded Balsillie when his company was acquired by a Dutch business. He was not acquired with it.

Like Dan Doctoroff, Balsillie didn't always leave a great impression on co-workers and business partners. And like Doctoroff, he didn't have much trouble taking his talents elsewhere. Balsillie's now-former boss suggested he invest his severance in a business—maybe one of their suppliers, called Research in Motion. There, Balsillie met Mike Lazaridis, a born tinkerer who had tinkered his business to the brink of death. The company had gotten access to a network that could send wireless messages using data, but struggled to get the tech to work, let alone give it a real business model. Balsillie agreed to come on board as co-chief executive, shoring up the business side of the company, which was nicknamed RIM.

Just as they became encouraged that they could make their own revolutionary communications device, Balsillie and Lazaridis made an enemy: Illinois company USRobotics. Itself a modem maker, USRobotics had bailed on a $16 million wireless modem order that RIM had gone into debt to fulfill. USRobotics insisted the modems were defective. Balsillie thought the company was trying to kill RIM's cash flow and fought back, starting a long legal conflict that eventually forced USRobotics to pay RIM a couple million bucks. The experience forged Balsillie's skepticism of American corporate giants. "Jim believed everyone was out to kill us and he couldn't trust anyone," Lazaridis told the authors of the RIM history *Losing the Signal*.

Lazaridis wanted to uproot e-mail. Mobile companies were focused mostly on pagers, and phones that tried to do more than make calls were clunky or expensive, or had network trouble. He thought RIM could do much better. As his team raced to get the technology together, Lazaridis wrote a late-night screed that would change how the world communicated. Fuelled by the sweet vibes of a Joe Satriani record, he called for a device with a long battery life, thumbable keyboard and VCR-inspired track wheel that would "maximize adoption by minimizing complexity."

The next few years were RIM's most inspired. After its network provider saw Lazaridis's post-manifesto prototypes, RIM got its first $50 million device order; just weeks later, Balsillie negotiated a second deal with a competitor, frustrating RIM's earlier partner but backing them into a corner nonetheless. His ruthlessness crystallized into a business strategy. And it worked: RIM's revenue began doubling annually. Technology can frustrate busy people, but RIM made its technology invisible. Wall Street, then Main Street, began filling up with tilted necks. And after bringing in professional branding help, RIM finally gave its product a name that stuck: BlackBerry.

Canadians didn't have a spectacular track record in building or supporting technology companies. They were conservative and loved their blue chips, anchored by six big banks with little differentiation beyond signature colour schemes and abbreviated names forced into acronyms. (BMO? "Bee-mo.") Striking out in Canada, RIM had to list on New York–based Nasdaq in the late nineties to shore up investor interest.

By that point, some of Canada's blue chips—Bee-mo included—had gotten into the tech game, too, plowing money into nine-figure venture funds and incubators, hoping that their fortunes would rise with the tidal wave of personal computing. The chair of one incubator, called EcomPark, boasted in the *Globe and Mail* in February 2000 about funding five companies, insisting that "there is absolutely no shortage of good Internet ideas."

Yet just as in the United States, the business models behind the would-be stars of this era were flimsy. There were Amazon rip-offs, domain-name hoarders, a publicly traded astrology advice website. The dot-com bust killed many of them. Consider the fate of North Vancouver's ClipClop.com Enterprises Inc., once valued at more than $32 million for a business model that depended on horse-related companies paying to list services in an "online mall." When management filed their last report with regulators in mid-2002, the company admitted it had lost hundreds of thousands of dollars over its last three months, as well as a key source of side income: running the Ricky Martin Fan Club's official website.

The most spectacular crash of all was Nortel. The Northern Electric and Manufacturing Company was spun out of Alexander Graham Bell's namesake Canadian business in 1895 to manufacture telephone equipment. The same internet frenzy that had whipped up the dot-com boom and bust had also inflated the value of telecom equipment manufacturers who'd been building out the physical infrastructure behind the information highway. By May 1999, three-quarters of American internet traffic was running through Nortel gear. At its height, Nortel was Canada's most valuable company by a factor of ten.

But by mid-2000, it turned out the market had too much equipment, and the Nasdaq's telecom index fell 62 percent. Executives had seemed not to notice the world changing around them. Worse, those executives had given themselves compensation packages worth hundreds of millions of dollars in recent years. Nortel went into a free fall, unleashing what the author James Bagnall called "one of the deepest and fastest downsizings in Canadian corporate history." It shed nearly sixty thousand people in a process so rushed that the paperwork for nearly a billion dollars in liabilities went missing. An investigation followed that emptied out Nortel's top ranks, as competitors ate so deeply into its market share that it later filed for bankruptcy protection.

As the new century progressed, regular Canadians returned to their blue chips and resource stocks. The tech companies who survived, like Waterloo's OpenText and Descartes, didn't build the next great device; they mostly created software that made running businesses easier, pulling in steady, reliable revenue and adding value through acquisitions to make shareholders happy. The dot-com bust left the country fearful of anything that might be a fad.

Except for RIM, which bounced back from its 75 percent share price collapse from the dot-com bust within months.

Within a few years, RIM was shipping millions of BlackBerrys. Jim Balsillie revelled in the newfound attention, befriending fellow billionaires like George Soros. His reputation for strong-arming people grew over the course of the 2000s, too. He tried three times, unsuccessfully, to buy a National Hockey League franchise to locate near Waterloo. The league's board of governors admitted in a court filing that they didn't like his character and didn't want to "be associated with him." The league that let Dan Doctoroff own a piece of the New York Islanders wouldn't let Balsillie in.

His battles grew larger and more complicated. A Virginia-based company called NTP launched an infringement lawsuit against RIM after seeing Balsillie quoted in the *Wall Street Journal* about RIM's wireless patents. A judge first awarded royalty payments to NTP that could have cost RIM more than $250 million a year, but Balsillie kept fighting, and RIM redesigned its U.S. services to circumvent its patent problems, settling with a $612.5 million payment to NTP. Fending off the NTP claim unleashed an obsession within Balsillie, and he began pushing for patent reforms in both the United States and Canada. He'd learned the hard way that so much of the benefit of technological innovation hinges on the value of intellectual property.

Lazaridis's relentless focus on e-mail set RIM up for a much more existential battle in the second half of the decade. RIM gave customers a device that delivered convenience; in 2007, Steve Jobs revealed a device that delivered the entire internet—the iPhone. At the same time, Google sped up its own efforts to build an internet-driven phone with help from the mobile-tech start-up Android, which it had bought two years earlier.

Lazaridis and Balsillie didn't at first perceive these new devices as a threat: they had faith in their keyboards, their battery life and their customer base. But Apple and Google changed the market. Their new class of smartphones became people's personal lifelines, and consumers followed fun, useful apps away from their old BlackBerrys. In one remarkable illustration of this sudden shift at the decade's turn,

RIM's devices went from making up 95 percent of Verizon's smartphone sales to 5 percent in just two years.

In public, RIM executives projected confidence, but Balsillie and Lazaridis had quietly fallen out over another corporate conflict: a years-long battle with Ontario's securities regulator over backdated stock options, which had forced the two CEOs and another executive to pay one of the biggest-ever sanctions of its kind in Canada—C$83 million. The pair hadn't been allowed to discuss the case between themselves, and frustration coloured their relationship. The divide only widened as they tried to figure out RIM's future while Apple and Google clawed away the market that RIM had helped create.

By 2011, in some of their last-ever conversations, Balsillie and Lazaridis had decided to quit as executives, together. In early 2012, they passed the torch to one of Lazaridis's deputies. Months later, Balsillie became furious that the new CEO wasn't prioritizing the messaging system Balsillie felt was RIM's future. He stepped down from the board and sold his shares.

RIM reinvented itself as a much smaller, data-security-focused company in the 2010s, finally changing its corporate name to BlackBerry just as its devices began to disappear from people's pockets. At the same time, Canada began warming up to homegrown technology companies again. In BlackBerry's shadow, a start-up culture thrived in and around Waterloo. Some of those start-ups had old-school flame-outs. Kik Interactive, a messaging company, and North, which made glasses that showed e-mail and calendar reminders in a lens, both flirted with valuations near a billion dollars on the private market, then collapsed. Across the country, a Vancouver social-media managing company called Hootsuite tried to sell itself for nearly a billion dollars as it struggled to adapt to the changing internet as quickly as its competitors, but couldn't find a buyer.

The Toronto Stock Exchange became home to a growing number of survivors, however—none bigger than Shopify. Shopify was started by a German snowboarder named Tobi Lütke, who fell in love with a woman from Canada's capital city of Ottawa. What began as a snowboard shop developed into a service to help merchants launch, finance and promote their own online stores. For a while in the early 2020s, Shopify was Canada's most valuable company.

After Justin Trudeau became prime minister in 2015, he and his top deputies often spouted talking points about Canada becoming a home to as many more Shopifys as the country could create. Trudeau appeared keen to make Canada's economy less dependent on natural resources, particularly as the environmental consequences of pulling oil out of Albertan dirt grew more obvious and controversial. But his government's approach to technology was not particularly nationalistic. Tech grew faster than any other sector in Canada during Trudeau's first four-year term. But many Canadian tech entrepreneurs were privately frustrated with his government's regular photo ops with Amazon and Google executives—especially when those companies were expanding their operations across the smaller companies' turf, with Ottawa's encouragement. Few of those entrepreneurs would speak out against the prime minister on their own, but just before he and the Liberals swept into power, they found a voice to speak on their behalf.

During the Toronto International Film Festival in September 2015, John Ruffolo gathered the CEOs of all of his pension fund's start-up companies at the Drake Hotel on Queen Street West, just far enough from the festival's core to avoid the glitz, glamour and traffic snarls. He brought Jim Balsillie with him.

Ruffolo and Balsillie had known each other for nearly two decades. They'd been working-class kids who rose to riches in separate segments of Canada's tech economy. As Balsillie was building RIM into a global powerhouse, Ruffolo began running Deloitte's Canadian tech practice, becoming the gatekeeper of its rankings of fastest-growing companies, which RIM regularly topped.

Balsillie hadn't jumped to another tech company when he left RIM. For a few years, he'd decompressed, relishing the luxuries of a rich man's retirement, shuttling between his home in an old mill and his waterfront cottage, traipsing elsewhere by private jet, and investing in an effort to recover a pair of century-and-a-half-old shipwrecks in Canada's Arctic. He eventually started seeking newer, quieter ways to exert influence. He co-founded a group to bring together Canada's most influential chief information officers and lobby for industry standards, and he tried to prod policy-makers into viewing innovation with urgency. Years after RIM's $612.5 million battle with NTP over patents, he still obsessed over protecting intellectual property. Trying to educate governments about how technology companies could influence economies—for good or bad—soon became a fixation.

As Ruffolo's start-up execs sipped cocktails and beers at the Drake Hotel, Balsillie warned them of the consequences of Canada's lack of protectionist technology policy. This absence, in his view, could put their companies at a disadvantage against the increasingly hegemonic influence of American tech giants, who were shaping markets for new ideas, then cornering them first, which created barriers to entry for start-ups. The country's tech sector, Balsillie continued, needed to get in front of governments to explain the policies Canada needed in order to remain competitive. "At first I didn't know what the fuck he was talking about," Ruffolo says. "But he was always ten steps ahead. That day, he was magical. He just rattled all of us."

As they moved to the rooftop patio to mingle, Ruffolo told Balsillie he wanted to fix all these problems, but he didn't know how. Any chance Balsillie would be interested in helping lead an organization that could make it happen? "Not really" was Balsillie's initial response, until he told his partner, Neve Peric, about the suggestion. "Jim," she told him, "you can't rile people up and not follow through."

So Balsillie and Ruffolo launched an advocacy group called the Council of Canadian Innovators. It began regularly pointing out that Justin Trudeau's approach to Big Tech hurt the companies the prime minister claimed he wanted to help. Representing dozens of fast-growing start-ups—the candidates to become Trudeau's next Shopify—the council was free to say what its individual members were sometimes afraid to say on their own. Though some its founders' peers on Bay Street quietly felt the council fostered a culture of government dependence rather than of genuine innovation, dozens of companies suddenly found themselves with a voice.

Balsillie still used a BlackBerry, and he kept flinging missives from its physical keyboard at politicians and senior bureaucrats in Ottawa and Toronto. In his second act, he'd come to see himself as a voice of reason on a national scale: he believed intellectual property could raise the fortunes not just of individual companies, but of the whole economy. He insisted Canadians should be furiously patenting their ideas and racing to get them in front of customers. He was living proof of what happens when you don't come prepared to outwit American giants.

Canada had always struggled to understand tech companies. When Sidewalk Labs and its BlackBerry-loving CEO came to town, the man who had sold BlackBerry to the world decided it was time to show his country how quickly power can fall from your hands.

By the time Ruffolo had given up on his bid for Quayside in 2017, he was regularly talking with Balsillie about the project. Sidewalk's involvement, they agreed, was too mysterious for anyone's good. The pair had questions about the company's business model, about its intentions with the little L-shaped site on Lake Ontario, and about how the company found its way to Toronto. Ruffolo wondered if Sidewalk might have had an influence over the bidding process. Justin Trudeau's comment at the project announcement that he and Eric Schmidt had been talking about it for years only fuelled this concern. Not knowing that a tamer version of this fear was true—there was no interference, but Schmidt had stoked Trudeau's excitement—Ruffolo and Balsillie began to wonder how much power Sidewalk was already exerting in Canadian governments, thanks to those governments' uncritical embrace of companies like Google.

Balsillie had already been pushing government officials that summer to come up with a strategy for how the country handled data collection. It was a matter he felt carried more urgency with each passing year. In the early years of the twenty-first century, collecting data was considered a part of doing business. The rise of smartphones and sensors in the world around us in the 2010s only accelerated the rate of data-harvesting. At first, few questioned the consequences of letting companies operate like this—until later in the decade, when a critical mass of people began to wonder why tech companies got all the financial benefit from data they were generating.

Data reveals patterns about things, which can show how those things can improve. Those ideas can become valuable intellectual property, such as patents, to sell elsewhere. That boring formula has built billion-dollar companies. Google, then Alphabet, grew into one of the world's biggest businesses by squeezing money from ideas—ideas like the algorithms that underpinned Google Search and the easy functionality of Gmail. Intellectual property helped Google make its billions. The idea of a neighbourhood built by Alphabet in Toronto only made Balsillie and Ruffolo more nervous about where its next few billions would come from.

During the 2010s, Google and Alphabet had also invested heavily in people and companies who excelled at certain branches of artificial intelligence (AI) that could make Google even better at processing data. Two of its highest-profile acquisitions were a London AI company called DeepMind Technologies and a three-person AI shop from Toronto called DNNresearch, led by University of Toronto professor Geoffrey Hinton, who'd been called the godfather of the AI field of deep learning. More AI power meant more processing power, which meant more power to find patterns or solutions to problems, which meant more ideas to patent.

Even though Sidewalk promised to minimize the collection of data that could identify people—and though it promised to take extra steps to remove identifiable information from data that it *did* collect through various sensors—the information the company gathered about the way people lived their lives would still be extraordinarily valuable to its parent company. Alphabet's routine investments in AI made this kind of data processing easier with each passing year.

In offering to host a test bed for Sidewalk technologies, Waterfront expected a share in whatever intellectual property would come out of the deal. If Canadians were generating money-making data as they moved around their city, Waterfront wanted its three government shareholders to get a cut of the eventual cash. But because the contract for Quayside wasn't public, Balsillie worried that citizens would be handing over valuable data to Alphabet that they didn't even realize they were generating.

Google often said that its chief motivation was to change the world for the better, playing with ideas until they were perfected. But once those ideas reached consumers and were given a business model, the company tended to outwit incumbents and whole industries, be they BlackBerry smartphones, car navigation systems or the entire advertising sector. How would economies be reshaped, Balsillie and Ruffolo wondered, if Alphabet scaled this domination strategy up to cities?

Sitting at a makeshift desk in an attic on the west side of Toronto's downtown, someone else was asking a different version of the same question: What would happen to democracies?

CHAPTER 9

That Don't Impress Me Much

BIANCA WYLIE WAS trying to hold her tongue. Across the table from her, an architecture critic was insisting that Waterfront Toronto's decade and a half of development experience would prepare it for dealing with Alphabet. Too much of the conversation about the Sidewalk Labs project centred around fears of Quayside being a "Trojan horse" that would allow Google to "take over the city," said the critic, Christopher Hume—and such fears, he added, were exaggerated. Waterfront would guard the public interest "because they are urbanists before they're technologists," he said. "This is something we should be excited about."

They were in a blue-and-purple-lit TV studio in midtown Toronto in early November 2017, taping a segment for *The Agenda*, one of Canada's most wonkish current-affairs shows, for a public broadcaster called TVO. The host turned to Wylie, who seemed to be trying to appear collected but looked anxious to speak, hands folded in front of her. "Are you as confident?" the host asked.

"No," Wylie said before he finished the last syllable, her expression turning to a smirk.

"Tell me why."

"We're talking about how we can do this with Google," Wylie said. "And personally I would prefer we were talking about different ways to do this, period." She looked a little out of place on the panel.

Dressed in a black sweater, she came across as more of a Torontonian-on-the-street than the blazer-clad talking heads across the table. She'd never even been on TV before. But she thought more than most people about how power gets exerted in cities.

Even if Waterfront served as a watchdog overseeing Alphabet's development of the lakeshore property, Wylie said, Toronto was missing a bigger chance to talk about how the city interacted with technology companies. "Big Tech, particularly—they are way ahead of the government in terms of understanding technology, working on technology, coming up with solutions and, most importantly, figuring out ways to make money using technology."

Wylie had been invited to the TV panel because she'd raised questions about the Sidewalk project's business model and public engagement for a blog called *Torontoist*. She was the segment's biggest Sidewalk skeptic, a position that was informed by a career spent raising questions whenever technology, democracy and advocacy intersected. She'd spent a long time thinking about how cities procure from technology companies. It was dry stuff, but something she saw as a battle for the future of democratic city life. She'd long had a hunch that technologists would make a play for cities by selling them on ideas so new and shiny that no one would mention the need for regulation; it would be a natural extension of Big Tech's hubris. And now one of the biggest companies in the world was trying to do this very thing in her backyard.

When she clicked through Sidewalk's bid for the project, she noticed that the company wanted to create a "cycle of ongoing improvement driven by the feedback of residents and the energy of entrepreneurs, rather than prescribed by planners and designers." That sounded a lot like how tech companies grew their market capitalizations by changing day-to-day life with smartphones: make apps that people can't live without, insist that the business model needs a ton of data about those people, then make it nearly impossible to regulate this kind of data collection—or at least make enough money to never have to worry about financial penalties for skirting regulations.

Plus, she wondered, if Sidewalk was going to collect information about people in one of the fastest-growing cities in North America, how exactly was the city going to benefit? Not only was there no regulation yet governing this kind of data collection, it wasn't even clear what Waterfront Toronto had signed up to build with this partnership.

Wylie had helped to found both the Open Data Institute Toronto and Civic Tech Toronto, volunteer groups that became a big part of conversations on how data about people and public services could be used for the public good. But she gradually took a more critical stance. Governments, she felt, tended to share data they'd collected only when it was either useless or beneficial for the vague purpose of "economic development." Wylie wanted people to learn about the *consequences* of data and technology. Governments used those things to make decisions, but the decisions weren't transparent just because "data" was mentioned. Traffic-monitoring sensors could guide a city to install new traffic lights at unexpectedly busy intersections, but how did that guidance work? Whose advice helped the city interpret the data? And were there other ways to interpret that data? *This* was the kind of thing Wylie wanted the public to learn more about.

The way technology companies and governments interact can have profound consequences on such decisions. Money, strategic meetings and personal connections have long shaped how politicians make decisions, but governments hadn't quite caught up to the speed with which Big Tech's influence was growing. Wylie felt that everyday people—everyday democracy—stood to lose from that power imbalance as tech companies began selling themselves to cities. At a minimum, taxpayers' money was at stake. At worst, citizens' rights and freedoms in the face of ever-increasing data collection could become forever constrained.

Governments have budgets to spend, she reminded the *Agenda* panel, and tech companies have products to sell. But their goals are different, because tech companies want to maximize income to maintain their enormous momentum and keep shareholders happy. "The challenge with that, to me, is—is this the best way of spending money, with a firm like Google?" she asked. "So they're going to be thinking of ways to keep money and services moving within their very large company." Wylie pressed further: Was Toronto ready to be a test bed for this kind of tech company?

Hume tried to slip a word in, but the third panellist quickly spoke up. "Smart cities have led to surveillance," said Ann Cavoukian, who'd spent three terms as Ontario's privacy watchdog. Tiny but outspoken, she'd learned long ago how to needle everyone in a room into coming to her terms on privacy. And she was all-in on Sidewalk Labs: "This is going to set us apart from the way that smart cities have been developed elsewhere."

Cavoukian was on Sidewalk's payroll. The company had contacted her when it was honing its bid for the Quayside project and told her they wanted to embed her vision of privacy into it. This was the kind of thing she liked to hear. She'd created a renowned framework for developing technology called Privacy by Design. It was based on the idea that people's privacy should be protected in new tech from its conception—any later than that, and the tech's capacity for surveillance could be irreversible. Cavoukian's framework had been embraced the world over as a human-rights-centred approach to tech, and Sidewalk had even referred to it in the Yellow Book.

Both Cavoukian and Wylie had built their reputations by taking a critical lens to how corporations change the way people live. But they approached those corporations from different perspectives. Wylie preferred to critique from afar, staving off any influence that might compromise her judgment or even create the perception that her thinking was compromised. Cavoukian, on the other hand, was more willing to work hand in hand with companies, especially when they embraced Privacy by Design.

"We can shine, at least in terms of the privacy and data protection associated with this smart city," Cavoukian told the panel. "We can be a leader in that."

But for Wylie, privacy was a red herring. It was an issue Sidewalk could defend against with public-relations moves like hiring world-renowned experts such as, say, Cavoukian. Wylie's greater concern was the privatization of public services, and what the public lost when shareholder-driven companies took on government responsibilities.

Wylie could hardly hide her frustration throughout the thirty-five-minute debate. When host Steve Paikin pointed out Toronto's obsession with appearing "world class," the architecture critic responded that Sidewalk had done its homework on the city.

Wylie rolled her eyes. "Of course they have," she said, shaking her head repeatedly before being interrupted by Hume.

"My impression is they're urbanists first and technologists second," Hume said, reiterating his earlier talking point. "That's an important consideration."

The camera returned to Wylie, who was nodding smugly. "And that's what I would call PR." The vertical nodding turned back to a horizontal, disappointed head shake. "The whole thing is a PR exercise. Because technology is a tool." She paused dramatically. "It's not a *way*; it's a thing. It's the people who bring technology into spaces and use it, right? If there's one thing I want us to think about more, it's other ways to do this. . . . There are underfunded but super-committed public servants who know how to use technology to deliver public services. Why are we going to even consider—it's like, why constrain them with this kind of stuff? Why don't we give them some chance to help define how this is going to look? . . . It is not innovative to be partnering with, basically, a monopoly. How is this an innovative strategy?"

The TVO segment turned out to be a homing beacon. Sitting a couple of hours away in London, Ontario, a retired data analyst named Paula Henderson-Johnson was watching *The Agenda*. She thought the Sidewalk segment might interest her son—the way Bianca Wylie talked about how government and technology interacted sounded a lot like the way he talked about his job. She called him up and encouraged him to watch the segment.

His name was Patrick Searle, and he was a former provincial Liberal spokesperson who ran communications for the Council of Canadian Innovators, the lobbying group Jim Balsillie and John Ruffolo had set up two years earlier to boost the homegrown tech sector. When he watched the segment, he was impressed by Wylie's ability to describe complex technology issues in ways that anyone could understand. He sent her a message on LinkedIn, and they agreed to meet for coffee.

At a patisserie on Queen Street West, they dug into how they felt about the Sidewalk deal. Searle thought the meeting would take half an hour, but it lasted nearly four times that. Wylie walked Searle through how she felt about the power imbalance between cash-strapped governments and rich tech giants. Searle explained that her concerns dovetailed with what his council was pressing governments about: for local start-ups to have a fair chance at success, they needed a seat at the table when big decisions were made, which would help Canada's economic sovereignty. It was another way to view the relationship between technology and democracy—one that could be interpreted conservatively as nationalistic, or progressively as anti-imperialistic. Although she and Searle weren't waking up in the morning for the same reasons, Wylie thought, they were moving in the same direction.

So Searle introduced Wylie to Balsillie. They talked about many of the same things: she about democracy and the privatization of government services, he about sovereignty. Though some of the early local media coverage of the Sidewalk deal focused on what privacy problems might arise in a community that had as many sensors as Sidewalk wanted to install—collecting data on traffic and pedestrian patterns, energy usage, noise levels and more—both were surprised by how little scrutiny Quayside was receiving on bigger-picture matters. They wondered if that had something to do with the local luminaries attached to the company, such as Ann Cavoukian and Ken Greenberg, whose reputations and relationships could help Sidewalk win over Toronto's close-knit circles of power. Though Wylie and Balsillie didn't know it, this was exactly the type of strategy Sidewalk had considered using in a city-building campaign since at least 2016, when staff wrote in the Yellow Book that the company "may reinforce its messaging" with the help of "top urbanists, technologists, architects, policy makers, and planners praising the project's approach and goals."

If someone was going to build a smart city, Wylie and Balsillie wanted to know who would make the decisions about how that city should be built and operated. For Wylie, this had consequences for democracy. For Balsillie, the consequences were economic. However you looked at it, the ways that data was collected and privacy was protected were a downstream consequence of how those decisions would be made.

Wylie's work wasn't exactly aligned with what the council was doing, but Balsillie had also co-founded a think tank called the Centre for International Governance Innovation (CIGI) years earlier. By the time he met Wylie, the think tank was focusing on how technology, policy and governance intersected. That interested the government-transparency advocate. Balsillie offered her a chance to write for CIGI as a fellow. It was a freelance gig, one that would allow Wylie's writing to reach more people. Though they came from different worlds—the erstwhile business magnate who changed how people communicate and the skeptic from progressive Parkdale—they each wanted to escalate the public's understanding of how technology companies influence governments and economies.

As soon as Balsillie had got wind of the Sidewalk announcement, he'd reached out to Waterfront Toronto and arranged a meeting. He wanted them to know how afraid they should be.

Within days of the announcement, Balsillie headed to Evergreen Brick Works, an event space in a repurposed brick factory at the edge of the Don River. The non-profit that owned the space shared a chair, Helen Burstyn, with Waterfront Toronto. She, Kristina Verner and Will Fleissig joined Balsillie, and he launched into his list of worries.

"What's the rush?" he began.

He pleaded with the Waterfront crew to slow down. Taking on a technology project like this, he said, would be as risky as him performing heart surgery. But he *was* surgical, he felt, in matters of data and intellectual property. So he urged the Waterfront team to be cautious, and to put guardrails around the project to protect Canadian interests. Some of those guardrails, especially concerning what Sidewalk could do with data collected at Quayside, would probably need to be enshrined in law, which he was already pressing the federal government to establish.

Fleissig explained that they were bound by process: that there was a one-year timeline to draft a plan with Sidewalk. Furthermore, it was no more than that—a plan. Sidewalk had been characterizing the Toronto project as closer to a done deal than its contract actually stated, but the Waterfront executives were planning to sort out the fine details over time. They weren't that worried.

Fleissig and Balsillie seemed at odds, so when they said they'd keep in touch, Burstyn agreed to manage any ongoing relationship between the agency and Balsillie.

The two sides left the meeting with very different takeaways. Verner took Balsillie's cautions to heart, deciding to brainstorm ways to scrutinize Sidewalk's ideas more quickly than she'd planned. Balsillie, however, walked away thinking that Waterfront was afraid of losing the chance to work with Alphabet.

The growing suspicion around Waterfront's Sidewalk partnership might have been quelled if the two organizations had made their agreement public. But they hadn't. Not even city councillors could see it. In January 2018, three months after the project was announced, Bianca Wylie went to City Hall to fight for the contract's release.

Nearly nine hours into a marathon meeting in a fluorescent-lit meeting room deep in City Hall, Mayor John Tory invited her to have a seat across from him. The city council's executive committee was hearing from citizens about a newly public report from a deputy city manager who'd been granted access to the confidential agreement. Though the report acknowledged that the finer points of the Quayside partnership would be worked out in subsequent agreements, it also noted how complex some of those negotiations might be, including around data collection and privacy, and outlined the city's concerns about how much power Sidewalk seemed to want over land and technology.

Wylie was the first member of the public to speak about the report. She looked anxious to get started. She'd recently co-founded an advocacy group called Tech Reset Canada that wanted to shed light on the importance of maintaining the public interest in technology projects. She scrolled through notes on her MacBook and told councillors she and Tech Reset had two requests: one, that data from Sidewalk's project be publicly owned; and two, that the Quayside contract be made public. Sidewalk had promised to spend $50 million on drafting its ideas, including on public consultations—nearly a million dollars a week ahead of a one-year deadline to publish a master plan—and the public deserved to know how that money would be spent.

"This is not an urban-planning project. It's a technology project," Wylie said to the room. "As far as a technology project goes, the biggest issue with this is not privacy. It's governance." Control, in other words. Anything related to data collection in cities, she said, should be a civic asset. But laws and regulations hadn't caught up to the internet age, and she worried about the public losing control of this asset to a company whose shareholders expected it to maximize profits. The city report raised concerns that Sidewalk's proposed tech pilots were too exclusive for Toronto's procurement policies, which were supposed to give a fair deal to taxpayers when the city spent money. And it highlighted the possibility that in order to deploy some of Sidewalk's new technologies at a scale that made financial sense, the company would need to expand its project beyond the 12 acres in the deal.

All of this raised red flags for Wylie. "This staff report shows what happens in a governance vacuum," she said. "This consultation is not a consultation. It's product development." She looked up from her notes and directly at the mayor and councillors. "We need to understand the terms of this deal, and the way this money is being spent. We must protect our digital infrastructure and data, and the immense value of our public assets."

Larry Page's first creation, Google, had become one of the world's biggest companies by collecting as much data on users as it could in order to figure out how to improve its products—and its revenues. The massive growth of internet-connected sensors presented the same opportunity in the physical world. The word *sensor* or *sensors* appeared seventy-one times in the portions of Sidewalk's bid for Quayside that it had made public. "This city must protect the immense value of our civic data," Wylie said. "Fifty million dollars is a rounding error in the conversation about the value of our data. We cannot undervalue this asset. Innovation is born of openness. As owners of data, we retain the power to open our data under our terms. At that point, true innovation can happen under our terms." She then translated this into language that the councillors were used to: "That's when broad and wide economic development can take place."

To make that happen, the city needed to design something else councillors were familiar with: policy.

Only one councillor spoke up in response. "There's a risk, then, if this data, these products, are monetized, that we don't have control over those things," said Janet Davis, who represented a neighbourhood just east of Quayside.

"Correct," Wylie said. "We need to be clear about something here. If we, the city, and our residents own this data, we can make a decision to open it. But that's *our* decision to make. It's not a decision for a vendor or an investor to say that *they* will open data. We have to think about this as critical public infrastructure. Our data infrastructure is like our physical infrastructure."

The councillor seemed to understand Wylie's chief concern. "If you have these fixed boundaries," she said, drawing a box in the air with her fingers, "in which you're going to pilot some new technologies, how are you going to contain those? And then run the risk that you've opened the door and there will be a desire to extend it?"

"As far as I understand, that's already happened."

"Oh."

"On the third page of this staff report, there's talk of using pilot projects in the city of Toronto proper," Wylie said. "This is why we need to see the contract."

Sitting across the room at this never-ending meeting was Denzil Minnan-Wong, the right-wing deputy mayor who sat on Waterfront Toronto's board. He actually *had* seen the contract, and he was furious about it.

Minnan-Wong had been fed up with Waterfront for years. He saw the agency as a money-waster that didn't add value to the city. In 2014, he showed up in the *National Post* grimacing at a C$11,565 pink umbrella that Waterfront had installed at Sugar Beach, just west of Quayside. Waterfront was trying to improve the image of the once-industrial dead zone that it had turned into a public park, but Minnan-Wong felt that the custom-made permanent fibreglass umbrellas were an example of reckless spending. Combined with two candy-striped granite rocks the agency had hauled in from the Laurentians, the new amenities had cost nearly C$1 million. He thought Waterfront's life-extending flood-protection plan was a sham, too—another example of money that could be spent elsewhere.

Why was he even on the Waterfront board, then? "I thought they needed a lot more scrutiny," he says. And at the marathon executive committee meeting in January, he did everything he could to point out that the Sidewalk deal also needed more scrutiny.

Minnan-Wong was caught in the middle of a project he didn't like. He'd actually voted in favour of the Sidewalk deal, but only because he was the mayor's representative on the board and voting yes was what the mayor wanted. Now, as a board member obliged to keep the details confidential, he still couldn't say what he felt: that the agreement left too many openings for Sidewalk to have sway over more than just Quayside. He believed its language meant that more of the eastern waterfront might be up for grabs than councillors realized, including nearly 500 acres of city-owned land. This was some of the most valuable underdeveloped land on the continent, and he worried that an agency he didn't respect was at risk of handing it away. He was constrained from saying much, but hoped to convey enough urgency that the debate would be moved to the full city council, where he felt his colleagues would ask hard questions and steer debate and decisions into more serious territory.

"I'll just ask a couple of questions," he told the committee that evening, looking tired in a dark-grey suit and striped yellow tie. "Do you think this deal is big enough that you need to know what the terms are, when we're dealing with Quayside plus 500 acres of land?"

Another councillor, David Shiner, piped in. "Councillor Minnan-Wong, aren't you on Waterfront Toronto['s board]? So aren't you able to get your questions answered there?"

"I know enough about the agreement"—Minnan-Wong paused, his eyes darting—"that I think you would like to know more about the agreement."

CHAPTER 10

In Undertow

EVEN BEFORE THE deputy mayor issued his cryptic warning in January 2018, city staff were scrambling to make sure Sidewalk wasn't asking for more than Waterfront Toronto could give it.

"We were very clear with WT and Sidewalk Labs Thursday that we did not want the proposal document to be released this week and that it needs a thorough review by City staff," wrote David Stonehouse, who ran the municipal department that oversaw the city's relationship with Waterfront Toronto, in an e-mail to municipal staff in October 2017. It was just five days before Justin Trudeau had joined Dan Doctoroff and Eric Schmidt to announce the project, and Stonehouse and his colleagues were already trying to do damage control. The pieces of Sidewalk's proposal that the company planned to make public included at least one graphic of the company's plans extending eastward into the Port Lands, which would have upended Toronto's official plans for the area.

"Just to reiterate, it is 'not now, not ever' because of how different it is from the City's vision of what is to happen in the Port Lands," economic development manager Michael Williams wrote in the thread.

Sidewalk included maps in the bid to make it clear that its intentions for Toronto were far bigger than Quayside. The company needed to think bigger, given how much money it would be required to sink into all the ideas it was promising. It didn't need all of the Port Lands, but it did want more space to make its investment in Toronto worthwhile. Its staff had the impression that they could get *some* more land to work with, and they began a campaign to make that clear to the government officials.

City staff were proud of their plan for the Port Lands, however, and didn't want any meddling. So when Sidewalkers began lobbying City Hall within weeks of announcing they were coming to Toronto, they were frequently rebuffed. It turned out that the eagerness they had encountered at Waterfront Toronto did not extend to everyone in the city's government. In fact, officials were surprised Sidewalk was asking for more land at all: Sidewalk's contract was with Waterfront, and Waterfront's request for proposals made the 12-acre boundary of its partnership clear.

In the city's view, Sidewalk needed to succeed at Quayside before anything else would be considered. Some officials found their conversations with Sidewalk lobbyists almost comical—one of the reasons it had taken so long to develop Toronto's waterfront was that this was a city that followed due process to a fault, even if some of those processes were faulty. The city was also less than two decades removed from one of its most controversial spending scandals, and the scandal was over a tech contract for computer equipment that cost tens of millions of dollars more than it was supposed to. The situation forced Toronto to set up a lobbyist registry, update its procurement policy and hire an integrity commissioner. So municipal staff and councillors were especially sensitive when it came to technology.

Waterfront Toronto caught some flak for allowing this misunderstanding thanks to the vagueness of its RFP. The agency had "overplayed their hand a bit," an unnamed councillor told the urban affairs magazine *Spacing* in December 2017. But it was ultimately Sidewalk, not Waterfront, that was taking the RFP's ambiguity and running with it. Bureaucrats and politicians marvelled at how equivocal the Sidewalkers could be—how they would discuss a contract that guaranteed them the chance to plan only 12 acres in such a way that presumed access to hundreds of acres more. One city official described

Sidewalk's approach as like "answering a question that wasn't on the exam." Others felt that Sidewalk's brashness was at odds with its track record: as much as the Bloomberg administration had built in New York, Sidewalk itself hadn't undertaken a project like this before. "They waltzed in and acted like every door should be open to them," says another person who dealt with the company. "They never wanted to hear the word *no*."

The land questions came up in Sidewalk's occasional meetings with Mayor John Tory, as well. His staff repeatedly warned them of due process. Offering a subsidiary of one of the richest companies in the world hundreds of acres of extremely valuable real estate wouldn't just break procurement rules. It would be political suicide.

It was Sidewalk's political animals, often Dan Doctoroff, Josh Sirefman and Micah Lasher, who dealt with Tory. They were expert navigators of politics and bureaucracies who felt they spoke the same language as the mayor. Tory was a former media executive and chair of a professional football league, and later led Ontario's right-wing Progressive Conservative party. He saw Sidewalk's tactics as, well, tactics. Sidewalk had a shareholder—Alphabet, controlled by Larry Page and Sergey Brin—whose needs its staff had to fight for. "The private-sector Dan Doctoroff shone through, I think, oftentimes more than the public-sector one," Tory says. The mayor respected the company's need for scale and return on their investment, but he had more than three million shareholders of his own to fight for. He was a stickler for following rules, and the rules said Sidewalk could plan for 12 acres. But Sidewalk persisted in courting his influence, so much so that the mayor's staff had to set up an "alarm system" for when Doctoroff or other Sidewalkers contacted the mayor to speak or meet alone. His senior staff wanted to avoid any appearance that Tory was being swayed by the company, and they demanded to be told when Sidewalkers reached out to him.

Doctoroff and his team were having greater luck winning over the business community. The former deputy mayor of New York had built his reputation by charming the private sector and pulling it into city-building projects. He was used to making the rounds at executive offices and working hotel ballrooms to drum up support. Sidewalk began making friends in the skyscrapers just south of City Hall, in the Bay Street financial district. Some of those friends eagerly bolstered an illusion of inevitability around Sidewalk's grander plans.

The Toronto Region Board of Trade was the most visible business lobby group in Canada's biggest, most businessy city. By November 2017, the board was embracing the idea of "smart cities," sending a delegation to the Smart City Expo World Congress in Barcelona. Finding the organization already open to grand city-building ideas, Sidewalk quickly turned it into a supporter. The board's chief executive officer, Jan De Silva, flew to New York with Tory that month for a series of meetings with companies that had operations in Toronto, hoping to encourage more local investment. Dan Doctoroff and his team invited them to tour Hudson Yards, where he walked them through what Sidewalk wanted to achieve. Upon De Silva's return, the Board of Trade didn't let the details of the actual Quayside agreement get in the way of its excitement. "The solutions advanced in the Quayside project will be brought to scale in the multi-billion-dollar development of the Port Lands, an underdeveloped area of more than 325 hectares along Toronto's eastern waterfront," read the board's recap of the New York trip. Asked about this years later, De Silva pointed out the enormous opportunity Toronto would have had if Sidewalk had built beyond Quayside, including a massive Google headquarters. "It was exciting to think that Google was that invested in Toronto's talent and ecosystem," she says.

Sidewalk made efforts to win over people who lived near Quayside, too, including Suzanne Kavanagh, an active community member who at one point ran the nearby St. Lawrence Neighbourhood Association. Many were invited to working groups to discuss Sidewalk's ideas, and some forged relationships with company executives. Doctoroff and Sirefman, for instance, would routinely bounce ideas off Kavanagh.

At one point, Kavanagh managed to snag a midday meeting with the Sidewalk CEO. Doctoroff seemed hungry, and she offered to treat him to lunch. It was her neighbourhood, and she appreciated his time. As she paid for the food, she noticed Doctoroff trying to stifle a yawn. "I said, 'Oh, you must be exhausted. I'm sure they book you with back-to-back meetings,'" Kavanagh recalls. "And he turned to me and said, 'Oh, but I've been looking forward to this one all day.' Well, I just burst out laughing. And I said, 'Oh my God, you *are* in sales.'"

Though Sidewalk had left Waterfront out of some of this closed-door campaigning, much of the sales job involved getting in front of the public, and there Doctoroff and Will Fleissig put on a united front. They issued the project announcement press release together, co-wrote an op-ed in the *Toronto Star* and soon hit the fireside-chat circuit to promote the potential partnership. In a two-day span in early November, they wound up engaging with very different ends of the socioeconomic spectrum.

At Evergreen Brick Works, just days after Jim Balsillie's showdown there with Waterfront, Fleissig and Doctoroff spoke on the same stage as Justin Trudeau, Eric Schmidt and DeepMind CEO Demis Hassabis in front of more than 650 tech-sector workers at a Google event. Schmidt wore Canadian-flag socks and thanked Canada for supporting the artificial-intelligence work of pioneers like Geoffrey Hinton, whose talent and intellectual property Google had bought the rights to years earlier, infuriating economic nationalists like Jim Balsillie. Fleissig and Doctoroff spoke about developing a new way to build neighbourhoods that could be replicated elsewhere without appearing too cookie-cutter or sterile. "It is not about doing cool things," Doctoroff said. "What it's about is staying focused on quality of life. And I think if you start with that as your north star, you really risk much less of that sort of [cookie-cutter] thing happening."

The two executives had faced a more existential question the evening before. At a packed community meeting in a performing arts space downtown, they were greeted by fifteen or so activists from a local branch of the Association of Community Organizations for Reform Now (ACORN). The group was frustrated at not being consulted on Quayside and worried about who, exactly, a Silicon Valley–affiliated company would be making affordable housing for.

"What is affordable for rich people?" one of the activists, Alejandra Ruiz Vargas, asked from the crowd.

Doctoroff praised her for pushing him on the issue and explained that he'd made significant efforts to build affordable housing in New York. (Weeks earlier, however, he'd published a memoir about his time as a deputy mayor there in which he bragged about lowballing the city's affordable housing commitments in rezoning negotiations with communities and city council.) The crowd was getting riled up. "Welcome to Toronto, the city of many strategies and very little money," one woman told Doctoroff. The meeting had an anxious energy: as much as Sidewalk and Waterfront were getting pushback, the room was filled with people who wanted to get involved. Torontonians had tripped over themselves for tickets to the event, and there was even a line for last-minute rush tickets. The city had held meetings on technology and data issues before, but none had ever drawn this kind of excitement.

People far from Toronto were also watching this Google-funded city project. One of the world's foremost critics of the social implications of technology, Evgeny Morozov, wrote that Alphabet wanted to be "the default platform" for municipal services. He argued that the Quayside project was the beginning of another data play for the massive company at the expense of public control. "Its bet is to furnish cool digital services to establish complete monopoly over data extractivism within a city," Morozov wrote, marking a corporate effort that might "be an attempt to privatise municipal services." Sidewalk had more of a voice in an *Atlantic* feature in which Lasher insisted that "we're not going to gather up all Torontonians' data and sell it, we're not building Sensorville." But the story's author, then-Toronto-based Mols Sauter, still insisted that if Sidewalk built out what it promised in its RFP, "the area would become some of the most heavily surveilled real estate on the planet."

The work to protect the Port Lands from flooding had already begun, marking the biggest undertaking in Waterfront Toronto's history, but the 12 acres it was plotting out with Sidewalk was consuming far more of the public's imagination—and, Waterfront staff began to realize, far more energy than they'd expected. Every time someone said in public that they were worried about what Sidewalk might do with Quayside, Fleissig tried to remind them that all the company had won was the chance to draw up a plan. "We have the right to walk away from this if we don't like it," he told the crowd at the performing arts centre.

The spotlight on the project, and the mounting pushback, changed the cadence of Waterfront's work. They'd hoped to sign a more detailed agreement with Sidewalk Labs within six weeks, but that timeline quickly melted away. Public consultations were drawing larger audiences, the board and staff needed to be briefed more often, and media were knocking down the door. So as Sidewalk was working boardrooms and backrooms to drum up support for a plan much bigger than Waterfront could actually permit, on land the city didn't want to hand over, Waterfront's to-do list kept getting longer.

What had begun as a legacy project for Will Fleissig—to build a sustainable, forward-thinking neighbourhood that future generations of Torontonians could be proud of—was by early 2018 growing into something that was increasingly more complex and contentious than his agency had expected. Waterfront was gradually losing control. It turned out that control was hard to come by when dealing with platform companies.

When the *Guardian* broke the news in late 2015 that Texas senator Ted Cruz's presidential campaign had paid at least $750,000 to a company that "harvested data on millions of unwitting Facebook users," it provoked discussion about how social media companies collect information about people—but not a particularly lasting debate.

The investigation found that a political consulting firm with ties to Cambridge University researchers had built highly detailed models of Facebook users' personalities based on what they'd "liked" and shared about themselves on the website. The data collection had begun with personality questionnaires, but some of the participants complained that the process handed over access to details from their Facebook profiles and information from the profiles of their friends. Consent was not paramount. Yet the researchers were combining their findings with other data sets, such as gun-ownership databases, and selling the personality models to political campaigns for hundreds of thousands of dollars to help them target different voters on different issues. People's social media histories had been hijacked to influence voting choices, potentially changing the shape of politics and policy.

The consulting firm was named Cambridge Analytica. It was later revealed to have ties to a variety of personality quizzes circulating on Facebook, and was connected with the U.K. Conservative Party, major Republican donor Robert Mercer and a right-wing news editor named Steve Bannon, who went on to manage Donald Trump's 2016 presidential campaign. In March of 2018, all the pieces came together in the public eye, and the blowback began. An investigation by the *New York Times* and London's *Observer* reported that Cambridge Analytica had collected information from more than fifty million Facebook users. The *Times* alleged it was "one of the largest data leaks in the social network's history" and had helped boost Trump's 2016 election victory. The U.S. Federal Trade Commission later tried to fine Facebook $5 billion.

By the time the full extent of the scandal was made public, Facebook was already under fire for its role in hosting the spread of "fake news" and propaganda in the 2016 election, and for being weaponized as a force against political critics in countries like the Philippines. Already slowly fading, trust in Big Tech began to plummet, and the rebuke from the public was dubbed "the techlash." It went beyond just a grassroots campaign urging people to delete Facebook. Google got caught up in the fray. Especially since buying DoubleClick in 2007, Google had developed sophisticated ways to understand how people clicked around the internet, aiming to boost its advertising business. Both companies took pains to show users how they could control the data they were collecting, but had to fight back against increasing accusations of non-consensual surveillance.

Over the course of nearly two decades, Google, then Facebook, had ushered in a new kind of economy in which the traces people left behind while navigating the internet became a form of currency. Advertisers got richer by using this data to target potential buyers with ever-increasing precision, the platforms themselves got richer selling the ad space, and a whole lot of middlemen got rich, too. It was generally assumed that users benefitted by seeing ads more accurately curated toward their interests—that their role in this new economy was, effectively, to spend more money. As more data was collected with each passing year, everyday people were confronted with both a growing power imbalance and a potential privacy disaster—one that a small but vocal group of advocates had long tried to warn could become a nightmare. European lawmakers had been pushing back for years, developing the progressive, bloc-wide General Data Protection Regulation, which came into effect in May 2018. Still, it took Cambridge Analytica's role in the election of Donald Trump to force most people to take data collection seriously. Suddenly it wasn't just critics like Bianca Wylie and Jim Balsillie warning about the unchecked influence of technology companies. It was your neighbour, your desk-mate, your mother-in-law. The battle for digital privacy became a *cause célèbre*.

Torontonians were no exception. As the techlash unfolded, more people watching the Quayside project began to understand the urgency in the voices of Sidewalk critics. "What was once viewed by most people as a positive, and I think continued to be viewed by many people as a *potential* positive, became questionable," says Mark Wilson, the former Waterfront chair who was also a retired IBM executive. "The context of the world changed dramatically."

The secretive way in which Waterfront Toronto and Sidewalk Labs were working out the details of the partnership wasn't helping. Though Sidewalk's winning pitch for Waterfront offered nearly two hundred pages of ideas for Quayside, both organizations understood that they'd need to spend months or years figuring out what would actually get built on those 12 acres, while promising to keep digital privacy top of mind. But it was easy for the public to look at Waterfront and Sidewalk and see a deeply opaque relationship. After all, the two organizations were describing the project in very different ways: one affirmatively, one tentatively. In the context of the Cambridge Analytica scandal, more people approached Sidewalk with skepticism, if not outright fear. "Because they weren't very specific for such a long time, the vacuum was filled with speculation about how bad they might be," says Cynthia Wilkey, a local resident who routinely consulted with Waterfront.

The agency, meanwhile, was largely left on its own to deal with the suddenly wary public. Sidewalk "couldn't disavow" its sister company Google, Waterfront vice-president Kristina Verner says, but Waterfront was forced to spend more and more time trying to assuage people's concerns. "Sidewalk Labs really left us to be the ones to answer for those issues."

Waterfront Toronto's staff and stakeholders were supposed to be negotiating the details of its partnership with Sidewalk, but just months into the process, they found themselves fighting a war on three fronts. They had a Sidewalk problem: the company was pushing for more land than Waterfront could give it. They had a public skepticism problem, made worse by a global scandal totally out of their control. And some of them also felt they had a Waterfront problem. He was sitting in the corner office.

CHAPTER 11

The Other Shoe

BACK IN THE rushed days before the Waterfront Toronto board voted to partner with Sidewalk Labs, the two organizations were struggling to come up with a brand for their efforts. A handful of executives and marketing experts convened in a small meeting room at Waterfront and tried to agree on a name. Will Fleissig and Kristina Verner were there; Dan Doctoroff and Micah Lasher were sharing ideas from Sidewalk. But they were hitting a standstill.

Waterfront staff didn't want the kind of Project Sidewalk–style branding that had marked Sidewalk's experimental Yellow Book ideas. Likewise, some Sidewalk staff didn't want to call the project Quayside, because that would imply it was constrained to a much smaller plot of land than they wanted. People flung innovative-sounding ideas like "Open City" around the room, but Verner thought they sounded too technical. The Waterfront staff weren't huge fans of "Sidewalk Toronto," which sounded like a local version of a standardized product—Sidewalk Singapore or Sidewalk Bangkok might be next. For a while, they danced around ideas that paid tribute to Jane Jacobs. She was connected to both their worlds: a famous New York urbanist who spent the final decades of her life in Toronto. Someone suggested J2 might work as a Jacobs tribute, but again, it was struck down as too technical-sounding.

As Sidewalk's envoys pushed the discussion back to the name Sidewalk Toronto, Verner recalls, the Waterfront team felt like they weren't being heard at all. Doctoroff and Lasher, two veterans of New York City politics, had a much more aggressive way of working a meeting than the more process-oriented Canadians. "The two of them can be a pretty powerful force to reckon with," Verner says. Fleissig grew impatient; he didn't think the project needed a name this early, and he had a mountain of other work to finish ahead of the project announcement.

"Just call it Jane!" he said sarcastically. He picked up his things and left abruptly, the frustration showing on his face.

The others eventually settled on Sidewalk Toronto. It was a name that created the impression Sidewalk was in charge, even fuelling speculation among project critics that secretive side companies might be involved. Like much of what unfolded during Sidewalk Labs' Canadian incursion, the name was a compromise forged in awkwardness that delivered little but trouble. This was a project with two CEOs, but only one of them was able to take control.

Fleissig is the German word for diligent and hard-working, and Will Fleissig largely lived up to the name at Waterfront. The people who hired him, the people who worked for him, and even people who had worked alongside him decades earlier in Colorado tended to describe him as visionary. The Waterfront CEO was a big-picture guy, and he was obsessed with building neighbourhoods that would enable a better life for future generations.

At the University of California's Davis campus, he'd pulled together the pieces needed to develop a 180-acre pedestrian-friendly neighbourhood that blended housing, school, research and recreation facilities. It billed itself as a "model in public-private partnerships" and, thanks to an intense investment in solar power and efficient design choices, it claimed to be one of the biggest developments in the United States that might someday produce all of the energy it used. In Toronto, he'd pushed for a similar vision for Villiers Island, and he hoped to do the same at Quayside.

But Quayside was no longer just Quayside. Sidewalk wanted more land, and its ambition to extend into the Port Lands now had people at City Hall, and even inside Waterfront, wondering whether Fleissig had overstated what Waterfront could deliver. ("No," Fleissig says when asked about this. "Period. Full stop.") And by early 2018, some of the people who'd admired Fleissig's vision began to worry about his execution. They felt that Waterfront wasn't displaying the kind of local leadership it had for years under former CEO John Campbell. Waterfront had long described its style as one of "relentless implementation." Its staff prided themselves on getting things done. In this context, some of the people who called Fleissig visionary did not always mean it as a compliment.

The board that had hired Fleissig—the one that admired his vision and encouraged him to seek partnerships with the private sector—was not the board he was answering to in 2018. Thanks to a coincidence of timing, the government-appointed terms of most of the directors had ended, and some members of the new board saw a CEO that didn't fit with their working style. The reshaped board inserted itself into Waterfront business much more than its predecessor had done, sometimes to the frustration of Waterfront staff, especially as the agency was thrust into the limelight with Quayside. Reputations were on the line. Vision would not be enough.

Some directors worried that Fleissig was being unduly protective of Sidewalk instead of more clearly defending due process when Doctoroff casually talked about building across the Port Lands. Directors also began to raise questions about Fleissig's relationship with Sidewalk, which was supposed to be nothing more than a "preferred partner" for Waterfront, and which would be subject to years of intense scrutiny. Yet Fleissig was making public appearances and courting influential people with Doctoroff in support of an only partially approved project.

"I definitely complained that there should be no joint public voice until due diligence was completed," says Julie Di Lorenzo. Since the board vote to bring on Sidewalk as a partner, she had often been the lone voice of dissent, but the sight of Fleissig joining Doctoroff and his deputies when Sidewalk was courting support had drawn wider criticism. "The board was concerned that we didn't have a good sense of what meetings were taking place, or that we were receiving a full enough account of what was being discussed," says Helen Burstyn.

Directors were not unanimous in their views of Fleissig's style. Some thought their colleagues were overstepping their bounds in critiquing his management or bypassing him to take concerns to other staff. Others sympathized with Fleissig over how he had to grapple with the intensity of the Quayside project and with the skeptical factions of the public who wanted answers *right now* about a deal Waterfront was still just negotiating. Nonetheless, the board urged Fleissig to be more mindful about how he communicated project details. This request had the unintended consequence of Fleissig's voice retreating from the public conversation, leaving Sidewalk to control more of the project's narrative.

Nothing ever seemed to be enough for Sidewalk Labs. The agreement-to-agree that Sidewalk had signed with Waterfront Toronto in October was growing stale, and by the spring of 2018 the two organizations were overdue to hash out a more formal contract. After months of meetings, the contract negotiations had led to a showdown in the conference room of a Bay Street law firm, with Waterfront and its counsel on one side facing Sidewalk and its lawyers on the other.

Even simple matters like how they'd talk about Quayside became the subject of intense debate. Waterfront had taken the lead on project communications throughout its history, with both the public and governments. Sidewalk was, in procurement jargon, a vendor: an organization that Waterfront had granted the privilege of working with it. Waterfront expected Sidewalk to follow its lead, and it planned to ensconce that dynamic in the new agreement.

Sidewalk's policy and communications chief, Micah Lasher, refused to accept that. A veteran campaign staffer from New York's

Democratic circles, he could be as charming as he was cutting—though the latter trait tended to get him more attention. Reverend Al Sharpton alleged that Lasher was "bigoted" because of his involvement in a campaign flyer for Mark Green in the 2001 New York mayoral Democratic primary that included a political cartoon depicting a Green opponent kissing Sharpton's behind. Lasher was still apologizing for it twenty years later, saying, "I was part of an ugly campaign operation that was wrong." With Sidewalk, Lasher was happy to wade into arguments on Twitter and in meetings. At one 2018 event, he told Bianca Wylie that he believed her criticism of Sidewalk was an act of selfish brand-building. Yet when it came time for him to push Sidewalk's brand, he was willing to peddle a selective version of the truth—like claiming in a tweet that the company "NEVER" wanted taxation powers when the Yellow Book's urban vision said the company "will require tax and financing authority."

And he was stubborn. At City Hall, one person who worked with Lasher described laughing with colleagues about how often his argument for more land boiled down to *but we want it!*

He took the same tack with Waterfront as they debated the details of their next agreement at the law firm. "We're not going to ask permission to talk about the project," he said.

Kristina Verner tried to explain that this was Waterfront's standard practice. "Partners have to work *with* us," she said. Lasher, Doctoroff and others were already scheduling meetings with government staff that Verner and her fellow Waterfront executives felt they should attend. The Waterfront staff felt that Sidewalk taking these meetings alone was disrespectful. It would be akin to Waterfront setting meetings with Alphabet executives without telling Doctoroff—a strange, if not clandestine, way to do business.

Lasher persisted, and he and Verner began arguing. The idea of having to follow Waterfront's rules when dealing with governments made him visibly upset.

He's like a moth to a flame, Meg Davis thought: unable to see any light but the one he was focused on. Verner tried to convey to him that if this was going to work, Sidewalk needed to work on Waterfront's terms. It wouldn't be until after the meeting that Sidewalk finally agreed the two organizations could work together when they had to discuss the project in public or with governments. Not that the agreement was consequential. Sidewalk would keep having those meetings, at one point gathering dozens of influential Torontonians together behind Waterfront's back.

But there were more pressing matters to deal with in the meeting than public perception and co-operative communications. Sidewalk staff wanted to break ground quickly—and on as much land as they could get their hands on. As Sidewalk and the agency's had been exchanging drafts of their new agreement, Davis says, Sidewalk kept inserting language that would allow the company to more easily expand its development into the Port Lands. Waterfront executives kept pulling the language out. This back and forth continued in person at the Bay Street law firm.

Sitting at the conference room table, Doctoroff insisted once again on access to more land. Will Fleissig and Meg Davis said it wasn't possible. Not only would it break procurement rules, but developing the Port Lands wasn't what Waterfront had asked Sidewalk to do.

But Sidewalk had been dreamed up by Larry Page to circumvent rules in order to make cities better places to live, and he'd hired Doctoroff for the task. So, as Doctoroff had done so many times in New York's City Hall, he raised his voice, saying, yet again, that he wanted more land.

The increased volume did not deter Davis. "It's Quayside only," she told Doctoroff. "If you do well on Quayside, then we can talk about more things with the city down the road."

Doctoroff refused to demur. Getting louder still, he said that wouldn't work.

Unwilling and unable to give the CEO what he wanted, Waterfront staff said they were willing to give Sidewalk the chance to draft a *plan* at a larger scale, but the agency wouldn't just hand over a chunk of the Port Lands. "Not only do you have to pass the hurdles, but we didn't procure you for that," Davis said. "We didn't say you could be a developer, or that you could have any geography beyond Quayside." The contract between the organizations, she explained, "doesn't give you that pathway."

Doctoroff then launched into a series of moves he'd used endless times before. He yelled angrily that 12 acres wasn't enough, then slammed his fist on the table, stood up and left the room abruptly. The performance was as cartoonish as it was predictable—he'd described this kind of behaviour in his memoir as a negotiating strategy, even though the business world had begun to move past the heyday of the domineering white male executive using anger to get his way.

The parties, slightly stunned, took a break. When they launched the meeting again a few minutes later, Doctoroff returned, much calmer.

Eventually, Sidewalk gave in. The new agreement gave them even fewer grounds for expansion than the original. It didn't mention the "eastern waterfront" once. Instead, it offered the chance to draft "plans at scale" that would get built only if Waterfront and governments said so. But this didn't resolve all of Waterfront's headaches.

Helen Burstyn may have stopped short of finishing her Ph.D. dissertation in English and comparative literature, but her academic research experience opened doors for her in the 1980s at Queen's Park, the seat of Ontario's government. After slipping her resumé under the door at the legislature's non-partisan research department, she got a job offer almost immediately. Thus began two decades of working in government, where she crossed paths with several generations of premiers, governing parties, bureaucrats and consultants, including her eventual husband, David Pecaut, a civic-minded businessman who was sometimes referred to as "the best mayor Toronto never had." Armed with a wide smile and a knack for details, Burstyn rose to the senior ranks of Ontario's public service.

After taking a leave of absence when Pecaut was diagnosed with cancer in the early 2000s, she returned to Queen's Park and convinced the Liberal government to let her oversee a provincial grant-giving organization whose chair's term had expired. This appointment inadvertently launched a second career—one spent serving the public through boards, and which would eventually help earn her the honorary title of Member of the Order of Canada. In 2016, in the waning years of the Liberals' long reign over the province, the government made her a director at Waterfront Toronto. By the end of the year, Mark Wilson had left the board, and she was named Waterfront's chair.

There aren't many organizations that report to three levels of government in equal measure, so Waterfront proved a difficult vehicle to steer. But Burstyn had a lot of respect for what John Campbell had accomplished and saw a chance to make the neglected shore of Lake Ontario a better place. She liked the new board that was filling out; the directors who'd appointed Will Fleissig as CEO before her arrival were gradually leaving, but in recent months, they'd brought in new directors like University of Toronto president Meric Gertler and the development executives Stephen Diamond and Julie Di Lorenzo. It was a good group, and Burstyn was excited to work with them.

As Fleissig got to work, Burstyn was surprised by some of his decisions. Board members and key politicians were brushed aside in the summer of 2017 when the prime minister, the premier and the mayor joined Fleissig for one of the biggest photo ops in Waterfront history to announce more than a billion dollars to reroute the Don River. When Waterfront's management picked Sidewalk as the winner for Quayside, Fleissig's excitement felt disorienting for some directors: at times, he seemed to want to lock the partnership down as quickly as possible, despite the board having final say. (This was first due to media leaks, and later, in part, because Waterfront originally wanted to negotiate a more formal agreement within six weeks of the first one; the formal agreement was later delayed by nearly eight months.)

Despite the key role she played for Waterfront, Burstyn wasn't asked to meet Dan Doctoroff or Josh Sirefman until shortly before joining them for a meeting with the mayor. At the meeting, she couldn't help but notice how collegial Doctoroff and Sirefman were with Tory's staff, even though this was before the board had approved the Quayside partnership. In the months following the vote, she found herself among the ranks of people wondering if Sidewalk had been promised more opportunity than Waterfront could deliver. "It's not like I reeled back in horror, but I certainly did have this feeling that they were creeping outside the boundaries of the agreed-upon area," Burstyn says. Worried, she began asking to join any future meetings Fleissig took with Doctoroff and other Sidewalk executives.

Directors also sometimes found themselves being briefed about the project by Doctoroff or other Sidewalk staff, rather than Fleissig or his executives, which made them worried that Sidewalk's presence might put a chill on open discussion in those meetings. Burstyn learned that some Sidewalk staff were even working out of Waterfront's offices. She asked them to leave. By late spring, some members of the board found Fleissig scrambling to provide details at the last minute before—or during—meetings, making it hard for the board to make well-informed decisions.

As Waterfront's staff watched this unfold, they began pushing Fleissig to hire a chief operating officer to take on some of his work. Directors, meanwhile, were worried enough that they instituted what's called a 360 review of Fleissig's performance. Working with outside consultants, a board committee began interviewing just about anyone the CEO had worked with, including both Waterfront and government staff—though not Sidewalk. The results, they found, justified the need to take drastic action.

The board mulled what to do next. Thanks to Sidewalk's connection to Google, the world was watching. Waterfront's biggest project, and its reputation, was on the line. All signs pointed to needing a new CEO. Even the mayor, John Tory, weighed in, telling Burstyn that Toronto didn't have time to wait for Fleissig to change his ways.

At a tense meeting in the early days of the summer of 2018, the board came to the same conclusion.

Two directors asked to meet with Will Fleissig at a legal office—neutral territory, away from his Waterfront colleagues. They presented him with a letter outlining the findings of the review and gave him the opportunity to resign.

It was strange timing. Fleissig was on the cusp of hiring a chief operating officer, which could have helped him bring calm and focus to Waterfront. But he also realized it would be difficult to keep leading Waterfront when he and the board, including the chair, were incompatible. He acknowledged, too, that he was a big-picture guy, and that it made sense that someone else might be better suited for the day-to-day work of executing the massive Port Lands Flood Protection and Quayside projects. So he agreed to resign, feeling confident that the staff he'd leave behind would keep working for a better Toronto, even without him. "I still believe in that mission," he says.

In the midst of tense negotiations with one of the world's biggest companies over a city-building project under extraordinary scrutiny, Waterfront Toronto lost the CEO who had brought them the project in the first place. In a press release on July 4 announcing his departure, Burstyn praised Fleissig for his "innovative vision."

CHAPTER 12

Left and Leaving

WHEN THE ONTARIO government appointed Michael Nobrega to join Waterfront Toronto's board in 2018, he had no idea about the firestorm happening behind the scenes. The offer sounded exciting. Waterfront had been, for much of its life, an infrastructure-building organization, and financing infrastructure had been his specialty.

Nobrega was a soft-spoken man who liked to work in the background on the best of days. He was in his mid-seventies, four years into retirement, but could easily be mistaken for twenty years younger. His short mop of straight dark hair looked like it belonged on the star of a nineties boy band, not the former CEO of a pension fund with nearly C$72 billion under its control—the Ontario Municipal Employees Retirement System (OMERS). He had taken that job in 2007 after running the fund's infrastructure division, steering OMERS through the financial crisis and steeling it for the future. Nobrega had opened new international outposts and taken chances on young Canadian tech companies at a moment when few mega-funds would.

Nobrega had two investment specialties that appealed to Waterfront: there was infrastructure, of course, but he'd also helped establish OMERS Ventures and remained enthusiastic about the tech sector. He'd spent time in retirement chairing two different government agencies that helped Canadians bring new inventions to market, too. Sidewalk, he felt, could bring global ideas to downtown Toronto, and he had experience and connections of his own that could help Waterfront build even more ambitious plans.

When Nobrega joined the board, its 360 review of Will Fleissig was under way. Nobrega hadn't been expecting that. Nor did he expect, when he walked in late to a meeting a few weeks later, that the review would have such a stark conclusion that directors were planning to present Fleissig with a letter offering him the chance to resign.

Soon after that meeting, Nobrega sat at Waterfront's long boardroom table with his fellow directors trying to figure out the agency's next steps. It needed a new CEO, but more urgently, it needed an interim CEO first, to run the show until a permanent executive was hired. He looked around the table. There were other experienced business operators there, but most were still running organizations of their own. Nobrega, on the other hand, was retired. He had experience financing hefty infrastructure projects and forward-thinking tech companies. And he was a meticulous executive, an accountant by trade who could be a steady hand at a moment when Waterfront really needed one. The other directors encouraged him to step up. If no one did, Waterfront could fall apart. So Nobrega agreed to run the agency until it found a permanent CEO. He thought it would be, at most, a five-month gig.

Nobrega had some conditions. Because he saw the role as a public service, he would refuse any payment—even a parking pass. And for continuity, he would bring Will Fleissig back on as a management consultant as part of his settlement with Waterfront. (Fleissig agreed to do so, spreading his severance out like a paycheque.) The ex-pension-fund boss also wanted to refine Waterfront's chain of command so staff could better grapple with problems before they grew unwieldy, and to present a common front to Sidewalk Labs. The interim CEO role was a chance to be both surgeon and personal trainer, cautiously reconstructing Waterfront and making it stronger in the process.

Just as Nobrega was stepping into the role, Sidewalk was beginning to share some of its early digital ideas with Waterfront—and Nobrega's own protege was making things even more complicated for the agency.

Nobrega had helped turn John Ruffolo into a start-up sector star in the early 2010s by giving him free rein with OMERS's chequebook. By 2018, Ruffolo commanded significant influence over the Canadian tech sector as founders fought for his time and money. Because of his expertise—and his clear interest in Quayside, thanks to his consortium's failed bid—Ruffolo was one of the first people Kristina Verner had invited to what Waterfront was calling its Digital Strategy Advisory Panel.

In the past, Waterfront had put massive effort into ensuring that new developments along Lake Ontario had world-beating internet speeds, both to entice tech companies to the lakeshore and, thanks to a clever broadband pricing structure, to ensure that all residents could afford the price of entry to the digital economy, regardless of income. It called this its "intelligent communities" effort, which was a couple of stretched synonyms away from the phrase "smart cities," but different by definition. What Sidewalk wanted to build was consequential not because of high-speed internet cables, but because of what would be attached to the ends of those cables. Even Sidewalk's garbage-sorting plans depended on "smart," data-hungry sensors. But as many of the people who were angry that Waterfront had picked Sidewalk for Quayside gleefully (and truthfully) pointed out, the government agency didn't have enough in-house experience with technology to scrutinize Sidewalk's plans for all that data.

Waterfront's chief strategy officer, Marisa Piattelli, had encouraged staff to find a way to parse Sidewalk's proposals in a thorough, methodical way. Soon after the partnership announcement, Jim Balsillie had pressured Waterfront to do the same—but with great urgency. So in early 2018, Kristina Verner began contacting what she hoped would be a dream team of experts. Two of Canada's best-known scholars in internet law and intellectual property, Michael Geist and Teresa Scassa, signed on. Kurtis McBride, the CEO of the traffic management technology company Miovision and an advocate for fair data collection, also agreed to join. There were urban-planning experts, like Pamela Robinson, a professor who would soon become the director of Ryerson University's planning school. Some crossed disciplines: former Waterfront chair Mark Wilson knew the agency inside and out and had also been an IBM executive; Charles Finley's world had bridged entrepreneurship, coding and strategy. The inclusive-design scholar Jutta Treviranus joined, as did Saadia Muzaffar, an entrepreneur, writer and advocate for immigrant women in science and technology.

Verner also reached out to people who were critical of the Sidewalk partnership. She wanted skepticism at the table. Bianca Wylie was one of those critics, but turned down Verner's request, feeling that she'd be more productive studying the project from the outside. Ruffolo, however, said yes. He was hesitant at first, but figured he might be able to help. He'd been warning friends like Balsillie about Sidewalk in private; if he learned through the panel that something was going wrong, he'd have a chance to sound the alarm loudly and publicly.

Their first meeting took place over an entire day in early June. After some initial onboarding with Sidewalk and Waterfront staff, the new panellists turned to how Alphabet might use data collected at Quayside. A handful of Sidewalk's top and middle managers were there. Head engineer Craig Nevill-Manning told the panellists that any data collection had to be tied to improved quality of life—that collection itself wasn't the end goal. The company would ensure privacy was protected in the data it collected, but it would be managed through a platform that would let others access and benefit from that data. Keeping the platform open would mean those others could learn how their city worked, and maybe build technologies of their own from the patterns they found.

Next came Alyssa Harvey Dawson, who oversaw privacy and data matters for Sidewalk. She shared some examples of the kind of data they might collect. The ideas were in their early stages—what got built would depend on what Waterfront and governments allowed—but they had a direction. Trash can sensors could alert city staff when the cans were full. Outdoor sound level measurements could help people find a quiet spot to read. Streetcar locations could be mapped.

Harvey Dawson and Nevill-Manning both routinely returned to privacy protections: anything that collected data in public that could identify someone would strip away identifying details. Lauren Reid, a consultant hired to work alongside Harvey Dawson to focus on privacy in a Canadian context, told the panel it was important to develop data-handling policies that reflected the country's pride in developing technology responsibly.

Sidewalk didn't have all the answers the panel was looking for. Andrew Clement peppered the company's staff with questions and got little that he liked in return. The retired University of Toronto professor with wispy hair, a sneaky smile and a raspy voice had spent his career studying surveillance and privacy through a progressive lens, and was thrilled by the chance to hold this mysterious new company to account. When he asked why Alphabet was interested in the Quayside project, Sirefman's response seemed vague and high level, frustrating the professor. He couldn't imagine a situation in which Google or Alphabet wasn't interested in the enormous amounts of data at Quayside. The professor launched into a tirade about the risk of "function creep," a phenomenon in which corporate promises around privacy are gradually eroded, with the collection and usage of data shifting over time, sometimes expanding into surveillance. Facebook's experience with Cambridge Analytica showed how easily it can happen.

Ann Cavoukian, the former privacy commissioner that Sidewalk had hired on retainer, tried to assuage Clement's concerns. Scrubbing personal information from data as soon as it was collected, she said, would drastically reduce the chance of identifying someone if the data got leaked. Clement argued that wasn't good enough—the spectre of Cambridge Analytica was still fresh, and it should serve as a warning about how any data that companies collect could be misused.

The panellists were left uneasy. They pointed out that Canadian policy and regulation were far behind on data issues, and warned that they might need to be revised to keep safeguards around the Sidewalk project. Some walked away feeling frustrated that they hadn't really discussed anything of substance.

Something else was making the digital experts nervous: Waterfront wanted them to sign a confidentiality agreement. A handful of panellists, including Clement and Ruffolo, spent the next few weeks pushing Waterfront to include language that would ensure they could publicly discuss as much of what they learned as possible. It was Waterfront's standard non-disclosure agreement, but the agency agreed to change some of the wording to accommodate those worries.

This wasn't enough for Ruffolo. He demanded that Waterfront's management explain why they needed the agreement. He saw the panel as a group of public advocates; if they saw something going wrong, he wanted the public to know. Management replied that most of what the panel discussed would be public, but sometimes they might need to give advice on contracts or ideas that Waterfront was actively negotiating with Sidewalk.

Ruffolo still felt like he was being strong-armed into silence to keep Sidewalk comfortable. And he was not a quiet guy. He mulled it all over for a couple of weeks, and on July 4 he e-mailed Verner. He couldn't stomach signing the agreement. He was quitting instead. "I am still unclear what exactly the nature of the confidential matters might be," he wrote, "and I cannot sign such a vague and broad agreement as I do not even know exactly what I am agreeing to."

Verner contends that Ruffolo's resignation was one miscommunication away from being resolved—that Waterfront wanted everything to be public. At first, non-disclosure agreements weren't even going to be necessary for the panel. But soon, an irony emerged. At a board-related meeting held before the panel first met, Julie Di Lorenzo had insisted that the panel give feedback on the data- and digital-related segments of the more detailed partnership agreement Waterfront was negotiating with Sidewalk. The cautious director wanted the agency to have the assembled experts inspect and review every word of its contracts with Sidewalk. But because the draft agreement was confidential, Waterfront needed to ensure that anyone who saw it would keep it that way. This meant that a director pushing Waterfront to scrutinize Sidewalk in the name of public interest inadvertently created a situation in which a high-profile adviser quit in the name of public interest. Worse, Ruffolo quit the same day it was made public that the CEO who had brought them the Sidewalk project had resigned. It was shaping up to be a rough summer.

The image of a progressive dream city that Sidewalk and Waterfront had promised Toronto was being progressively chipped away. The deputy mayor was making cryptic, concerned comments with cameras rolling as his City Hall colleagues pushed back against a brazen attempt at a land grab. Negotiation meetings featured blow-ups on both sides. People were leaving the project, some by choice, some less voluntarily. Opponents were getting louder. The instability reached the point where, in its briefings with government, Waterfront promised it would focus a summer public consultation for the project exclusively on "non-controversial" subjects like building and street designs.

Julie Di Lorenzo couldn't see anything but controversy when she looked at drafts of Waterfront's new agreement with Sidewalk, called the Plan Development Agreement. Since the fall, she'd racked up tens of thousands of dollars in personal legal fees consulting with lawyers about how to deal with the Sidewalk project, and her mind constantly lingered on her responsibilities as a director. At board and committee meetings, she raised her hand whenever she could to ask for more scrutiny on the deal. She'd pushed for Waterfront's financial statements to incorporate the 2.7 million square feet of potential development that Quayside had been zoned for, so that the public would know how much value could be unlocked across 12 acres. She would later pay for a land appraisal at her own expense, finding it was worth C$570 million even before development. The director had also grown frustrated with what she felt were infeasible ideas that Sidewalk was putting forward—things like heated pavement tiles and tall wood buildings seemed more about hype than serious development plans.

But what bothered her most was the language that Waterfront had struck with Sidewalk around communicating about the project. The final version of the agreement said the two organizations would have "clear, consistent and coordinated communications to the public and government stakeholders." Waterfront's executives were quite satisfied with this part of the agreement: it blocked the company's push for Sidewalk to make announcements and speak with governments in Waterfront's absence. But as Di Lorenzo viewed it, the new language meant she might not be able to bring her concerns about Sidewalk to governments without coordinating *with* Sidewalk. This, to her, was inexcusable. She'd already dissented once before. As July progressed and she once again felt backed into a corner, she realized she could be much freer to criticize the project from the outside, as a Torontonian who paid taxes to Waterfront's three government stakeholders.

Waterfront spent the final days of July preparing to release both the original Sidewalk agreement and the new one to the public, which staff hoped would assuage their critics' greatest fears. Just in case, however, communications staff prepared a list of fifty-three questions that their colleagues might be asked by media and the public. The questions ranged from why a partnership with Sidewalk was needed to develop Quayside in the first place, to "Is Waterfront Toronto in trouble? Why did the CEO depart before the [agreement] was signed?" (The answer to that one began "We're doing great as an organization.")

But as Waterfront prepared to get in front of any negative coverage, Di Lorenzo was agonizing in front of a word processor. She was writing a resignation letter. She completed it just hours before the new agreement was released on July 31, sending it to Helen Burstyn and Waterfront management. "Circumstances have prevented me from performing my fiduciary duties in the interests of Waterfront Toronto and the tri-level government stakeholders," she wrote. ". . . The time has come where I am unable to support the new business of Waterfront Toronto as predicated under the [new agreement] which i do not believe is in the interest of the Corporation and our Country."

There was another irony here, just as when Di Lorenzo had unintentionally caused Waterfront to issue the confidentiality agreement that led John Ruffolo to resign. Unconvinced that Waterfront could retain its autonomy with its new agreement, Di Lorenzo chose to leave the board. But the language she feared most was intended to protect Waterfront's interests.

And so in July 2018, two of the most critical voices inside Waterfront departed the project due to misunderstandings. This had unintended but far-reaching consequences. They began taking their harsh words public just when their quiet scrutiny might have helped Waterfront the most.

Julie Di Lorenzo's resignation cast a shadow over what Waterfront Toronto had hoped would be a celebratory release for the Plan Development Agreement. After months of negotiations and resignations, the agreement was meant to shed light on a partnership that was routinely accused of too much secrecy. It attempted to quell fears about a land grab, too: as they worked toward a draft master plan, the new agreement made it clear that Sidewalk had to meet Waterfront's goals at Quayside, present a detailed business case and get government approvals before the public agency would consider an expansion. "Any proposed options at scale shall be subject to and closely tied to the achievement of the [draft master plan] Targets and supported by robust business planning and financial analysis," the agreement said. ". . . Both Parties recognize the value in strong collaborations: the success of the Project depends on the support and confidence of government and the public."

An entity with Sidewalk in its name wasn't called the "master developer" anymore. The organizations clarified some confusion over the $50 million Sidewalk promised for the project—it wouldn't be considered an equity investment, but would be dedicated to drafting their master plan and consulting the public. They broke the funding down into individual tranches, including $11 million for consultations and to "ensure support" among "key constituents." Waterfront's new temporary CEO, Michael Nobrega, used the opportunity to remind the public that the agreement had plenty of exit ramps in case the project "goes sideways."

Di Lorenzo's resignation coloured much of the agreement's coverage just as Waterfront was hoping, rather belatedly, to create the appearance of control. Others began to go public with their anger, too, as they started looking for gaps in the agreement that could hurt the public. The new agreement didn't have much to say about data, for one thing. Beyond a few progressive-but-vague principles and promises to create "citizen-centered" policies recognizing privacy as a human right, and similarly vague assurances that the project would not "foster monopolies" around data or tech, there was still little information about what kind of data would be collected if Quayside were built—or what Sidewalk might do with it.

Jim Balsillie and a group of like-minded intellectual-property hawks were thinking many steps ahead of the new agreement. Even if Sidewalk allowed other companies to install sensors at Quayside, it looked like nothing could stop the company from installing the *most* sensors, collecting the *most* data about how residents went about their lives, and turning insights fuelled by that data into patents. Google had made many of its billions that way.

Larry Page had filed for patents in Google's early days that would form the basis of its search engine and ad products, including one called "Method for node ranking in a linked database." That patent was originally "assigned to"—effectively owned by—Stanford University. Universities often host, fund and provide materials for the research that leads to patented ideas. At U.S. universities, the rights to inventions made with federal funding are usually, by law, held by the school. In Canada, the policy depends on the school, but in many cases the university will split the rights, and the lucrative money from licensing the patents, between the school itself, the department, and the lab and inventors.

The rights to patents are often split like this—with a share of the licensing money going to organizations that helped create the conditions for the invention to be invented. That's why the data that Quayside would have generated mattered so much. If cities all over the world wanted to use inventions that had been developed with the data harvested at Quayside, the patent owners could have made as much, if not more, money as the hundreds of billions of dollars that technology companies were raking in from learning how people behave online. In offering to host Sidewalk's test bed, Waterfront was expecting a share in whatever intellectual property (IP) was developed as a result of the deal. If Canadians were generating money-making data as they moved around their city, Waterfront wanted its three government shareholders to get a cut of the cash.

George Takach, the lawyer Waterfront hired to guide its IP negotiations, had helped write details into the new Sidewalk agreement that he believed would go even further than a university's contribution to a student's research: that Waterfront should get a share of IP from giving Sidewalk the "social licence" to install technology and collect data. Through Waterfront, Sidewalk would get advice from governments, subject-matter experts and the public, on top of the opportunity to collect data and develop technology in a thriving city. The July 31 Plan Development Agreement included a commitment to establish how Waterfront and Sidewalk would ultimately share patents, other intellectual property and the money all of it generated.

But in the agreement, the IP promise was still just that—a promise. It was vague. Jim Balsillie was, predictably, angry about this. Not only had he fought through patent disputes at Research in Motion, but he routinely liked to point out that he'd brought IP to market in more than 150 countries. He felt Canadians could easily be left behind by patent-hungry Alphabet. Their stake, he felt, needed to be clarified swiftly. "As it currently stands, someone experienced in commercialization can drive a truck through the language on IP and data in the new agreement," he said the week of the new Quayside agreement's release. "Between Google and [Waterfront Toronto], I think we can guess who is more experienced on matters of IP and data."

Natalie Raffoul, a patent prosecutor from Ottawa who would later work with Balsillie on a government panel to shape Ontario's approach to IP, said Canadians could be left in the dust by the agreement's vague language. "Nothing in this agreement spells out monetization from our own data," she said. "The benefit to Google is going to be enormous."

But for the third time that summer, a curious twist would come to shape a major moment in the Quayside saga. For all the hand-wringing about data and intellectual property, it turned out that Sidewalk was making plans for IP in ways that didn't even touch on data.

In late August 2018, I came upon a confidential document. It was a call-out from Sidewalk Labs asking architects and designers to submit ideas for a study it was conducting about small, affordable apartment units that could be built at Quayside that would maximize usefulness with minimal space. Many of Sidewalk's ideas were about finding better ways to build the physical pieces of cities: to make construction cheaper and less carbon-intensive, to make rent more affordable, to use less energy. Original designs can be their own form of intellectual property, and can be safeguarded with patents, too. But the document I found made it clear that whoever shared their ideas would do so on a work-for-hire basis: "All rights, including in the intellectual property, will vest in Sidewalk Labs and Sidewalk Labs will have the opportunity to incorporate it into future phases and other work." In cases where full-on IP ownership wasn't possible, Sidewalk wanted an exclusive, royalty-free, worldwide licence to use it, giving the company maximum benefit either way.

Sidewalk had signed an agreement with Waterfront weeks earlier that promised the two parties would develop a system for sharing IP that was developed during their project. Even if the work-for-hire designers Sidewalk tapped for the study were fine with handing over their IP to Sidewalk, Sidewalk's new agreement with Waterfront said the Canadian agency likely deserved a stake in that IP. The agreement was supposed to guarantee that the public benefitted from a public-private partnership that sought to improve how cities were built everywhere.

From my fifteenth-floor desk in the *Globe and Mail*'s downtown Toronto newsroom, where I could see the mud of Quayside if I stood up and leaned the right way, I started making calls about the document. "The taxpayer won't derive any benefit," Natalie Raffoul told me. Jim Hinton, another IP lawyer who would later fall into Balsillie's orbit, said Waterfront needed to assert itself for a fair cut of IP rights. "The governments need to be [saying] . . . 'We're a partner, a co-investor—we want to be paid out when this is implemented in Singapore or wherever else this will be done,'" he said. Balsillie put Waterfront in his crosshairs. "Here we have Google [systematically] building their IP and data assets by doing cartwheels around the incompetent Waterfront Toronto board," he said by e-mail.

A colleague got hold of Julie Di Lorenzo. "This is saying, 'Give me what you have and it becomes mine,'" she told him. "If you ask people to relinquish their creative work, that's not conducive to innovation and collaboration." She also questioned why the procurement document was issued by Sidewalk Labs alone. "I don't understand how that can not have Waterfront Toronto participating."

Sidewalk referred me to Waterfront for comment. George Takach, Waterfront's IP specialist lawyer, said there was nothing to worry about. "The real question about what happens to that IP, who gets to own it, use it, make money from it . . . will be determined later," he said. "But retroactively, they'll apply to all the IP that's been developed."

The *Globe and Mail* story about the design call-out was published in the final hours of August. Sidewalk and Waterfront had very different ideas about how fair it was. Many Sidewalkers were frustrated with the story, but some were downright angry at their bosses. The company hadn't explained or defended the design contract on the record. Leaving comment to Waterfront only deepened the impression that Sidewalk had something to hide. If Sidewalk management had just reinforced that the contract was a standard work-for-hire agreement common to many organizations, it would have softened the blow.

Waterfront had a very different response. Kristina Verner was aghast. It didn't matter that the design contract was a generic work-for-hire agreement. As with all of the IP brainiacs I'd spoken to, it was the *spirit* of Sidewalk's confidential design call-out that offended her, not the scope. Collaborating on IP was supposed to be central to the partnership with Sidewalk—written right into the latest agreement—yet she found herself doing damage control over news that Sidewalk was looking to control some IP on its own. Left unchecked, the consequences could ripple throughout the entire partnership and everything it stood for when it came to the future of cities. And it didn't help that she didn't know about the call-out until I asked Waterfront for comment.

Verner spoke to Waterfront's planning and design team. It turned out they *had* seen the design contract but weren't aware how much trouble the IP terms could stir up, given Waterfront's hope to get a stake in the ideas Sidewalk deployed there. Verner insisted that Sidewalk never use such heavy-handed language again. Waterfront had to start scrutinizing every contract Sidewalk issued even more carefully than it had before.

The optics and implications were terrible. Just weeks after John Ruffolo and Julie Di Lorenzo resigned from Waterfront out of worry that they couldn't criticize Sidewalk, Sidewalk was violating the spirit of its agreement with Waterfront, and Waterfront only found out through someone else. And the optics for Sidewalk were just as bad. With its reputation on the line, the company was making questionable decisions that were starting to spill into the public eye. Even its own staff struggled to understand why.

CHAPTER 13

Blue

WHEN PEOPLE WALKED into the blue-splashed building at 307 Lake Shore Boulevard East for the first time, they were bound to find something to fill them with awe. Depending on the day, as they passed through the lake-facing doors and into Sidewalk Labs' central showroom, they might have zeroed in on the floor laden with wooden prototypes of its hexagonal paving tiles. Maybe the vast models of a neighbourhood of wood-framed buildings would make them smile, or the walls plastered with city-building ideas scribbled by members of both Sidewalk and the public.

It conferred the impression of a welcoming oasis in a part of town that was not particularly welcoming. Sidewalk's Toronto headquarters sat at the long-ignored post-industrial foot of Parliament Street, wedged between Lake Ontario and a raised expressway, right in the elbow of Quayside. Natural light flooded through windows onto gleaming white walls, which were occasionally obstructed by wooden-beam prototypes that looked like what the company someday wanted to use at the core of innovative "tall timber" skyscrapers. The 307 Lake Shore space was designed to be a symbol of optimism and opportunity. It was supposed to be a glimpse of the bright future Sidewalk wanted to build for Toronto, starting on that very site.

The blue paint made the office stand out among the monochromatic low-rise commercial buildings that lined the block. This deliberate symbolism pulled in optimists from all over Toronto. But it was just a coat of paint. Underneath, it was just as old and grey. Some people cried with joy when the company offered them a job. Some cried when they realized Sidewalk wasn't the company they'd thought it was.

The Toronto office opened in the middle of 2018, and Sidewalk filled it with fresh-faced new recruits. They had quit careers at investment banks, tech companies and governments, and often marvelled at how freely money flowed as they flew back and forth to Hudson Yards to plot out an inclusive neighbourhood of the future. Some staff were surprised by what they viewed as arrogance, sometimes verging on aggression, from some of the New Yorkers who'd fly into Toronto. Their confidence struck the Canadians as bordering on a saviour complex, with Quayside as their plan to turn some backwater Canadian village into the next Hudson Yards. Even the New Yorkers' mannerisms could feel abrasive, such as when they started meetings or e-mails by going straight to business instead of with friendly small talk. The local staff weren't sure if it was a New York thing, an American thing, or a Bloomberg administration thing, but the culture clash grew exhausting.

Tensions became so severe by the late summer that the company agreed to hold presentations on Canadian culture for each office. Dozens of people showed up to them. They were led by a former City of Toronto media relations officer named Giannina Warren, who had gone on to do a Ph.D. in the branding of places. She walked the staff through the countries' divergent histories, and how the city's angst over the Quayside project was rooted, at least in part, in Canadian resistance to American hegemony. Giants of corporate America had failed in Canada before, she told Sidewalkers, pointing to the department-store chain Target, which had pulled out of the country after only a couple of years in the market. She also explained Canada's affection for the gravy-covered, french-fry-and-cheese-curd delicacy of poutine, a Molson beer ad campaign titled "I Am

Canadian," and Justin Trudeau's apparent friends in the Tragically Hip—a shining example of a pop culture phenomenon that succeeded without America's warm embrace.

Though some Toronto staff thought it was a cringe-worthy exercise, others were glad to finally have someone explain to their superiors that they were not in New York. When Warren then gave the presentation to the New York office—whose staffers probably needed the education more—people voiced their appreciation but flitted in and out regularly, and most left abruptly at the end. Some people who were there don't recall Doctoroff attending; a company spokesperson says he attended "parts of these sessions" (and insists that they were not a response to any tensions within the company).

Building in Toronto depended on gaining the trust of Torontonians, but secrecy was baked into Sidewalk's DNA, stretching back to the Javelin days of exploring Larry Page's fantasy of a regulation-free city on the sea. Not only did this culture make it hard for some employees to trust the company, it left them struggling to do their jobs.

As staff received the public at 307 Lake Shore, they were often asked what the company's business model was, or how the project would be funded. But responding to those questions was a problem: Sidewalk hadn't given its staff all the answers. Even a map was hard to find, leaving some staff to speculate that the company didn't want to remind people they had the right to only 12 acres. All of this put the once-enthusiastic Canadians in an awkward position. In some instances, they were told to answer simply: "It's way above my pay grade."

Some employees began to doubt whether the company itself knew what its business model was. In interviews and public appearances, Dan Doctoroff had begun to rhyme off combinations of real estate investment, infrastructure investment and, sometimes, tech development—the company hoped to make money from one or all of these. After a number of staff complained about the uncertainty, the company sat them down to explain the model, only to give them a variation of the same answer. This was not what staff were looking for.

They needed more to win Toronto over, especially as opposition to the project rose throughout 2018.

The public still cycled through the showroom, leaving ideas on cards that would be affixed to a wall, and Sidewalk counted every single person so they could say how many Torontonians the company had consulted. Sidewalk said that number eventually exceeded eleven thousand, though one person at 307 Lake Shore says it couldn't have topped five thousand. And all those people's ideas—just like the brainstorming of the experts consulted for the Yellow Book—didn't appear to matter. "None of that information went anywhere," one former Sidewalker says. "It didn't mean anything."

Few people see the basement of the Sony Centre for the Performing Arts, the downtown theatre a little west of 307 Lake Shore. This lower level is a cavernous place—big enough to host a mini stage and a handful of meeting and dining areas. And for some reason, during a tech conference called Elevate in September 2018, I was allowed into the VIP "creator stage" section of the fluorescent-lit, low-ceilinged space. I am not sure why my conference badge allowed this. Other media had trouble getting in. But it was a great coincidence: Eric Schmidt was scheduled to speak. And though he'd recently stepped down as Alphabet's chair, he was still a technical adviser to the company.

I wandered into the audience and sat down as Schmidt chatted with the head of the Canada Pension Plan Investment Board. Schmidt was onstage in a bright-red chair wearing a grey suit and Canadian-flag socks, which had become a running gag during Canadian appearances since he'd shared a stage with Justin Trudeau the year before. Schmidt and the pension executive praised the Canadian start-up sector and talked about the future. Data, Schmidt said, would be what cements the leaders of the new economy. When they opened the floor to questions, I put up my hand.

Nearly a year had passed since Schmidt had joined Dan Doctoroff in Toronto to announce the Quayside project, but Sidewalk's parent company hadn't said much since about its big foray into cities. I'd been talking with start-up executives about the project for months, and many were worried that the plans for Quayside might prevent them from getting a fair chance to play in the economic future Schmidt had just described. Alphabet was a world leader in capturing, analyzing and extracting new ideas from data—and in cornering markets. So I asked Schmidt how he could reconcile Alphabet's data-driven market power with the struggles of the start-ups he'd just said he was encouraged by.

He called the premise of my question "completely wrong"—even though it directly built on comments he'd just made—then talked about the same economic spinoffs that big corporations always bring up when people question their market power. "Employees of ours quit and do start-ups," he said. "The economic growth that comes from Google's participation in a city or a country is well measured. We've got all that data."

That data, he said, contradicted start-ups' concerns about . . . data. Quayside, he declared, "is going to build an awful lot of space that can be used by these sorts of people." Then he took someone else's question.

As Sidewalkers fanned out to win over Torontonians, they reached out to many of those start-ups. With the help of Nicole LeBlanc, a well-known tech investor Sidewalk had poached from the government-run Business Development Bank of Canada, staff met with as many interesting local companies as they could find, especially if they were exploring technology that could be applied to cities. LeBlanc helped build a promising relationship with InnerSpace, a start-up that used sensors to help companies better understand how employees use their office spaces. Though InnerSpace had already been taking a privacy-first approach, tracking office workers anonymously, Sidewalk staff actually helped the company improve its privacy policies, pointing out ways that, even when location-tracking data is anonymous, patterns can help reveal a person's identity. InnerSpace founder James Wu now calls it "the CEO problem": repeated anonymized smartphone pings from the corner office could easily be reverse-engineered to reveal that the CEO was there—a problem that could extend to any person in any space. InnerSpace reconfigured its systems so this kind of information couldn't be accessed. "They helped us see something we hadn't seen before," Wu says of Sidewalk.

The CEO of one of Canada's most city-focused innovators, the traffic measurement company Miovision, also kept up a conversation with LeBlanc. In both the physical and digital worlds, the ways data gets collected and stored are directly tied to who benefits from it. Kurtis McBride was following the Quayside project closely—and had joined Waterfront's panel of digital advisers—for precisely this reason. His company used video cameras at intersections and artificial-intelligence systems to monitor traffic patterns, allowing cities to respond to problems quickly and adjust traffic-light patterns to make travel smoother. Before Miovision, this kind of data collection was usually analog, even counted by someone standing at a street corner. McBride wanted to make traffic measurements more efficient, but he also wanted traffic data to be more shareable, so that more people could benefit from it. It used to be hard for cities to share data between departments; at Miovision, McBride saw a chance to share data, patterns and congestion solutions with cities the world over.

Given the complexity of his work, McBride likes to bring in relatable comparisons when explaining it. Cities, he says, are facing the same kinds of decisions as the pioneers of the early internet. In the mid-nineties, many internet users went online using services like America Online and CompuServe, which kept users in their own limited, proprietary ecosystems. With the World Wide Web, however, "anyone could contribute to this loose network of systems that now gave you access to an ever-growing set of content and experiences," McBride explains. This was enabled thanks to global open standards: choices about how the internet's many parts link together are generally not set by any entity that might benefit disproportionately from that design. Without a constant push for open standards, today's generation of digital titans would be free to build their own versions of the internet—say, a Google internet or an Amazon internet—that users couldn't send messages between because data moved differently in each one. "The economics of that world would be very, very different," McBride says.

The arrival of smart-city technologies brought the same promise, and the same threats, to the physical world. Whoever created the standards could design them to their unique benefit. As McBride says, if you design the pipes through which data flows, you can design them in a way that makes you lots of money. Take the number of cyclists going through an intersection as an example: in a world with truly open standards, cities and companies should be able to share that data just as information flows across the web, as opposed to each city and company handling the data differently, like AOLtown versus Compuberg. McBride had made his name building forward-thinking technology, but he was just as interested in convincing people to help make the market for data in the physical world a fair one.

Sidewalk Labs consistently said it wanted to design technology for Quayside that would work with open standards. Its staff were also very interested in working with McBride. Miovision was headquartered only an hour and a half's drive from Quayside, in the start-up-filled Waterloo region. It already had a strong presence in Toronto, where its technologies had helped convince the city government to make a streetcar-priority pilot program on a major downtown street permanent.

Sidewalk wanted to use Miovision's tech to help it run a pilot at one of Toronto's airports that would monitor passenger pickup and drop-off spots for rideshare drivers and alert them to open spots. In one August 2018 discussion, Sidewalk staff told McBride that it wanted data to be open, but that Sidewalk would have the rights to any decisions made using the data collected. McBride thought they were having the wrong conversation. If he was going to work with

Sidewalk, he wanted to help set fair standards around data collection and use, so that if Sidewalk took the technology it made in Toronto and sold it around the world, smaller firms like Miovision had a real shot at working with Sidewalk, or even a shot at out-innovating the company. "It's the only conversation that matters," McBride says. "Whether a garbage truck robot is driving down a tunnel is not the point, right? The operating system that all this stuff is going to get built on top of—we're not having that conversation." Data confers power, and McBride wanted to build a market for urban technology where that power was distributed fairly.

Over the course of a few meetings, McBride struggled to get his points across. Sidewalk staff also asked him to sign a non-disclosure agreement, which he worried might compromise his ability to fully hash out Sidewalk's ideas on Waterfront's digital expert panel. He gave up and walked away.

Between the growing public backlash and Sidewalk's inability to clearly describe its plans for Toronto, a few other start-up CEOs were wary when the company came knocking. Sidewalk invited a handful of Waterloo-area executives to dinner at a local hotel in October 2018, which at least one invitee interpreted as meaning Sidewalk was struggling to rally support in Toronto. When Hongwei Liu, the CEO of indoor mapping company Mappedin, asked Dan Doctoroff to explain Sidewalk's business model, Doctoroff rattled off the same rehearsed points he was giving to staff and media, including infrastructure, real estate and technology investment. Facing a room full of people who liked to solve problems, Doctoroff didn't seem to know what problem his company wanted to solve.

Back at 307 Lake Shore and 10 Hudson Yards, Sidewalk was reaching out to professors, executives and communications firms—even students. The company sponsored a fellowship for students and researchers who flew to Amsterdam, Copenhagen, Malmö, Boston, New York and Vancouver to study waterfront developments and new technologies. A trip to Hudson Yards included what one person interpreted as a strange joke about needing to conceal a copy of the

Yellow Book from view. It ended with a boozy event where, according to independent accounts from multiple people who were present, Sidewalk staff, some very senior, drunkenly told the fellows about some of its plans—including for land beyond Quayside, the pursuit of which was not widely on the public radar.

Staff also put together working groups of people from around Toronto to advise on matters like data governance and community services. The most cynical factions of Quayside opponents felt that this outreach was a superficial public-relations move to win over influential backers; some of the people that Sidewalk courted in this way would eventually agree. "They did not understand the nuances of the stakeholders they were engaging," says Lekan Olawoye, the founder of the Black Professionals in Tech Network, who gave up on Sidewalk's community services working group after two meetings.

Enough opposition began to rise against the Quayside project that the company at one point assembled a team to begin doing political-style opposition research, compiling information on dissenters like Julie Di Lorenzo and members of the media. "This was not run as a project," says one person who watched it all unfold from inside. "It was run as a political campaign."

Doctoroff liked to make people feel important by giving them a seat at the table; he could get his own work done more easily that way, even if he didn't always pay attention to the people he'd invited. Before he was New York's deputy mayor, as he drew up the Olympic plan that included a stadium on the land that would become Hudson Yards, he included staff from several city agencies on his facilities advisory board. "As a result, the territorial jockeying that can occur when public officials believe an outsider is attempting to do their jobs for them was minimized," he later recounted. Toward the end of his time at City Hall, as New York developed a long-term sustainability and economic plan, he perfected his seat-at-the-table scheme. Reveal a project's goals, get feedback from people and politicians, then add details. That way, he said, "people would feel some ownership of the plan," lowering the likelihood of dissent.

This was a tactic he'd used at Sidewalk, too, in 2015, splashing the names of dozens of luminaries like Richard Florida and Janette Sadik-Khan across the back page of the Yellow Book. Doctoroff wanted to do the same in Toronto, but Waterfront management had warned him against it: those kinds of meetings were both too exclusive for the government agency's comfort, and possibly broke the terms of the agreement signed in the summer of 2018. Sidewalk decided to disregard Waterfront's warnings.

Doctoroff and a close circle of staff, including a former long-time Liberal staffer named John Brodhead who'd joined Sidewalk earlier that year, began compiling a list of influential names who could help spread Sidewalk's gospel. It totalled nearly seventy people: urbanists, not-for-profit champions, executives and at least one former mayor.

They were invited that fall to a private lunch at 307 Lake Shore, where Doctoroff was planning to pitch them personally, with quarterly meetings to follow. The "advisory council" was kept a secret until Bianca Wylie got her hands on the invitation e-mail. She published its flattering appeal on her blog: "As a thought leader in Toronto, your input is essential to helping us as we develop this bold new vision," Doctoroff wrote. ". . . We know we still have lots to learn about this great city, and if our vision is going to succeed, we need help from those already shaping its future."

Wylie couldn't believe it. The company claimed it was working toward a neighbourhood for everyone, but was secretly buttering up the city's elite for guidance. "Sidewalk Labs is using Waterfront Toronto's brand and reputation to engage with people in the context of what should be a democratic conversation and using their advice for product development," she wrote that October. "This is a corporation that says it's working on 'a vision for the future of the city.' That's for our city government to do."

As stressed as this made Wylie, the Sidewalk staff who had to prepare for the private lunch were put in an even more stressful situation. Staff were tasked with transforming 307 Lake Shore as the CEO seemed to sweat every detail: the chairs, the tables, the place settings, the floral centrepieces. His obsession with pulling off the perfect day seemed to carry once-in-a-lifetime stakes. Behind his back, some of his employees called the meeting "Dan's wedding."

On a cloudy October 17, a fleet of cabs, Ubers and luxury cars delivered the CEO's hand-picked group of luminaries to the doors of bright-blue 307 Lake Shore. They were greeted by a sea of familiar faces—the same crew of power brokers that usually showed up to rubber-chicken dinners hosted by the Toronto Region Board of Trade and the Canadian Club. The guests sat down at their tables, each with a Sidewalk staffer there to collect their ideas.

Doctoroff soon stood up, called Toronto his "second home," then launched into scripted remarks, which included a joke about how recreational marijuana becoming legal in Canada—that very day—made Toronto an even more enticing host city. He ran through Sidewalk's origin story, calling the company an "essential catalyst" for Waterfront Toronto's dream of a better, more sustainable future. The CEO tried to add some clarity to the business model questions that had plagued the company. Sidewalk would invest in infrastructure—an advanced electrical grid, stormwater management—and new technologies to manage traffic, buildings' energy use and the delivery of social services. The company wanted to prove it was possible to build skyscrapers with wood-based structures. The ideas were so wide-ranging that he sounded like he was playing start-up whack-a-mole. He also hinted at his grand vision beyond Quayside: "To create this new community will require substantial development on parcels of land that have laid fallow for decades," he said.

The CEO got defensive, too, hinting at the Quayside pushback that had ramped up over the past year. "There are times when the prejudgment of what we will propose—and assumptions of ill intent—have taken me by surprise," he said. "But I can tell you there is nothing more behind the curtain than what I have shared with you today."

Doctoroff was trying to create the perception that everything was under control, even when it clearly wasn't. People kept seeing his choices, and the choices of the company he'd built, much differently than he did—including in the room around him.

As one attendee took his seat, he couldn't help but notice that the rented chairs, centrepieces and room arrangement seemed familiar. There even seemed to be a head table for VIPs. "Hey," he whispered to the person next to him. "This feels like a wedding."

CHAPTER 14

In Too Deep

EACH IN THEIR OWN WAY, the Cavoukian siblings devoted their lives to helping others find order in a disorderly world. As a portrait photographer, Onnig "Cavouk" Cavoukian managed to capture the essences of cultural giants like Queen Elizabeth II and Oscar Peterson in a single frame. His brother, Raffi Cavoukian, soothed generations of children by distilling life's lessons and eccentricities into iconic songs like "Bananaphone" and "Baby Beluga." But while their younger sister moved in a less creative world than her brothers, her work may end up having the most significant legacy.

Ann Cavoukian made her living protecting people's privacy. For three terms, she served as Ontario's privacy commissioner, but her influence extended far beyond Canada's most-populous province. In the mid-nineties, she and a group of Dutch experts began to develop a set of guidelines to help businesses and technologists reduce the risk of compromising people's personal information. The guidelines required organizations to take a proactive approach, often working beyond the scope of existing laws or regulations, to ensure that, from their inception, new products and practices protect people's privacy. Following the guidelines meant guaranteeing data security at all points where it was collected, "cradle to grave." It meant being transparent about the fact that data was being collected, and ensuring that users consented to how the data was collected and used. It meant always keeping the user's experience and privacy top of mind when designing products and practices. If a user didn't consent to data collection, any information that might identify them would need to be stripped from the data as early as possible.

Cavoukian's guidelines were called Privacy by Design. Sidewalk Labs had been interested in following them since the Yellow Book days.

In 2010, the International Conference of Data Protection and Privacy Commissioners declared Privacy by Design a global standard. When the European Union put the landmark General Data Protection Regulation into force in 2018, it embedded the framework into its rules for data-collecting organizations. By then, Cavoukian had finished her final term as Ontario's privacy commissioner and launched a new career spreading the gospel of her work.

Part of her new career was spent studying the work of high-profile smart-city projects around the world. Cavoukian didn't like most of them. Dubai called itself "smart" by installing thousands of closed-circuit TV cameras across the city, then setting up a system to scan the footage with artificial-intelligence and facial-recognition software for the benefit of law enforcement. In Shanghai, officials had the power to capture more than twenty million images daily through a mix of drone, satellite and close-range cameras, many of which had facial-recognition capabilities. Cavoukian wasn't entirely against smart cities, but she considered Dubai and Shanghai examples of "smart cities of surveillance." She hoped to see more municipalities try to become "smart cities of privacy."

Sidewalk called her in 2017 with an enticing offer: Privacy by Design would be its default mode of operation, and if she wanted, Cavoukian could hold the company to its word as a paid adviser. She accepted, and soon became a helpful ambassador, too. While some critics had warned about the privacy implications of lacing Quayside with sensors, their concerns were mostly speculative—they raised important questions, but those questions were about technology that hadn't been built yet. Hiring Cavoukian gave Sidewalk an instant, credible line of defence against the skeptics. She was a sound bite machine, pointed and direct, alarmist while never quite hyperbolic. If someone uttered a warning about what could go wrong with data collection at Quayside, she'd be on their tail, reminding people on Twitter, in news media and on TV that she wouldn't let Sidewalk do anything without following Privacy by Design.

In private, she reviewed ideas and policy proposals—including with Sidewalk's data governance working group, which pressed Sidewalk to add more details and privacy protections in its early data-use policies, especially in the aftermath of the Cambridge Analytica scandal. The group also discussed ways to "de-identify" or remove personal information from collected data, such as using artificial-intelligence software to scrub faces from camera footage. One of the biggest issues Cavoukian recalls discussing with Sidewalk staff surfaced around garbage collection. The company wanted to divert more residential waste to recycling. By monitoring what was being thrown out, such a system could easily be used to track individuals' private conduct in intense detail. "I said, 'Absolutely not. You do not do that,'" Cavoukian recalls.

Her relationship with Sidewalk was going smoothly. In early September 2018, she exchanged e-mails with policy director Micah Lasher and privacy head Alyssa Harvey Dawson, hoping to confirm that the company still planned to immediately strip personal information from any data that might be collected at Quayside without people's consent. They gave Cavoukian the assurance that the company would do so, and as summer turned to fall, she didn't think about Quayside much. She didn't need to.

That was about to change.

In the mid-2000s, Google was at the centre of the biggest and fastest redistribution of information in the world's history. Because information had become easy to find, information about *people* was easy to find, too. People's pasts crept up on them—they were just a Google search away. But the way that Google made this information easy to find led to another historic change. Its tailored advertising system created a market for intensely specific data about people's behaviour that became intrinsic to how whole industries worked: not just technology, but media, retail and many others.

The sheer volume of data was both unwieldy and controversial. In 2006, when the United States Department of Justice subpoenaed Google, Microsoft, Yahoo and AOL for what likely amounted to billions of search records, only Google refused to comply, claiming it would violate their users' privacy and reveal trade secrets. But the Electronic Frontier Foundation, a leading privacy and digital rights organization, still warned that Google and other search engines' trove of data was a "honey pot" of information ready to be exposed.

When it bought the ad company DoubleClick in 2007, Google said the acquisition would make its ad targeting better. It could collect even bigger troves of information about Google users as they surfed the web. It also put the company in the crosshairs of competition regulators and made the honey pot even more enticing for bad actors. In that era, however, Larry Page and Sergey Brin thought many concerns about privacy breaches were overblown.

In time, even Google's biggest boosters were having a hard time grappling with the sheer amount of data the company collected. Richard L. Brandt, the author of *The Google Guys*, a swooning hagiography whose 2011 edition is subtitled *Inside the Brilliant Minds of Google Founders Larry Page and Sergey Brin*, drew the line at data. "The question of whether Larry and Sergey can be trusted with all that data can never really be answered in the affirmative," he wrote. In the wake of the 2018 Cambridge Analytica data misuse scandal, Google made numerous public efforts to show the public they were caring for data responsibly. The company even published a paper outlining how to delete data from its cloud platform. But these efforts were undercut by Google's own admissions of how the digital world works. In his 2013 book with Jared Cohen, *The New Digital Age*, Eric Schmidt, then the chairman of one of the world's biggest data collectors, acknowledged that "the option to 'delete' data is largely an illusion."

Dan Doctoroff and his deputies tried to distance Sidewalk from Google and its hunger for data. They tended to mention the company only when referring to it as a source of "patient capital" through Alphabet—the money with which Sidewalk could build the city of its dreams—while taking pains to say Sidewalk wouldn't use Torontonians' data to sell ads. But scrutiny of Google was on the rise. In August 2018, the Associated Press reported that the company was tracking people's movements even when they'd expressly asked it not to. And the lines between divisions of Alphabet were beginning to break down, with public trust eroding alongside them. The health division of DeepMind, the pioneering artificial-intelligence company Google had acquired in 2014 but kept as a separate unit, like Sidewalk, had once promised that "data will never be connected to Google accounts or services, or used for any commercial purposes like advertising or insurance." Then, in late 2018, DeepMind's health team, which was collaborating with the U.K. National Health Service (NHS) in its research, was absorbed by a new Google division called Google Health. Though both Google and DeepMind said that the NHS would remain in control of its data, this led one law researcher to speculate to *Wired* that the move might give Google "unprecedented access to the best repository of health information on the planet—that of the NHS—in a way that patients have zero control over."

By October 2018, Sidewalk Labs, and by extension Waterfront Toronto, could no longer isolate itself from these heightening concerns.

Months earlier, Toronto's MaRS Discovery District entrepreneurship hub had sent Waterfront an unsolicited proposal about storing data from Quayside in a trust. In principle, a trust allows one person or entity to possess and manage something given to them by a second person, often on behalf of a third person or entity. A banker, for example, might manage a trust filled with a parent's money on behalf of a child. Data trusts can operate similarly: information collected from a specific place—say, a neighbourhood—can be stored on behalf of the people who provided that information and managed by a trustee or trustees for the benefit of the public. Whoever creates the trust can, in effect, create the rules: how the data is chosen, collected, stored, managed. Sidewalk had floated the idea of setting up an independent data trust as early as 2016, and Doctoroff was mulling the concept publicly by early 2018.

But Waterfront was the Quayside partner in charge of shaping and overseeing anything that might get turned into official government policy, and on October 4, the agency took up the offer from MaRS. They began discussing trusts in earnest. Then, on October 9, Bianca Wylie published an article about data trusts through Jim Balsillie's digital governance think tank, the Centre for International Governance Innovation, with Sean McDonald—whose research had helped popularize the concept. "Data trusts can play a role in ensuring governments guarantee that data required for public services isn't captured by commercial interests or held hostage for shareholder value," they wrote. ". . . When it comes to maximizing the value of our data and technology, data trusts are a new version of one of history's most useful legal tools."

October proved to be a busy month. Three days later, the *Toronto Star* published a story about Sidewalk software called Replica, which created "synthetic" population models using anonymized data to help planners understand how people move about their city. Replica would take cellphone location data, which Sidewalk said was stripped of identifying information, and adjust it with public census data to match a city's demographics in order to help planners run projections of realistic travel patterns. Aggregated location data was already for sale, and some governments were already buying it, but Sidewalk promised more accuracy. The Canadian Civil Liberties Association issued a warning. "Do people even know their data is being collected?" asked the association's privacy director, Brenda McPhail. "Is it reasonable for that data to be used by a for-profit vendor to sell back to the government?" she continued. "It's a morass of ethical issues." Jim Balsillie weighed in, too, arguing that Replica was further proof Waterfront shouldn't have signed the Sidewalk deal without a clear policy on data use. But he'd already used his angriest words in a salvo of his own.

Balsillie had finally had enough. He'd spent nearly a year criticizing Sidewalk to friends, executives, politicians, bureaucrats and, occasionally, reporters. He had little faith in Waterfront Toronto after his Evergreen Brick Works meeting. Having watched John Ruffolo's and Julie Di Lorenzo's resignations, and the slow start of the agency's digital strategy panel, Balsillie decided to take his frustrations public. On October 5, as the one-year anniversary of the Quayside partnership loomed, he published an op-ed in the weekend *Globe and Mail.* Editors teased the piece on the front page. The headline read: "Sidewalk Toronto has only one beneficiary, and it is not Toronto."

Balsillie tied together everything that made him feel the Quayside process was out of control: the prime minister fawning at the announcement a year earlier and Eric Schmidt's comment that someone should give Alphabet a city; deputy mayor Denzil Minnan-Wong's cryptic quotes about Waterfront's contract with Sidewalk; Di Lorenzo resigning out of frustration; and Bianca Wylie's description of public meetings as a masterclass of "arrogance and gas-lighting." He drew a line from Big Tech's massive historical data collection to Sidewalk and the future of Canadian rights. "Any data collected can be reprocessed and analysed in new ways in the future that are unanticipated at the time of collection and this has major implications for our privacy, prosperity, freedom and democracy," he wrote. "As long as Waterfront remains clueless about IP and data while deferring to Sidewalk on all the critical decisions, Canadians will continue to be treated to glitzy images of pseudo-tech dystopia while foreign companies profit from the IP and data Canadian taxpayers fund and create."

The community that might one day be called Sidewalk Toronto, he continued, "is not a smart city. It is a colonizing experiment in surveillance capitalism attempting to bulldoze important urban, civic and political issues. Of all the misguided innovation strategies Canada has launched over the past three decades, this purported smart city is not only the dumbest but also the most dangerous."

The Balsillie op-ed marked the apex of a shift that had begun quietly over the summer. Sidewalkers' friends were asking why they worked at the company, and some staff began having doubts about what they'd been sold on. Even local Google employees were facing such questions—under all this pressure, Sidewalk couldn't distance itself from the more recognizable sister company. With public trust eroded, Sidewalk's leadership began second-guessing their relative silence on data collection and privacy.

"The change in tone came after Jim Balsillie's criticism," Ann Cavoukian says. "They really felt that. I can't tell you what a huge impression that made."

For much of 2018, Sidewalk privacy chief Alyssa Harvey Dawson and her Canadian counterpart, Lauren Reid, had been sharing ideas about privacy and data governance with Waterfront's digital panel, their own in-house working group (which included Cavoukian), and officials from both the national and provincial privacy commissioners' offices. They'd discussed de-identifying data, with checks and balances to ensure that anyone who collected or used data did so responsibly. Reid, in particular, worked to incorporate the Canadian context into Sidewalk's ideas, hoping to embed caution, care and respect into ethical processes for data use that could be replicated worldwide.

The company's plans often shifted, just as the unbridled ideas of the Yellow Book had given way to more realistic proposals in its Quayside bid. For all the accusations lobbed by opponents that the terms of the Quayside project needed to change, Sidewalk had been flexible with its ideas, if not reactive to criticism. And as both data trusts and privacy concerns flooded the discourse around Quayside that fall, Sidewalk was preparing to introduce a pair of concepts to address both—to the surprise of some of its own privacy advisers.

First was a take on a "civic data trust" that would be fully independent: even Sidewalk would need to apply to collect or use data in the neighbourhood, and provide proof each time that it planned to use the data responsibly and ensure privacy risks were minimized. The trust would review the applications, approve any sensors or devices that would be used to collect data, and have the power to audit projects and shut down data access for some users. Any data that didn't identify a person would be freely accessible to the public. Because the trust would be run independently, any person or business could apply to collect or use data. Allowing this access helped Sidewalk distance itself from accusations that it might do something nefarious with data, or that it would corner smaller players out of the market for city-driven technology.

The *kind* of information the trust would store, however, is where Sidewalk's new ideas got murky. For its second new concept, Sidewalk proposed a whole new category of information for the trust to manage called "urban data." If images or information were captured by a camera or sensor in a public place like a park at Quayside, and didn't originate from a person's cellphone or other electronic device, Sidewalk wanted to designate it as urban data. But Sidewalk wanted to extend the definition of the term to include semi-private spaces, too: areas like stores, courtyards, lobbies and other privately owned spaces where the public was welcome to loiter. Even some data from individual apartments—such as temperature data that could help monitor energy usage—could count as urban data, so long as it was collected with consent and could help the community.

Because the concept of urban data was an attempt to formalize a new kind of thinking about corporate data collection, it had no legal definition outside of Sidewalk's own slide decks and documents. It surprised many of the advisers with whom Sidewalk management shared details that fall, leaving them to marvel at the countless reasons it wouldn't work. It certainly didn't fit in the context of Canadian law, and Canada was where Sidewalk wanted to build first. That was the whole point of Quayside as a test bed.

As proposed, the framework didn't seem to distinguish between data about people, such as the direction in which they were walking, and data about the environment, such as air quality. Each would require very different regulations and protections. Governments had their own, sometimes looser rules to guide data collection—they run the census and administer taxes, after all—and it wasn't clear if the City of Toronto would need to follow a different set of rules if one of its departments wanted to collect or use information at Quayside. Sidewalk also distinguished urban data as separate from that collected online or through cellphones—adding more confusion because some of its own software, in particular Replica, relied on cellphone location data. Some Sidewalk and Waterfront advisers who were privy to the urban data concept noted that exempting cellphone data would also mean that Google's own long-criticized data collection would exist outside the trust's regulations. So even though Sidewalk would follow the rules of a data trust at Quayside, there would be little stopping Google from collecting its usual cellphone app data over the same geographic area and, say, combining it with something that a sister company (such as Sidewalk) collected from the trust. Doing so could give Alphabet a serious advantage over rival urban-planning companies, and the upper hand in designing lucrative technologies that would cater to urban life.

Armed with all these concerns, the experts Sidewalk was privately courting, including staff from two different governments' privacy commissioners' offices, tried to tell the company that the term "urban data" was too ill-defined. Even though some of them were the very officials Sidewalk would need to work with if Quayside were built, Harvey Dawson and her team pressed forward with the urban data concept. On Friday, October 12, at least one person in Sidewalk's own data governance working group warned Harvey Dawson and her team about the ambiguity of the term. The group asked Sidewalk not to go ahead with an idea that could upend existing privacy regulations, but instead find ways to show that the company would uphold even better standards than those regulations allowed. Undeterred, Sidewalk spent the next few days fine-tuning a slide deck and blog post on urban data and a civic data trust.

When Kristina Verner and others at Waterfront finally saw the details, they couldn't believe Sidewalk would propose such a confusing way to distinguish what was effectively a made-up classification for data. The agency's lead privacy lawyer, who had once served as interim federal privacy commissioner, was astonished. Chantal Bernier felt Sidewalk was trying to establish some kind of exemption from data protection regulations simply because someone was out in public. "We really contested that from day one," Bernier says. "Personal information is personal information, whether you're walking on a sidewalk or at home." From the day Waterfront launched the Quayside RFP, the agency had been eager to see what kind of technological innovations a partner like Sidewalk could bring to Toronto. But this wasn't technology, it was policy, and policy was supposed to come from governments. "The mere fact that they would elaborate such detailed policy on the management of data was really an appropriation of policy-making that did not belong to them," Bernier says.

Timothy Banks, another technology lawyer who would go on to advise Waterfront, was similarly surprised that Sidewalk had actually put the proposal on paper. "They didn't fully understand the complexities of dealing with the city, the province, Waterfront Toronto and the private-sector institutions that would be participating," he says. Sidewalk's vision for urban data and its model for a trust "seemed both premature and potentially doomed to fail."

Saadia Muzaffar felt Waterfront Toronto had lost the plot. When Waterfront built its digital strategy panel, it sought out people who looked at tech from a mix of angles, including law professors, tech execs and urbanists. Muzaffar brought the lenses of both entrepreneurship and intersectionality to the oversight group. She was the founder of a not-for-profit called TechGirls Canada, which pushed for equity in science and tech for LGBTQ+, immigrant, refugee and

Indigenous women. She'd co-founded Tech Reset Canada with Bianca Wylie and several others, too, to advocate for technology projects and policies that focused on the public good. She thought a lot about the structures in society that determine who gets to prosper and how. She wasn't confident that Waterfront was thinking enough about how that kind of power worked.

The panel hadn't gotten much to review by early October, and Muzaffar was stunned at how she saw the partnership between Waterfront and Sidewalk developing. At a public meeting in August—the one that Waterfront and Sidewalk had said in an internal document would focus on "non-controversial" subjects—she was surprised that the two organizations sidestepped conversations about data and digital infrastructure.

After mulling the meeting over for weeks, Muzaffar submitted a resignation letter on October 4. "Waterfront Toronto's apathy and utter lack of leadership regarding shaky public trust and social license has been astounding," she insisted. "There is a growing list of squandered opportunities to take ownership of the narrative that would clarify the boundaries between who is in charge of how this 'partnership' unfolds." As the sole visible person of colour on a panel that was supposed to represent the public in one of the world's most diverse cities, she said her decision wasn't easy. But she didn't want to be affiliated with Waterfront's choices anymore.

Though Muzaffar had previously dropped some hints about her frustration, the decision caught Waterfront by surprise. The panel's work hadn't really started, since Sidewalk hadn't given them much to work with. Kristina Verner wished she'd had a chance to talk things over with Muzaffar, to let her know that Waterfront had been waiting on Sidewalk to share details with them that the panel could review.

There was little time for regret. The news that Muzaffar had quit broke later that afternoon, and the coverage that followed drew newfound attention to Waterfront's panel of digital strategy experts. What was the panel supposed to be doing, and who were all these people? So far, the public didn't know much more than the fact that John Ruffolo and Saadia Muzaffar had resigned. The resignations began prompting even more questions. What the hell was happening at Waterfront?

Reporters spent the next few days trying to figure out what might come next from the digital panel. Then, half an hour after most newspaper deadlines on October 15, Sidewalk sent media a copy of its long-awaited data plan proposal. Arriving too late in the day for reporters to ask independent experts to review and analyze it, the proposal outlined the company's ideas for a civic data trust and urban data, and argued that storing data on Canadian servers was neither necessary nor legally required, despite some Canadian privacy experts' long-standing fears that storing data in other countries could potentially make it subject to surveillance.

Sidewalk was scheduled to discuss these plans with Waterfront's digital advisers three days later. As I canvassed several members of the panel the next day, three told me that, like Muzaffar, they were on the cusp of quitting. Urban-planning professor Pamela Robinson was demanding more substance on Waterfront's plans. Andrew Clement, both a due-process geek and the most vocal surveillance hawk on the panel, said that Waterfront, not Sidewalk, needed to be developing data plans on behalf of the public. "It's fundamental of the consultation process that it be independent of the interests that are driving it," he said. Teresa Scassa, one of Canada's most prominent academics at the intersection of technology, privacy and policy, wondered whether she and the panel were "window dressing" for the project instead of genuine advisers.

Three top advisers had left the project—Muzaffar and Ruffolo from the digital panel and Julie Di Lorenzo from the board—and three more were considering it. The Quayside project was in the midst of an image crisis. If all these experts saw what was happening and walked away, how was anyone supposed to trust that Sidewalk and Waterfront knew what they were doing?

Dan Doctoroff spoke the next morning at a conference at Toronto's Ritz-Carlton, and a handful of reporters pressed him on the advisers' frustrations. He downplayed the controversy. "I think, to some extent, some of the criticisms were unfair, because they prejudged something before the processes actually unfolded," he told us. "I'm also somebody who believes, even when you disagree with something, you should try and make it better," he added. "I think we have done nothing but demonstrate that we are open and listening." When the miniature press conference ended, he headed to 307 Lake Shore, where he arrived in time for Sidewalk's first secret advisory council meeting.

It had been exactly a year since the federal, provincial and city governments had all sent their top politicians to the Quayside announcement to welcome Sidewalk with glee. With resignations and controversies mounting, however, politicians were now distancing themselves from the excitement. Federal innovation minister Navdeep Bains was at the Ritz-Carlton that morning, too. A *Financial Post* reporter and I caught him in the hall and asked if he was concerned about the state of the Quayside project. "We're closely monitoring the situation," he said, sticking to his talking points with a promise that his office was working on new legislation to protect Canadians' data. (It took twenty-five months for that legislation to arrive, and then his government let it die in Parliament.)

It was a rapid-fire day for the press. Mayor John Tory was up for re-election in less than a week, and he came to the *Globe* newsroom that afternoon to take questions from reporters. With the CN Tower and Bay Street skyline unfolding to his left, Tory shared less enthusiasm for the Quayside project than he had a year earlier. "There was some considerable zeal about announcing it," he said. "I think it was announced with too many unanswered questions." He had "acute concern," wanted more assurances around intellectual property and data before he could confidently back the project, and said Waterfront should probably have figured out those terms before announcing the partnership. "The process here was not handled the best way," he said.

But Tory was a perpetual optimist, and was willing to give Waterfront a chance. "I think this still could be very beneficial for the City of Toronto on the right terms," he told the crowded room. Sitting to his left, his communications director, Keerthana Rang, was keeping her eye on the time, ready to get Tory to his next meeting. Five days later, Tory would be re-elected mayor. Barely a week after that, Rang left politics to handle public relations for Sidewalk Labs.

Three dozen people—media, Sidewalk staff and members of the public—were packed into the tiny public gallery at the back of Waterfront Toronto's boardroom the next late-October day, some anticipating mass resignations from the digital strategy panel. Sidewalk, Waterfront and government staff lined the walls next to the board table; one senior civil servant from Ottawa was e-mailing a play-by-play of the meeting to her bosses.

"It has been an eventful stretch since our last meeting," said the digital strategy panel's interim chair, the University of Ottawa internet law professor Michael Geist, from the head of the table. Without naming them, he called Saadia Muzaffar's and John Ruffolo's departures a "real loss" for the panel, but said he welcomed the attention that their resignations drew to the group's work. He also made it clear that Waterfront's latest woes wouldn't be resolved at the meeting. "I'm not convinced that an initial proposal tabled about seventy-two hours ago means the vast majority of privacy and data concerns have been addressed," Geist said. That the panel had so little time to review the proposal was "unacceptable," he added. And he warned that Sidewalk's newly proposed data policies had done little to assuage confusion about who was actually in control of the Quayside project.

Meg Davis and Kristina Verner took the floor, insisting that Waterfront would be the partner in charge of determining how intellectual property and data would be guarded at Quayside. If Sidewalk made proposals, Verner said, they "should be seen as an effort to catalyze conversations around these topics. . . . Nothing will proceed without our approval and confirmation of adherence to the laws of the city, province and country."

After a handful of other discussions, Alyssa Harvey Dawson stood up and began to backpedal on Sidewalk's behalf—first for how little time the panel was given to review the data policy proposals, but then for the policies themselves. "They are merely ideas," she said. ". . . We're eager to hear where you think the proposals are good, where you think they fall short or go in the wrong direction. As a matter of process, I wanted to make sure that it's fundamentally clear to everyone that we defer to this panel, and to Waterfront Toronto, on where to go from here."

The tension in the room dissipated. The panellists began discussing Sidewalk's data proposals civilly. Andrew Clement warned that urban data would be the least-sensitive data collected at Quayside, given that its Replica travel-measurement software would use cellphone location data. Pamela Robinson pointed out that Sidewalk's new data ideas hadn't yet incorporated the public's perspective. Teresa Scassa told Harvey Dawson and her colleagues Micah Lasher and Craig Nevill-Manning that defining urban data wasn't enough—the public should know *all* of the data that could be collected or used at Quayside, not just what was in the trust. Further, privacy was just a subset of "governance" issues around that data, she said. The security of the data, where it was stored and who got to own and access it all had implications for how collecting data in the first place would affect human rights, how that data could be monetized and who got that money. All of this needed to be clarified, Scassa continued, especially given the outsized role a private company would have in overseeing the Quayside community.

Scassa, Robinson and Clement had voiced their frustrations, but they were no longer seething. The day's drama seemed to be over. But there was one person sitting along the wall next to the boardroom table who seemed visibly agitated. As the meeting wrapped up and I wandered toward the table looking for comments, I felt someone tug on my elbow. It was Ann Cavoukian. "Hey," she said, nearly whispering. "I think I'm going to quit."

The data trust proposal had irked Cavoukian. She'd heard about it only days earlier. If the trust was independent, Sidewalk wouldn't have the authority to guarantee that its users followed Privacy by Design—and therefore couldn't guarantee that it would use only sensors and other tech that stripped data of identifying information. When, minutes earlier, Alyssa Harvey Dawson had told the digital panel that Sidewalk would defer to Waterfront for decisions about data policy, Cavoukian felt the last of her influence slip away.

She mulled her future overnight, then sent Lasher and Harvey Dawson her resignation. Two hours later, Dan Doctoroff e-mailed her in astonishment: "Is this a miscommunication?"

It wasn't. The news of her departure went public hours before Cavoukian had planned to announce it. Within days, the story was making news around the world, adding yet another layer of public suspicion to Sidewalk's relationship with data. Overlooked by the likes of the *Guardian* and Gizmodo was the fact that Cavoukian's decision to quit Sidewalk wasn't all that consequential: she almost immediately agreed to consult for Waterfront instead. Toronto's city of the future would indeed follow Privacy by Design, no matter who enforced it. Nonetheless, Sidewalk had fumbled another opportunity to build trust. And not just the public's trust. The surprise urban-data gambit had cost Sidewalk its own privacy experts.

That's experts—plural. Lauren Reid didn't carry the same celebrity in privacy circles that Cavoukian did, but she had done much of the heavy lifting on the subject for Sidewalk during its first year in Toronto, guiding the company toward a way of thinking about privacy that would work in the context of Canadian law. The company's management sometimes bristled at advice, but she would take time to explain to her colleagues why Waterfront and its digital strategy advisers were so often pushing back.

When Sidewalk executives introduced their urban-data concept to Reid, she tried to convey to them, just as their other advisers had done, that such an amorphous data category didn't make sense from a legal perspective. But her expert opinion didn't deter them, and she soon found herself toeing the company line.

"You could see the frustration and how, even publicly, she would have to say things she didn't agree with," says Waterfront's innovation vice-president Kristina Verner, who worked closely with Reid. In the weeks that followed, Reid grew quieter and quieter. Verner describes her as turning "numb."

Reid resigned in November. Though her departure was much quieter than Cavoukian's, it was more significant. While Cavoukian had shaped Sidewalk's understanding of Privacy by Design on a part-time basis, Reid had been a full-time ethics watchdog. "Losing her voice was a massive hit to the project," Verner says.

By the time Reid left, Sidewalk had gotten a reprieve from its stream of negative headlines. Its parent company, however, wasn't enjoying the quiet. In late October, the world watched as a group of citizens drove an Alphabet project out of a different city altogether.

CHAPTER 15

Prelude to the Feud

MANY MONTHS BEFORE Sidewalk Labs and Waterfront Toronto grappled with mass resignations, a man in Berlin who used the pseudonym Sundar Panet found himself reading a book about insurrectionism by the Invisible Committee, a far-left group or person whose identity, or identities, was also pseudonymous. Panet was struck by a photograph about a hundred pages in. Seen through the front window of a bus in Oakland, California, the image showed two people blocking that bus's path while holding up a banner. *Fuck Off Google*, it read.

In 2004, Google had begun chartering a private, biofuel-powered bus to shuttle employees from across the Bay Area to its suburban office park headquarters. Back then, the service catered to about 150 people a day, who could relax or use its wireless internet to work during their commute. Its success had unintended consequences. By the eve of the global pandemic in 2020, more than a thousand private commuter buses were flooding the Bay Area's highways, forming an ad hoc transit network that was worth more than $250 million.

This had some benefits—namely, as many as fifty-two thousand fewer commuters sitting alone in cars inching through traffic every day, in turn preventing tens of thousands of tons of carbon dioxide emissions. But Google was accused of taking up space at public bus stops to load passengers into its private luxury buses. The company donated $6.8 million to San Francisco's bus system in 2014 to cover the cost of rides for low-income youth. Close observers, however, worried that the donation fell short of offsetting the far-reaching consequences of its transit network, which had sent rents surging anywhere near those bus stops. And with fewer riders, the Bay Area's transit operators produced less revenue to reinvest in the systems everyone else had to use. People went so far as to lie down in front of the buses to protest the power Google was exerting on the Bay Area.

Panet kept staring at that photo. *Fuck Off Google.* The image had been circulating for a few years, but now he saw it in a new light. Made with a few simple materials, the banner had been noticed around the world. The phrase it bore kept turning in his head. It would make a fine slogan in Berlin, too.

Just east of Berlin's core, the Landwehr Canal carved a thin line between Neukölln and Kreuzberg, a pair of communities filled with artist types and recent immigrants. In the last week of the summer of 2019, after two months of searching, I met one of the people who shared the pseudonym Sundar Panet—the one who was struck by the photo of protesters in Oakland—at a pizza joint on the edge of the canal. Earlier in the grey day, hundreds of thousands of Berliners had gathered at the Brandenburg Gate to protest government inaction against climate change. Many of them were now unwinding together here, sipping tea and beer along the canal, and at the tables that spilled forth onto cobblestone sidewalks from bars, restaurants and the convenience stores lovingly called *Spätis*.

Raising a glass of white wine to his mouth with one hand and gesturing toward the canal with the other, the man who called himself Panet described this corner of the world as "a diverse and joyful mess."

German reunification had turned Berlin into a global epicentre of creativity in the nineties. Young people flooded the city, taking over abandoned buildings and warehouses to build communities around art and electronic music. The city was fun, accepting and cheap. But it was not impervious to the gentrification and soaring costs of living that afflicted other cultural capitals such as New York, London and Melbourne—it just took a little while to catch up after the Wall fell. By the end of 2017, Berlin had the fastest-rising home prices in the world; between 2009 and 2019, the city's average rent doubled.

This was, at least in part, because of start-up culture. Young tech companies flourished after the global recession that struck in 2008. Instead of money flowing *away* from tech, as it had during the dot-com bust a few years earlier, companies like Google, Facebook and Amazon became seen as beacons of the future. The promise of future successes had investors salivating in the 2010s. Venture capital investment surged. Berlin's burgeoning tech sector had a handful of superstars, like the fashion retailer Zalando and the music-hosting company SoundCloud, and the start-up community there earned the nickname Silicon Allee—a German play on New York's Silicon Alley, which itself paid tribute to Silicon Valley.

Investors need to buy a stake in many start-ups to find eventual winners, so start-ups benefit from running with low overhead costs. It so happened that Kreuzberg's real estate was, at least at the start of the decade, relatively cheap. When Panet first landed in the area, start-ups were even finding homes in off-the-beaten-path office spaces in the interior courtyards of Kreuzberg's apartment complexes. Tech was making its home there. And soon after his arrival, Panet was walking down the *Straße* and saw Google's logo on a poster.

Alphabet was all-in on start-ups, too. It wanted to set up a campus for them in an old electrical substation—an *Umspannwerk*—along the Landwehr Canal. The unassuming red-brick complex, built in 1926, was half a block long and had survived the Second World War. It was already home to an event space, a Red Bull–branded recording studio and a restaurant, and it was right on the Landwehr Canal, where Kreuzberg blended into Neukölln. In late 2016, Google announced it had secured a fifteen-year lease from the building's owner, a West End London property investor, to take over 32,000 square feet in the basement and soon-to-be-former event space. As we spoke, Panet still remembered his immediate reaction: "Oh, fuck no."

—

Panet was a computer guy, a self-described nerd and free-software advocate who had helped organize campaigns in the past to protest the power that Big Tech held over the internet. "We genuinely thought that the internet could enable more participation, more democracy, and not become the shithole controlled by a few that it turned into," he told me. The dominance of platforms like Google and Facebook had taken away the web's sense of endless possibility. Even Panet's pseudonym was a protest: a portmanteau blending the name of Google CEO Sundar Pichai with ARPANET, the U.S. military's experimental pre-internet information-sharing network. For Panet, the internet was something still worth fighting for. "Free software that belongs to everyone, plus hardware you can control, decentralized services and end-to-end encryption—that may enable us to use the internet and media technologies in an emancipatory way," he said.

The poster he'd seen after moving to the neighbourhood wasn't in support of the proposed Google campus; it was for a community meeting to stop the project. Panet showed up and was surprised at the range of people he saw. Nerds like him were sprinkled among anarchists, lawyers and concerned neighbours. There was a full spectrum of frustrations, from local gentrification to the global consequences of Google and Alphabet's concentrated power. Some worried that, as was feared in Toronto, a Google campus might herald the encroachment of a smart-city development and all the surveillance technology that had historically followed.

The concerns aligned with Panet's own. And when he later visited a meeting hosted by local anarchists, he was struck by the meeting's structure: like his vision for the ideal internet, the gathering was decentralized and open for anyone to present ideas.

Google had established start-up campuses to host and mentor local entrepreneurs in a half-dozen or so other cities, including London, Tel Aviv and Warsaw. The company marketed these campuses as resource centres for local tech communities, but skeptics saw them as opportunistic—thinly veiled attempts by one of the world's biggest companies to bring together local talent and ideas, then cherry-pick the best to buy for itself. The ragtag band of Berliners took that skepticism a step further. They saw Google as parasitic: accelerating the arrival of more start-ups and real estate investors into Kreuzberg, further pricing out the people who made the neighbourhood special. Google had long held sway over a massive swath of the internet. Now it was exerting power over the physical world—over their home.

Dan Doctoroff's failed dream of a New York Olympics was much less of a catastrophe than Berlin's attempt to secure the 2000 Games. Grassroots opposition coalesced to the point that the final International Olympic Committee visit in 1993 was met with ten thousand protesters. Berliners forced the stewards of their city to focus their reunification efforts beyond a five-ringed, multi-billion-dollar marketing blitz.

In Kreuzberg, this kind of activism is part of the community's very fabric. It has been home to Tag der Arbeit labour protests each May 1 since clashes with police in 1987 prompted a massive riot that left a supermarket burned to the ground. With much of it boxed in on three sides by socialist East Berlin during the Cold War, Kreuzberg had become something of a slum, with hundreds of buildings abandoned or slated for demolition. By the early eighties, squatters occupied more than 275 buildings as it became a home for students, artists, punks, anarchists, LGBTQ+ runaways and the often-Turkish working class. When West Berlin officials tried to raid and clear the buildings beginning in 1981, riot police and tear gas were met with hurled paving stones. Though some squats were cleared out, many others were legalized, cementing a culture in Kreuzberg focused on tenants' rights.

Stefan Klein wasn't living in Kreuzberg then, but after arriving from Frankfurt a couple of decades later, the music industry lawyer became one of the closest things to a leader in the anti-gentrification wing of Google's opponents. Klein, who does use his real name, is tall and a bit grandfatherly, and favours twilly blazers. Like many in Germany, his English comes through with a British flavour. In 2019, rolling a cigarette as he sauntered down one of the May Day riots' main thoroughfares on Reichenberger Straße a block from the canal, he explained what he called the "typical Kreuzberger mix": "A mixture of little businesses, people with not so much money, people with normal jobs, people on welfare, migrants, people here for fifty years, people who just moved here. Everybody comes together. You need just the normal things everybody needs."

Klein began fighting against gentrification in the mid-2010s when he found out a beloved local French bakery named Filou was under threat of eviction. The landlord wanted something "fashionable," Klein told me. "But you cannot come to a place and say, 'I want to change it after my plan without asking the people living there,'" he added. "We took action against that."

He and his neighbours formed a neighbourhood group called GloReiche, named after the intersection where Filou was located: Glogauer Straße and Reichenberger Straße. In what would become a theme in Kreuzberg, the pushback against the bakery's landlords grew and became decentralized. Different people and groups used different tactics to make their objections known, sometimes coming together only for larger protests, with one nearing two thousand demonstrators. It was only after regular protests, a smashed window or two and a plea from the owners' daughter to the landlords, that they gave in and signed Filou to a ten-year lease, Klein said.

Soon, GloReiche learned about Google's plans for the *Umspannwerk*. At first, the organizers were intimidated. Filou's landlords were "two little speculators from London town," Klein said, but Google was a multinational corporation that was many orders of magnitude more powerful. And there was another issue, Klein said: "Everybody loves Google."

The campaign started with brochures, in German and English, about Google's financial clout and how it could be wielded for gentrification. Next came flyers, laying out the same frustrations. One of the arguments was about the follies of venture capital. Normal businesses need to profit or at least break even to survive, but start-ups don't. Klein summed the argument up as we spoke: "They can pay high prices and don't look for income. And after they leave, the *kiez*"—the neighbourhood—"is devastated. But they don't care."

As GloReiche and other groups in Kreuzberg began to hold meetings, neighbours gobbled up the flyers and brochures. The different factions that came together each began to fight back in their own way, under names like Counter Campus and Google Campus & Co Prevent. After a while, Sundar Panet showed up and connected with other self-identified nerds. There were plenty of them in Berlin, home to the Chaos Computer Club, a hacker network focused on making the digital world safer, fairer and more secure.

At community meetings, the faction of nerds explained to their counterparts that Google wasn't a typical force for gentrification. "It is mass-surveilling, mass-sensoring, standard-defining, centralizing and tax-evading," Panet said. And they brought their fight with the internet giant back to its own turf: online. They made a wiki, a collaborative website, that augmented the decentralized nature of the campus opposition, sharing information about mass data collection and Big Tech's other city-changing projects. The wiki bore the motto from the banner Panet had seen in the photo from Oakland: *Fuck Off Google.*

As the nerds got word of similar fights in places like the San Francisco Bay Area and Rennes, France, they added details of those confrontations to the wiki. Soon enough, they were communicating with protesters in those cities, too.

Some protesters took tech in the opposite direction, using one of the oldest tools in communications history to spread their word: the printing press. They churned out thousands of copies of an opposition newspaper titled after Germany's favourite English metaphor, *Shitstorm*. Someone also made at least thirty thousand stickers exclaiming *Fuck Off Google.* The catchphrase became the decentralized movement's unofficial moniker. People tagged anti-Google slogans on dozens, if not hundreds of buildings, and protest signs sprouted from countless windows.

Google tried to talk with the protesters, but this was Kreuzberg. The protesters were going to protest. Even moderate branches weren't interested in engaging, worried that the conversation might be warped into creating the appearance of an endorsement. "The only talk we would accept would be that you would stop the project," Klein said.

At one meeting, someone brought up the idea of a "noise party": showing up and making a racket, both to draw attention to the cause and to frustrate the parties being protested. The first one was small, maybe thirty people banging pots, pans and whatever else they could find. Soon, some of them decided to throw a noise party the first Friday of each month.

The decentralized factions started to discuss the implications of regular in-person protests. Should they call the police to officially register the events?

No way, said the anarchists. That would mean recognizing the police's authority.

No one else felt comfortable making the phone call either. But they didn't stop making noise outside the *Umspannwerk*.

By 2018, Big Tech was threatening to reshape more than just Toronto and Berlin. Though its influence had been seeping out of the U.S. west coast for decades, the 2010s heralded whole new consequences. The demand for workers in the San Francisco Bay Area helped send home prices there surging 77 percent between 2009 and 2019, revealing only a fraction of the digital giants' influence on the physical world.

After years of arguing against independent reports that suggested Uber and Lyft increased congestion when they entered a market, the rideshare companies commissioned a study of their own in 2018 with similar findings, prompting Uber to admit that its business was "likely contributing" to congestion after all. And in Uber's quest to get driverless vehicles on the road, a company vehicle also killed someone. Elaine Herzberg, forty-nine, died in Tempe, Arizona, after an Uber-owned Volvo SUV struck her while travelling at about

40 miles an hour. The Volvo's sensors didn't recognize Herzberg as a pedestrian because she wasn't at a crosswalk, and the person in the driver's seat who was supposed to catch errors was distracted, transportation officials later said. Uber's vehicle of the future had been optimized for an impossibly orderly society; the company's understanding of cities, like Larry Page's, didn't account for the messiness of everyday life.

Pittsburgh was a haven for Uber's research, too, and activists there have argued that the high-paying jobs this brought to the city tended to go to young white men, who in turn shelled out big money to live near work, raising the prices for everyone else, including those in historically Black communities. Cities across the globe are undergoing similar transformations as the new titans of industry set up shop in cheap neighbourhoods. In many places, tech services have shocked the entire real estate market. In 2018, Canadian researchers found that Airbnb rentals had probably erased thirty-one thousand units from the country's long-term rental market. When the COVID-19 pandemic hit Toronto, listings for furnished condos in the region shot up 52 percent as people stopped travelling and treating the units as hotels.

Soaring real-estate prices weren't the only thing the residents of Queens feared when the Seattle-based commerce juggernaut Amazon announced in November 2018 that it would open a twenty-five-thousand-employee secondary headquarters in the New York borough. City and state governments offered nearly $3 billion in grants, credits and tax breaks between them to entice the company during its "HQ2" search—a scheme that was hatched by Amazon CEO Jeff Bezos out of jealousy of Elon Musk, who'd secured $1.3 billion in tax incentives to build a Tesla battery plant in Nevada. In Queens, local residents were furious at Amazon. Progressive congresswoman Alexandria Ocasio-Cortez became the de facto face of opposition. She fought an uphill battle trying to point out not just what the HQ2 deal would mean for rental prices in Queens, but what taxpayers were giving away because Amazon had dangled twenty-five thousand jobs in front of New Yorkers. Amazon—which paid no U.S. federal income taxes in 2018 despite making more than $11 billion in profit—cancelled its massive New York plan a few months later in the face of growing opposition. But it didn't give up on its second expansion location in northern Virginia, near Washington, D.C., where it was offered more than half a billion dollars in incentives.

The HQ2 saga helped Amazon do more than shrink its tax bill. It also gave the company reams and reams of data. Amazon received 238 bids in its hunt for a second headquarters—238 compendiums of how people lived and how they were governed. Amazon was handed demographic information, transit plans, zoning codes and more. With all this data, Amazon knew more about cities than virtually any of its competitors, and could plot future offices and warehouses in places with the lowest costs and greatest profitability while shortening delivery times. "This was about crowdsourcing data," said the University of Toronto professor and urban scholar Richard Florida, who himself was a Sidewalk Labs adviser. "This was never about an individual HQ2."

None of these companies were proposing anything revolutionary to benefit urban living. Sidewalk's plan for Toronto was bigger than a new corporate headquarters, but even its proposals were a mixed bag: society wouldn't be filled with garbage-hauling robot trucks anytime soon, let alone Larry Page's dream of buildings on wheels. Technology's push into cities is usually incremental, and Sidewalk's plans would have accelerated it slightly. But like Amazon's 238 HQ2 bids, and like Uber's self-driving cars sensing the roads around them, Google's push into cities would give it a much more immediate benefit: data.

A city filled with sensors, even those that help automate building temperatures, can capture information about the people who live there. More land to play with means more data and more money. By the time Sidewalk's Toronto project was facing mass resignations in late 2018, even the Berlin opponents' more moderate anti-gentrification faction began recognizing that data collection was the company's end goal. Free coffee and Wi-Fi at a Google campus would be a natural extension of free e-mail, Stefan Klein said: gradual steps to make everyday people "data slaves."

As Google's Berlin enemies realized the scope of their battle was growing, the scope of their tactics grew, too.

Whispers of the plan floated through the ranks of Berliners during the summer of 2018. As the days got cooler, little pieces of paper were passed around with a date, a time, a meeting point. Those who accepted the paper also received a warning: don't bring a phone; its signals could be used to surveil you.

In a courtyard near the *Umspannwerk* one morning that September, people began to trickle in. They didn't trade names; that would be too risky. As the time on the little slips of paper arrived, maybe thirty-five people had gathered. That wouldn't be enough. Organizers worried they might have to cancel. But it was morning, and this particular demographic wasn't made up of many early risers. Within an hour, the group had reached critical mass.

Someone gave a speech. "We're going to occupy the building," they said.

The crowd split into two groups. About twenty-five would enter the *Umspannwerk*, while the remaining sixty or so would stand outside and block the entrance, preventing police and security staff from entering to press charges for the illegal occupation.

They headed to the building, and the first group burst through the door. The space was surprisingly underwhelming—the phenomenon they'd opposed for nearly two years was little more than a construction site. The ceiling was unfinished, and the floor was bare concrete. Plywood, various tools and a scissor lift were left where they'd last been used. Cables dangled everywhere.

Then they saw something else: three construction workers. The occupiers suddenly realized they might be accused of trying to take hostages. That was the last thing they wanted. Then one of the occupiers realized they had some beers in their backpack—so they approached the workers and made an offering. "Okay, guys, it's not your fault," they said. "You can say you've been kicked out. Take my beer and go." The workers left, but a nervous security guard remained. He started freaking out and firing his pepper spray wildly. The occupiers eventually got him to leave, too.

There was little opportunity to celebrate. Though they'd seen the building's blueprints, they discovered there was an additional way in that they hadn't accounted for. If they weren't careful, cops might get in and arrest them.

Officers had already surrounded the building, their shadows flickering past the floor-to-ceiling windows. Some of the occupiers' nerves frayed. They searched for security cameras and other sensors, but held back from vandalizing the place—they didn't want to risk any extra charges. Some had brought sleeping bags and food, prepared to turn the occupation into an overnight affair, but it wasn't clear how long they'd last. Adrenaline was distorting their understanding of time.

One of the occupiers began calling the press. A handful of others ducked outside to speak to reporters when they arrived, and to hang up a *Fuck Off Google* banner behind their colleagues who were blocking the entrance.

Dozens of police had arrived by then, some in full riot gear. Down the block, a confidant acting like a normal neighbour listened to the cops' plans, leaking details to the occupiers with a burner phone. A little over an hour after the occupation began, the confidant sent a message to the team inside: police were planning on sneaking in through the back. So the occupiers ran out through the front doors, pushing past the officers and into the crowd of supporters, and mingling as quickly as they could so officers couldn't identify who was an occupier. Some of them quickly changed clothes while hiding behind umbrellas. It was chaotic: as officers tried to arrest people, other protesters tried to free them. A handful of occupiers were taken away by police. The others ran away.

Stefan Klein, who usually used the law to fight his battles, was impressed when he heard about the occupation. "Sometimes escalation is not so bad if you stop it in the moment," he said. "It was heard all over the world. If nobody knows, it's like a ship with no lights in the dark."

A little more than a month after demonstrators occupied the Kreuzberg building—just days after privacy expert Ann Cavoukian made global headlines for quitting Toronto's Sidewalk Labs project—Google cancelled the Berlin start-up campus. Instead, the company would spend €14 million to kit out a new "House for Social Engagement" and cover overhead costs, letting two organizations run it like a campus of their own for charities and social issues organizations.

"We won against Google," Klein told me. "It was incredible—for the moment."

For the moment?

"We are aware we have to stay alert," he added.

The non-profits would get support from Google for five years, but the company's lease was for fifteen years. It could return and try something else. Other tech companies were encroaching, too. The family behind the Rocket Internet start-up investment hub had property a fifteen-minute walk south in Neukölln; neighbours saw that as a threat. And a thirty-five-storey tower being built to the north in Freidrichshain, past the Spree river and the stretch of the Berlin Wall that had been turned into the East Side Gallery, was expected to make Amazon its chief tenant. There was more work to be done to protect the creative, immigrant-friendly enclaves to the east of Berlin's core.

"Change is a normal process," Klein said. "But what you can stop is speculation. People coming here, trying to get a few million in a few years—after a few rounds of speculation, the houses are no longer places for living. They're assets."

Some local politicians lamented the sudden death of Google's campus, fearing it might cast the city as anti-business. But, Klein pointed out, politicians often jump for joy when a private company does their city-building for them. "You see it in Toronto, where the government was happy to offer all the seaside space," he said.

Sundar Panet feels the same way. "No one is naive enough to think that when we defeated them here, we altered the course of gentrification forever," he said. "This is what it is: the giant got kicked in the nuts by a bunch of punks, artists and nerds."

CHAPTER 16

Twist My Arm

THE DAY AFTER Google walked away from Kreuzberg, Sidewalk Labs doubled down on Canada. Dan Doctoroff and his colleagues had embarked on a vast influence campaign that month, meeting with tech CEOs in Waterloo, visiting the Fortune Global Forum at the Ritz-Carlton to pontificate on cities, and launching their secret advisory panel. On October 25, the company pulled off perhaps its most important meeting yet: a dinner with twelve of the most powerful unelected officials in the federal government.

It was co-hosted by the office of Innovation, Science and Economic Development Canada and the Privy Council Office, the nerve centre of Canada's government. Securing the meeting was a coup for Sidewalk. The country's corporate elite pitched meetings to public servants all the time, but rarely did a comparatively small American firm command such attention. Four top innovation ministry officials were on the guest list, as were the secretary of the Treasury Board of Canada, the president of the Public Health Agency of Canada, and a handful of deputy ministers, including Kelly Gillis, whose portfolio at Infrastructure Canada meant overseeing Waterfront Toronto. It was a chance to sell Sidewalk's brand of optimism to the very people who would advise legislators on any regulatory or policy changes that the company desired at Quayside, if Waterfront ever gave it permission to break ground.

It was also a chance for Doctoroff's team to get in front of the reputational damage that been piling up for months. As innovation ministry staff briefed meeting co-host Paul Thompson, an associate deputy minister, they warned him about Sidewalk's vague proposals around the concept of urban data, about rumours that the company was pushing for more land than Quayside, and about the resignations of Ann Cavoukian and John Ruffolo. Canadian intellectual property was at stake, too, and federal staff warned about the contract I'd discovered that showed Sidewalk wanted full control of design partners' IP. In the talking points they drew up for Thompson, he was told to press Sidewalk to ensure that Canadians would realize the benefits of their ideas, and to bring up the importance of strong privacy measures and good data governance.

But the associate deputy minister was also told to strike a balance—one rooted in a growing global insecurity. That same day, a Berlin politician declared that Google leaving Kreuzberg would be "fatal" for the German capital. Thompson's colleagues feared the same thing: that if Sidewalk left Toronto, it could send a message to other tech companies that Canada was anti-innovation. Thompson was given specific instructions on how to address this, by saying: "I hope that Sidewalk Labs' experiences with the Quayside project could act as a positive test for additional product development and investment in innovation by Alphabet in Canada." The country's official stance on the Sidewalk project was to attract as many new and shiny things as possible. Welcoming the idea of progress without considering the consequences was a governing strategy hundreds of years in the making.

It's easy to create new problems while trying to solve an old one. Two centuries ago, parts of Europe—beginning with Prussia and Saxony—began to approach their economies with a newfound scientific rigour, trying to maximize the value of any asset they could. Forests weren't producing timber as quickly as they could harvest and sell it, so they planted forests anew, focusing solely on the trees they needed, aligning them in careful rows. The first generation of trees grew with resounding success. The second and third generation, particularly of spruce, did not. That first regimented generation had sucked the remaining nutrients from the soil of the old-growth forest it had replaced, the ecosystem was thrown off balance, and pests proliferated to "epidemic proportions," as recounted by the political scientist James C. Scott. Treating trees purely as an economic asset ignored what they really are: part of a large, more complex whole. Ben Green, a city-focused data scientist and author of the book *The Smart Enough City*, drew a line between "myopic" German forestry techniques and the ways governments see technology today: "Quite literally, it missed the forest for the trees."

For decades, cars and highways have taken the blame for destroying the urban density that gives cities their close-knit communities and characters. Though highways bridged the distance between cities, they often made life within cities more distant. Robert Moses, the titan of twentieth-century New York planning, reimagined the city and its surrounding region for the car, razing whole neighbourhoods and starving the city's public transportation system of funding. Toronto's fate was somewhat different. Moses's adversary Jane Jacobs moved there in time to help a band of opponents convince the provincial government to stop a north-south expressway from cleaving her new city in half. "If we are building a transportation system to serve the automobile, the Spadina Expressway would be a good place to start," said Premier Bill Davis, Waterfront executive Meg Davis's father. "But if we are building a transportation system to serve people, the Spadina Expressway is a good place to stop." Today's tech executives, however, often take a different lesson from the rise of the automobile. Canadian sociologist and technology scholar Vincent Mosco wrote that executives like Elon Musk and Mark Zuckerberg follow directly in the tradition of Henry Ford, who believed that "market victories justify the disruption of established institutions, especially cities."

In 2016, the Massachusetts Institute of Technology proposed a system of "intelligent" intersections that would allow automated vehicles, or AVs, to flow through cities without stopping. Not only would vehicles in this model be driverless, but the model itself seemed to mark a significant step toward reducing congestion and getting people to destinations faster, ending traffic lights as the world knew them. Reading about the proposal in the *Boston Globe*, one question struck Ben Green, who had spent a year working for Boston's Citywide Analytics Team: How were pedestrians going to cross streets without lights? Many smart-city advocates, Green believed, see the world through "tech goggles," giving them a narrow view of problem-solving that assumes technology alone can provide a solution.

The failings of this blinkered perspective have been made clear through numerous other technologies that were marketed to cities over the past decade. Predictive policing companies like California's PredPol use machine-learning algorithms to figure out crime patterns, selling cities on the idea of more efficient police work. But the historic patterns used to feed these algorithms include the biases baked into that history—and the history of policing is rife with well-documented examples of racial injustice and further marginalizing the marginalized. In 2021, an investigation by Gizmodo and the Markup analyzed more than seven million PredPol predictions from thirty-eight cities and counties, and found that the company's predictions "relentlessly targeted" communities "that were the most heavily populated by people of color and the poor."

Using such software to augment police work in the first place represents a possibly deeper-rooted bias in how city governments respond to crime, too. If cities want to invest in predictive technology, scholars and urbanists tend to ask whether policing, which responds to symptoms, is the right endgame, when improving social services would address the root causes of the criminality itself. Progressives have been arguing this for decades, but they don't have the marketing budgets of technology companies.

As Sidewalk Labs began the work that would lead it to Toronto, Dan Doctoroff wrote on his blog that Sidewalk's strategy for cities would be to "lay the foundations and let people create on top of it." The company's staff had studied many of the city-building and urban technology efforts that came before them, and were eager to distance themselves from past errors, going so far as to consciously avoid the term *smart city*. As much as Doctoroff chided top-down city-building, though, his tech goggles remained in place. Digital technologies, the Sidewalk CEO proclaimed, will "help bring about a revolution in urban life."

Sidewalk's October dinner with bureaucrats was, by all accounts, pretty banal. Doctoroff gave his usual sales pitch, flanked by policy chief Micah Lasher and communications leader Lauren Skelly. They talked about the new data trust proposal that Sidewalk had shared with Waterfront the week before, and how Sidewalk's model for data would promote Canadian innovation. No one in attendance raised much fuss. "It's senior bureaucrats," says one person who was there. "No one really gets into fights."

To many people in Ottawa, Sidewalk Labs was a Toronto problem. Yet this was one of the highest-profile efforts in human history to rethink how cities were built. And because government responsibility fell to an obscure agency, it was easy for officials with actual power to absolve themselves of accountability for the outcome. That marked one of the project's greatest flaws. The most consequential decisions to be made around Quayside were about data governance: those choices could have set global standards for who prospered, how fairly wealth would be shared, and how privacy would work in the world's increasingly sensor-laced cities. But Waterfront's own governance was broken. Its allegiances were to the city, the province, and the country in equal measure. With no majority shareholder and three very different sets of interests, it was a fundamentally paralytic organization—a superficial symbol of government co-operation that lacked any of those governments' authority.

Waterfront was forced to wait for those governments to write policies that would regulate data collection at Quayside—policies they hoped would be world-beating and privacy-protecting, and would stand up to the scrutiny such a widely watched project would inevitably get. Rather than listen to Waterfront or to the project's opponents, who were begging for leadership around data and privacy protection, the lawmakers and policy advisers in Ottawa did extraordinarily little. After Cambridge Analytica changed the world's view of corporate technology and Quayside was beset by months of scandals, the political opportunity was gone. Officials who dealt with or closely followed the file now privately admit that Justin Trudeau's government had previously approached Sidewalk and other tech giants with naïveté. One says, jokingly, that Sidewalk became "inconvenient." Others are more blunt. "Why would you run into that fire?" says Matthew Mendelsohn, a former high-level privy council official. "You only run into the fire if you've got a hose."

The abdication of responsibility happened not just between governments, but within them. Because Waterfront was, at its core, an infrastructure agency, it reported to the federal infrastructure department. But big decisions about how Canada would handle the future of citizens' data and privacy fell to the innovation ministry. With the Prime Minister's Office by this time deferring to Infrastructure Canada for any public comments on the project that Trudeau had once so eagerly embraced, the infrastructure and innovation departments fell into an awkward dynamic that did little to answer the big-picture questions Quayside posed for the future of data and privacy rights.

The Personal Information Protection and Electronic Documents Act, Canada's mouthful of a privacy law for private-sector companies that's usually shortened to PIPEDA, was nearly two decades old. It was woefully underequipped to deal with the mass data collection practised by technology firms in the late 2010s. By the time Sidewalk came to Canada, federal officials were struggling to balance the duelling imperatives of data protection and their government's pro-Alphabet stance.

Infrastructure Canada was hesitant to take ownership of the project's big-picture questions; the department says it didn't believe the Quayside project necessitated policy change on its own, since the government had already promised to overhaul PIPEDA entirely. The innovation ministry was trying to make that happen. But for years, the overhaul was hit with delays, in large part thanks to territorial jockeying: numerous senior staff and politicians wanted a say in digital regulations, which would ripple through departments from the Treasury Board to National Defence. And innovation minister Navdeep Bains was a deeply cautious politician—a team player to a fault, willing to dodge meaningful questions about Quayside as they related to his department's digital policy work, to avoid stepping on the toes of his colleagues in infrastructure who actually had responsibility for overseeing Waterfront. As one person who watched it all happen describes it, Sidewalk had become a political hot potato.

Since they were unable to get meaningful privacy and data governance reforms through Parliament quickly, the two departments discussed helping Waterfront fight for privacy and data protections in their agreement with Sidewalk, which would reflect the modern world better than PIPEDA did. Effectively, they hoped to protect Canadians' interests contractually, rather than legally. Frozen by complex internal politics and afraid to take responsibility for a controversial project, a government with the power to write a country's laws was addressing one of the most complicated city-building efforts in its history by avoiding laws altogether. Yet it was a different government that nearly destroyed the Quayside project, and Waterfront Toronto with it.

CHAPTER 17

It's My Way

IT WAS NOT a secret that Ontario's auditor general, Bonnie Lysyk, had spent much of 2018 studying Waterfront Toronto's books for a value-for-money audit. These kinds of reviews were common for a government auditor, though it was unusual in this particular case: of the 268 audits her office had conducted since 1997, only two full audits before this one had examined arm's-length agencies like Waterfront that didn't deal directly with the public.

Staff at Waterfront spent months doing interviews and handing over documents. Quayside had reshaped their jobs in many ways, but Lysyk and her team seemed more interested in the agency's day-to-day. By the fall, Waterfront executives were under the impression that the audit was nearing completion, with the agency expecting to file a response to Lysyk's findings. Instead, they recall, her office suddenly began asking more questions. About Sidewalk Labs.

This surprised Meg Davis and Kristina Verner. They hadn't fielded many questions about the Quayside project during the earlier interviews. At the same time as the Waterfront executives were grappling with the losses of Saadia Muzaffar and Ann Cavoukian—and wrapping their heads around Sidewalk's urban data proposal and its secretive meeting with Toronto luminaries—Lysyk's staff began asking questions about the project's origins, its data governance ideas and more. The definition of *value* in Lysyk's value-for-money audit appeared to have vastly expanded.

Her team queried the project's origins along the lines many critics had been pursuing, sometimes focusing, as Julie Di Lorenzo had, on the 2017 rush to vote on the Sidewalk partnership. And they zeroed in on the gaps in government policy that the Quayside project exposed, which Jim Balsillie and John Ruffolo had voiced concerns about. It turned out that, just recently, Lysyk's team had interviewed all three of them. The auditors had listened intently.

"Waterfront Toronto's communications to the public gave the impression that it was playing an irreplaceable role in the world-class transformation of Toronto's waterfront, a total of 2,840 acres," Lysyk wrote in her 2018 report. "This was not our conclusion."

She took sixty-one pages to make her case. Waterfront wasn't autonomous or powerful enough to make the right choices for itself, she argued, yet the autonomy it did have with Quayside raised urgent questions. In particular, she worried about data collection and its twin consequences: privacy risks and the capture of intellectual property. And since the federal government wasn't exactly racing to write policy to protect Canadians' data privacy and intellectual property rights, Lysyk recommended that the Ontario government take the lead.

But it was Waterfront itself that was caught in the report's crosshairs, and Lysyk focused much of her criticism on the origins of the Sidewalk partnership. After months of studying how Waterfront worked, the auditor general believed that by giving Sidewalk Labs and other potential Quayside bidders early access to information about Quayside before bids opened, Waterfront had risked giving them "an unfair and unequal advantage." The bidding process, she argued, was curiously brief—the six weeks that applicants were given to respond to Waterfront's request for proposals was shorter than past Waterfront RFPs that she'd studied. Governments weren't "adequately" consulted during the process, either. Lysyk questioned the board's rush to vote on the Sidewalk partnership, raising concerns about how little time directors had to review materials after Julie Di

Lorenzo refused to approve the deal the Friday before the Monday board meeting. And she highlighted Sidewalk's intentions, including the vote to expand across the Port Lands, when the RFP outlined that the company would need to jump some big hurdles to build across more than 12 acres.

Lysyk's office sent a draft report to Waterfront for a response. Sitting thirteen floors up at the south end of Bay Street, the agency's staff stared at the document and wondered what to do. It put them in a powerless position. The auditor general was chastising the agency for not prioritizing new ways to bring in revenue within pages of chastising it further for doing exactly that with Quayside. The information they'd shared with Sidewalk and other potential bidders before releasing the RFP, Verner and Davis say, was mostly maps and links to municipal and start-up sector websites. As far as Waterfront was concerned, that took place outside of the actual procurement process while they were trying to figure out what the process would look like. By asserting that Sidewalk and some other bidders got more details than others, however, the report could create the impression that the procurement process was deeply flawed, if not corrupt. But if Waterfront's assertion that the extra information consisted only of easily accessible website links and maps is true, then the impression Lysyk created was almost certainly unfair.

The bidding process that Lysyk had found to be too short was, in Waterfront's eyes, much longer than the report acknowledged. Bidders had six weeks to apply, but more than 150 days in total to expand and refine their Quayside applications. Once Sidewalk was selected, both CEO Will Fleissig and chair Helen Burstyn briefed Mayor John Tory at a meeting with Sidewalk staff present. Yet Lysyk cited an "internal Waterfront Toronto email" from the same period that said the mayor's office had received "almost no information about the project," not mentioning the meeting, whether it assuaged the concerns of the e-mail's author, or the fact that mayoral staff were attending board meetings. And though the board had only a weekend in October 2017 to review all of the materials before its vote to confirm

Sidewalk as Waterfront's Quayside partner, the directors, including one of Toronto's deputy mayors, had already seen much of the material and received two briefings about it.

Armed with more documents and interviews than virtually anyone else could legally gather, in 2018 Bonnie Lysyk was in the best position to present a clear-eyed version of why the Quayside project kept running into controversy. But because her office did not make public the research underlying her report, or share even redacted pieces, it was impossible to bridge the gap between the report's assertions and Waterfront's subsequent pleas for nuance. All that was clear was that Lysyk's research had pivoted at almost the same time her office had begun conducting interviews with some of the project's loudest opponents.

Waterfront executives would feel some relief, two years later, when a series of *Toronto Star* stories highlighted a variety of provincial officials accusing Lysyk of "headline hunting" and "cheap shots." "The most frequent complaint from public servants and experts is that Lysyk not only rejects their explanations and expertise, but imposes her own subjective values on value for money audits," wrote the paper's provincial politics columnist, Martin Regg Cohn. But in 2018, Waterfront had little choice but to suck it up and write an official response thanking the auditor general for her work. Lysyk would soon win the headlines she was later accused of hunting.

On the morning of December 5, a handful of tech reporters crammed into a meeting room next to the political correspondents who covered Ontario's legislature in the Richardsonian Romanesque building called Queen's Park, just northwest of Toronto's core. We were given USB sticks with the auditor general's report on them and a few hours to read it before it went online. We filed our stories, moved from the cramped room to a more cramped press conference theatre and asked Lysyk and various ministers about the dozen or so audits the auditor had conducted that year.

Monte McNaughton, the minister who oversaw Waterfront Toronto for the province, said he'd take "decisive action" over Lysyk's findings. "We're going to ensure that taxpayers get value for money," he said, echoing both the reason for the audit and the kind of fight-for-the-little-guy talking points his government often used. "We will ensure that proper oversight is brought forward. And that privacy laws are followed to the letter."

As the theatre emptied out, I wandered around the high-ceilinged hallway outside, handing my business card to anyone in the crowd who might want to share gossip about the auditor's findings. I very quickly ran out of cards.

I showed up at Waterfront's office early the next morning to find the agency in damage-control mode. The audit coverage had not been kind, and Waterfront directors were holding a special board meeting to address Lysyk's allegations. "Management remains confident that Waterfront Toronto's procurement processes and, more specifically, the RFP used for the Quayside project were fair, open and competitive," Helen Burstyn told the directors and handful of media who showed up. ". . . We remain committed to broad public engagement on a proposed plan for Quayside, which includes consulting with all potentially impacted government departments and regulatory agencies."

Staff circulated a copy of a letter from a retired senior judge who had overseen the RFP process as a "fairness" adviser. "I am satisfied that Waterfront Toronto successfully took all reasonable steps to ensure that the Quayside procurement proceeded on the basis that no proponent or potential proponent secured any unfair advantage," he'd written. From there, much of the meeting was in camera, and media were dispatched to the hallway. Once it wrapped, I did a brisk interview with Burstyn, hoping to determine what the board had talked about behind closed doors. She didn't say much, but gave me her cellphone number for future stories.

I headed to the newsroom and tried to make a story about Waterfront's procurement policy seem interesting, but there wasn't much to work with. I stared at a blank Word document for most of the afternoon. Then, late in the day, I got a call from the very last person I'd handed my business card to the day before.

"How late are you working tonight?"

As a rookie city councillor in 2011, Doug Ford had tried to impose a very specific vision on Toronto's Port Lands. Waterfront Toronto wasn't moving fast enough, he felt, to build out the area. In just five or six years, he believed the city could turn it into a major attraction, with a monorail, a major hotel, a massive mall and "the world's largest Ferris wheel."

"We had fifteen people in the room and everyone's jaw just dropped" when he laid out his vision to one group, Ford told CBC, not specifying who those people were. "It is spectacular, just spectacular."

Waterfront and its supporters panicked. Ford had real clout on city council—he was the brother of Mayor Rob Ford, who spent much of his single term battling accusations of smoking crack that turned out to be true. Mark Wilson, then the agency's chair, remembers being astounded at Doug Ford's Port Lands proposal. The pitch felt like a cruel joke. It being the age of YouTube, "everyone was pulling out the monorail clip from *The Simpsons*," Wilson says, referring to the episode in which a huckster sells a town a single-rail train it doesn't need while the rest of its infrastructure crumbles. Waterfront worked with community members to bombard the rest of city council with pleas to keep the land under the agency's control. Three weeks later, they won enough support from council to cancel Ford's plan.

When his brother Rob pulled out of the 2014 mayoral race after being diagnosed with a deadly form of cancer, Doug Ford took his place as a candidate. He lost to John Tory, a centre-right politician who had himself lost a mayoral race years earlier, then put all his efforts into provincial politics, only to lose the chance to become premier, too. Then a funny thing happened: after the leading candidate to run the Progressive Conservative party in the 2018 provincial election dropped out, Doug Ford won the party leadership.

He quickly ousted the long-governing Liberals to become Ontario's premier. Toronto was left with a mayor who wanted to run the province, and Ontario was left with a premier who wanted to run the city. And with the provincial Liberals voted out and Doug Ford's PCs swinging further to the right than John Tory typically did, Waterfront no longer reported to three governments that were philosophically aligned.

The dynamic wasn't great. One of Ford's first moves as premier was to slash the size of Toronto's city council nearly in half in the middle of a municipal election. The decision threw the city's politics into disarray and forced Tory to try to save the council in the middle of his own re-election campaign. Ford also made a point of erasing the work of his Liberal predecessors in the name of fiscal responsibility, sometimes clumsily. Soon after he was elected, he fired the well-compensated head of an Ontario electrical utility that his campaign had painted as a symbol of unnecessarily high consumer costs and Liberal overspending, calling him the "six-million dollar man." It was later revealed that the executive's departure would cost as much as C$9 million as payment for stepping down, far offsetting the savings Ford had hoped to glean in the name of friendly headlines.

One remnant of the provincial Liberals' legacy was the Waterfront board of directors. Helen Burstyn, Michael Nobrega and University of Toronto president Meric Gertler had all been appointed by Ford's predecessor, Kathleen Wynne. (Ontario had a fourth position to appoint, but hadn't replaced Julie Di Lorenzo.) When the auditor general published her report in December 2018—with its opaque research and what Waterfront Toronto executives would later cautiously describe as a seeming unwillingness to understand the nuances of Quayside's procurement—it said what Premier Ford wanted to hear: Waterfront wasn't giving Ontario taxpayers fair value for their money. He gathered his cabinet and advisers, who began deliberating how to intervene. One of them had my business card.

—

On the evening of December 6, Helen Burstyn was driving from a Canadian Opera Company reception to an event in Toronto's west end when she got a phone call. It was from Monte McNaughton, Ontario's infrastructure minister.

"I can't talk right now," she said. "Can I call you back?"

He said yes, Burstyn recalls, and she turned her car around to drive home, bailing on the next event. She was a Liberal appointee to Waterfront's board who had once run for election as a Liberal; she had a feeling something bad was coming. All she could hope for was that she, rather than Nobrega and Gertler, would take the fall for Waterfront. Their careers had been far less overtly political. But as the agency's chair, and with a Liberal history, she was the perfect target for a symbolic firing.

Burstyn arrived home and took the call. McNaughton thanked Burstyn for her service and said it was a pleasure working with her. But the Ford government wanted to tighten its hold on Waterfront, the minister told her, and installing new directors was the first step. He didn't use the word *fired*, but that's what he meant.

"All three of us?" Burstyn asked.

"Yes, all three of you," McNaughton responded. He sounded like he would rather not be making the call. "I really, really did like working with you."

Burstyn called Gertler to give him a heads-up. Across town, Nobrega had already heard from McNaughton. The minister had sounded similarly uncomfortable on the phone with the acting CEO. Nobrega, worried the board would be gutted for a long time, asked him one question: Was there anyone to replace the directors he was firing? Yes, McNaughton assured him, the government had people in mind.

Soon after, I was racing to call all three board members. The source who had my business card had tipped me off to the firings, but I had only just met that source and needed to make sure I wasn't getting played—so I needed to confirm that the government had actually followed through. A little after 9 p.m., Burstyn picked up at the number she'd given me just hours earlier.

I asked how she was doing.

"I'm okay," she said. She confirmed the news before I'd even asked. "I was fully expecting the call from the minister," she said. It turned out she'd even prepared a statement. She asked to read from it rather than respond to questions off the cuff, but paused at several points to laugh at the absurdities that had led to this moment. What she really wanted to get across was how carefully she felt Waterfront had worked on the Quayside project—that despite the auditor's findings, its team had done everything with diligence and care. "The staff and the board members at Waterfront Toronto are second to none," she said.

After fourteen months of controversies and resignations, a government had finally intervened in the Quayside project. New governments replace the appointees of their predecessors all the time, but by firing a third of the board, including its chair and acting CEO, the Ford government had only further destabilized the agency.

Ford's team apparently hadn't told Waterfront's other stakeholders, either. Less than an hour after my story about the firings went online, the senior federal staff that oversaw Waterfront Toronto were forwarding it to each other and trying to understand what would happen next. They couldn't tell if the board had quorum to make decisions, or if Waterfront even had a CEO anymore.

With Quayside, Canadian governments had a chance to take ownership of one of the world's most ambitious city-building projects. But during its biggest crisis, the people the federal government entrusted to oversee the project couldn't figure out how the agency handling it worked.

Still, Michael Nobrega had been fired only from the board. As acting CEO, he was a staff member. It would take the whole board to fire him, and it was the board that had asked him to take the job in the first place. But he wrestled with his future: the flood of stories that followed his removal from the board carried a sense of shock and controversy, and he had some long talks with his family and friends about how he might be perceived after the news. He could easily walk away from Waterfront entirely, but the long-time pension plan leader wasn't the type to make a sudden, emotional decision.

In the days that followed, Nobrega began to consider the consequences for Ontario, too, if he decided to step down as CEO. Waterfront would be left rudderless, still searching for a permanent chief executive while its board seats were nearly half vacant. In the face of an embarrassing firing, Nobrega soon decided to stay the course with Waterfront—to keep serving Ontario, even though Ontario had fired him.

"I'm determined to remain," he told me when I cornered him at a public consultation the next weekend, speaking out on the matter for the first time. Yet remaining also meant having to deal with the barrage of opposition. By then, Bianca Wylie and others had marshalled a growing group of supporters who were arguing that the whole Quayside project should be scrapped and restarted, with better protections in place for the public good. Doing so could open up Waterfront to lawsuits, Nobrega said that day, but he also pointed to the exit ramps that the agency had baked into its agreements in case the Sidewalk partnership really did go sour. Until then, he said, Waterfront and Sidewalk would keep negotiating in good faith.

Nobrega was not alone in trying to salvage the Sidewalk deal after all the drama it had caused that summer and fall. John Tory called his long-time frenemy Doug Ford afterwards to make sure he wasn't trying to destroy Waterfront. Adam Vaughan, a federal politician who represented downtown Toronto, was more pointed. Ford "has never won the argument with serious city planners on his vision for the waterfront," Vaughan said. His own argument for sticking with Sidewalk was more publicly aggressive than Tory's or Nobrega's, pitting his political rival's agenda against a decades-long effort to make the shore of Lake Ontario better. "With a couple high-tech activists and a really flimsy, poorly written auditor's report, he's got the cover he needs to reverse twenty-five years of good, solid work and forty years of dreaming on the Toronto waterfront, and I'll be damned if I surrender that just because you don't like the way Google or Facebook or Twitter handles personal data."

There were five vacant seats on Waterfront's twelve-seat board after McNaughton fired the three Ontario directors. On top of the provincial firings and Julie Di Lorenzo's resignation, Denzil Minnan-Wong's term had just ended, with no replacement yet named by the city. As Waterfront struggled in an uncomfortable spotlight, another director considered leaving, too.

Stephen Diamond had been a bit player in the Quayside saga. He sat on the same board committee as Julie Di Lorenzo that oversaw real estate deals, and as the committee reviewed the original Sidewalk deal in the fall of 2017, he'd pushed for clear-cut exit ramps and financial provisions, but then didn't show up for the vote. As controversies mounted throughout 2018, he did what he did best: kept his head down, got the work done and stayed in his lane. He'd taken this approach to his work for four decades.

A slight man in his late sixties, Diamond looked unimposing but had little patience for dithering. He spoke matter-of-factly and expected the same of others. Doing so helped him bridge seemingly impossible divides as a lawyer and developer in Toronto. He was capable of finding compromise in complex projects that pitted governments, builders and citizens against each other.

Development was a family affair. Diamond's father, Allen Ephraim Diamond, was a mogul whose Cadillac Development Corporation had merged with a firm owned by the Bronfman family to create Cadillac Fairview, once among North America's biggest real estate companies and still a large-scale operator of malls and office space. After years as a development lawyer, Stephen Diamond launched his own company, DiamondCorp, in 2008, investing in more than twenty residential and mixed-use projects across the Toronto area. As Diamond settled into his second career, he also found a new interest in giving back to the city. When the mayor's office asked him to join the Waterfront board in 2016, he welcomed the opportunity.

It was, for a while, a routine board position, and he felt that Waterfront was doing good work. As the Sidewalk project derailed the once-boring agency, though, he found himself more and more distracted from his day job. Months of damage control had left him exhausted. In December 2018, just as Ontario's government took a sledgehammer to Waterfront's board, he called the mayor and said he wanted to resign. "I've really done my best over these last few months, but now that we're coming to the end of the year, I've done everything I can do," Diamond told Tory. His three-year term was only a few months from finishing, but he wasn't sure if he could cross the finish line. "It's time for someone to take my spot."

The mayor was already anxious after the board firings. Even with assurances that the premier wasn't trying to dismantle the agency, the future of one of Toronto's highest-profile projects with one of the world's biggest companies was on the line. "You cannot step off this board," Tory told him. He trusted Diamond and knew he had a steady hand. The mayor had another idea for the developer: "Not only do I see you staying on the board, I see you becoming chair." He went on. "We need somebody like you there. We need *you* there."

Diamond said he would think about it. He was a big believer in public service, so it was an enticing offer. But Tory was thinking bigger than just keeping Diamond as a regular chair. Tucked into the provincial legislation that gave Waterfront Toronto its powers was a clause that had gone unused for most of the agency's history. On top of the four directors that the federal, provincial and city governments were each allowed to appoint, there was room for a thirteenth director—a chair appointed by all three. Though the board had gotten used to voting for its own chair, its ranks were ravaged as 2018 gave way to 2019. If there was ever a moment for governments to take control, and maybe even reach a compromise, it was now. And if there was a person in the thick of it all who could make compromise happen, it was Diamond. Tory and his staff began making calls.

CHAPTER 18

Raise a Little Hell

WHEN DAN DOCTOROFF was deputy mayor of New York, the closest subway station to Hudson Yards was a fifteen-minute walk—hardly a convenient mass transit option for his ambitions of square footage the size of downtown Seattle. He and his team thought they might be able to extend the east-west 7 line, which at that time terminated near Times Square. There was a problem with that plan, however. As Doctoroff has told the story, New York state assembly speaker Sheldon Silver wouldn't release funding for the state-controlled transit authority to extend the 7 train on the west side of Manhattan until a long-awaited Second Avenue subway line was built through to the Lower East Side, which Silver represented.

Doctoroff's Olympic-bid leader, Jay Kriegel, called him with the bad news after meeting with Silver. "This is a disaster," Kriegel said.

But Doctoroff had an idea. "Shelly didn't say that we couldn't build a subway," he said. "Just that we couldn't use state funds, right?"

After studying how Chicago paid for public projects, Doctoroff had come to like a funding mechanism called tax increment financing. It was a way to tie together private profit and the public good, as he often sought to do. In one common form that Doctoroff embraced, the model diverted future public income to private investors who were willing to pool money into a project or community with the assumption (or at least hope) that it would boost development interest in an area. That, in turn, would raise property values. The subsequent increase in taxes—the tax increment—would eventually get paid out to the investors.

There are dangers with this kind of project financing. It can divert increased property taxes from public services. It has been connected with gentrification. And for investors themselves, there's a risk of the project getting cancelled, taking their money, or at least the gains they'd hoped for, with it. But as long as the city wasn't on the hook for the investment, in Doctoroff's view the financing model could give life to a project that "essentially pays for itself."

With some help from the investment bank Bear Stearns, Doctoroff's team figured out how to make tax increment financing work to extend the 7 subway to Hudson Yards. Including other expenditures, the city would wind up issuing $3.5 billion worth of bonds related to the project, making it one of the biggest uses of the funding mechanism in history. But an analysis by New School researchers found that the whole Hudson Yards project cost taxpayers $2.2 billion regardless, thanks in part to tax breaks for Stephen Ross's Related Companies and other developers, and hundreds of millions of dollars in cost overruns. Tax increment financing got Hudson Yards past the subway extension hurdle, but it paved the way for a project that left New Yorkers on the hook anyway.

That was ancient history by November 2018, when Doctoroff's office sat halfway up the tower of 10 Hudson Yards, overlooking his life's greatest project. That month, he and his team gave a progress report to their masters at Alphabet. Though Sidewalk was months away from its next milestone—submitting a draft master plan to Waterfront Toronto—Larry Page liked to keep track of his dream neighbourhood, and Alphabet CFO Ruth Porat needed assurances that the hundreds of millions of Google-generated dollars the company was giving its city-building division were being well spent.

Doctoroff's team presented a slide deck to Alphabet executives, outlining mountains of project details that Waterfront had neither seen nor endorsed. Had the agency's staff seen the presentation, they'd have stopped Sidewalk midway and told them to scrap the deck. Sidewalk was walking an impossibly shaky tightrope, balancing the expectations of its parent company on one side and the restrictions of the Waterfront deal on the other, sometimes relaying only what was palatable to each party in order to keep pushing forward. At the meeting, Sidewalkers outlined detailed plans for neighbourhoods beyond Quayside, with additional powers across more than 475 acres of the Port Lands—land they still hoped to access. They told the Alphabet brass they had numerous financial backers lined up for the project, including Goldman Sachs, and that they'd discussed the capital structure of a deal with the Canada Infrastructure Bank, a C$35 billion federal fund set up by Justin Trudeau's Liberals to encourage big public-private partnerships. Yet according to the Infrastructure Bank itself, Sidewalk had only had an introductory conversation with its staff; capital structure wasn't discussed. Sidewalk's mythmaking wasn't directed just at the public, but at its own benefactors. (Asked about this discrepancy, Doctoroff replied, "I don't remember that," and said he wasn't personally involved in Infrastructure Bank discussions.)

The Sidewalkers proposed to Alphabet that the company would be a developer at Quayside, which Waterfront had never wanted, and at another site to the southeast called Villiers West, which City Hall would have refused (though a Waterfront vice-president entertained the idea of "adding lands to the current Quayside parcel" from Villiers Island in an e-mail to Josh Sirefman that month). They also shared detailed plans about investing in light-rail transit and infrastructure through much of the Port Lands, hoping to become an "essential catalyst" for "the creation of the world's most innovative urban district." In return, Sidewalk wanted payments, largely from the city, tied to future increases in the value of more than 100 acres of land and development charges levied there, while taking a slice of the city's incremental tax revenue generated from that land—just as Doctoroff had arranged for Hudson Yards. The company believed all this could be worth $6 billion over thirty years.

By early 2019, rumours of a confidential slide deck began circulating among Toronto reporters. The *Toronto Star* and the news website National Observer got copies and published stories about it in quick succession, starting on Valentine's Day. Though Sidewalk had long said it wanted to eventually build more than 12 acres, Waterfront and city officials had usually tempered those expectations. Now Dan Doctoroff was forced to explain why the company had made such detailed plans to ask for powers across most of the eastern waterfront behind the backs of government officials. "We don't think that 12 acres on Quayside has the scale to actually have the impact on affordability and economic opportunity and transit that everyone aspires to," the Sidewalk CEO said. Investing in infrastructure and long-delayed light-rail transit would be necessary for anyone to build through the Port Lands, he added, and Sidewalk had plenty of Google's money to lend to the cause.

For Sidewalk's biggest critics, the slide deck offered vindication: the company's pitch for the 12 acres of Quayside now looked like a Trojan horse from a company they already didn't trust. The plan to fund transit infrastructure through incremental tax revenue, meanwhile, made some Torontonians worry that the city might lose control of part of the transit system. Even Waterfront's executives were taken aback. Though they'd seen some of the ideas that ended up in the deck while working with Sidewalk on a draft master plan, many other details caught them by surprise. "To see the brazenness of their aspirations shared like that, without us being aware of them, was alarming," says Kristina Verner.

Government staff were surprised, too—not just at the scope of Sidewalk's plans, but because it appeared that Sidewalk had told Alphabet it was having weekly meetings with them that didn't actually happen. Sidewalk was extensively lobbying politicians and bureaucrats across Toronto and Ottawa, but they weren't the ones who mattered most if the company wanted to partner with Waterfront. "I can tell you that we are not meeting with Sidewalk weekly nor in any regular frequency," wrote Kelly Gillis, the top civil servant at

Infrastructure Canada, in an e-mail to colleagues the morning after the deck leaked. Other government officials took their surprise public. One councillor said the city should pull out of the Sidewalk deal, while other municipal politicians, including former Waterfront director Denzil Minnan-Wong, promised to give it more scrutiny.

At Queen's Park, Doug Ford had been coming around to Sidewalk since firing nearly a third of the Waterfront board. But the blatant land grab in the leaked slide deck was simply too much. One senior official in his government told the *Star* that "there is no way on God's green earth that Premier Doug Ford would ever sign off on handing away nearly 500 acres of prime waterfront property to a foreign multinational company."

Two days after the slide deck story broke, Ana Serrano, the chief digital officer of the Canadian Film Centre and co-chair of a conference called DemocracyXChange, sent an e-mail to Bianca Wylie and Nabeel Ahmed, a smart-cities researcher who had written critically of the Sidewalk project. "I'm curious as to what you both think about mobilizing a citizen protest similar to Fuck off Google in Berlin against Sidewalk Labs now that the tide is turning?" Serrano wrote. "I think the timing is right and many people would be willing to come out to picket at the 307 location. Are you already working on this? If yes, I'd like to help. If no, would you be interested in starting a campaign?"

For months, a couple dozen critics, including Serrano, Wylie and Ahmed, had been meeting to discuss Sidewalk, sometimes braving snowstorms to cross the city for homespun salons where they unpacked what frustrated them about the project. Sidewalk's leaked intentions to make a play for the Port Lands galvanized them to do something more public with their swelling numbers.

Many of the problems they saw with the project were esoteric. It was hard enough to explain how decisions about data governance confer power in society, and harder still for those explanations to sway the swaths of the public that were happy Google was setting up shop in town. The phrase "surveillance capitalism" was being thrown around a lot, thanks to retired Harvard professor Shoshana Zuboff's book *The Age of Surveillance Capitalism*, which had just been released. Zuboff argued that Big Tech was sabotaging both democracy and traditional capitalism, and even devoted a few pages to the rising fear that Sidewalk was bringing Google-style data harvesting into cities in ways that would make their governments dependent on the company at the expense of their citizens. But with more than five hundred pages examining deeply complex subject matter, the book was not particularly inviting to the average reader.

Now Sidewalk's critics had a much more relatable concern to work with. In Toronto, every square inch of land is considered a commodity. Real estate is an emotional issue. By wrapping the land grab up with the pitch to siphon away taxes from a city that struggled to pay for necessary things—there were more than 102,000 families on a wait-list for affordable homes, far more than Sidewalk's plans could accommodate—the company had suddenly offered the public something it could really get mad at.

The budding organizers brainstormed names for their campaign, cross-referencing suggestions with available website URLs and underused hashtags, before settling on #BlockSidewalk.

Over the next week, nearly thirty people attached their names to the growing campaign, including Nasma Ahmed, who ran a charity called Digital Justice Lab that sought to reframe technology around inclusion; Saadia Muzaffar, the entrepreneur and writer whose resignation from Waterfront Toronto's digital strategy panel had set off a month of tumult for the project; and University of Toronto law professor Mariana Valverde, who was writing sharp-tongued screeds about how governance oversights could cause trouble at Quayside. Two seasoned campaign organizers also came into the fold: JJ Fueser, a long-time researcher for a hospitality labour union, and Thorben Wieditz, a researcher with a faint German accent *The Telegraph* once called "strong," probably because he's more patient with his words than North Americans. Wieditz had recently signed on to a campaign to fight short-term rental companies like Airbnb, and he had fought for the Port Lands before, a decade earlier, helping to stop a film studio from being torn down to make room for a shopping mall.

#BlockSidewalk announced itself to the world with a press release on February 25. Though Wylie was heavily involved in its creation, she leaned into the "co" of being a co-organizer and, over time, gradually took a backseat to other organizers. Fueser and Wieditz were veteran campaigners, nimble and loud, and eventually came to the fore as spokespeople as they tried to capture the same wide-ranging attention that the Berlin campaign did.

Though their rhetoric suggested that the organizers felt there was no other option for Toronto than to kick Sidewalk out of Quayside, Fueser and Wieditz had spent enough time with unions fighting powerful companies that, in private, they knew to temper the group's expectations. "I told them that there's no guarantee that you can stop it," Wieditz says. "The only certain thing is that if you don't engage, you don't have any agency to help shape or stop it."

So they engaged. They didn't go so far as to picket 307 Lake Shore, as Serrano had hoped, nor forcibly occupy the space like their counterparts in Berlin did, but #BlockSidewalk's organizers began to rally supporters. They'd found a surprisingly kindred spirit in Jim Balsillie, whose office kept in regular touch, but focused on meeting with community groups, unions—anyone with a pinch of philosophical alignment. They explained that the land grab and tax-siphoning revelations had reinforced what Wylie had long been trying to explain to the public: *privacy* concerns were largely a red herring with Sidewalk Labs; it was *privatization*, of all the things people in a democracy want governments to do, that was at stake. "It's a consent problem," Fueser says.

They approached union leaders, some of whom had heard there'd be jobs aplenty with all the construction Sidewalk would bring to the waterfront and had been given promises of a factory to prefabricate building pieces from wood. But this was an Alphabet company, the organizers liked to point out—wouldn't it try to automate as much of the work as possible? Public-sector unions could be swayed, too; with all of Sidewalk's promises, their jobs were also at risk of being automated. Not everyone was on board with opposing the project—the vision Sidewalk was selling was very optimistic and inclusive-seeming—but the #BlockSidewalk organizers still tried to make a case for skepticism.

Within two months, #BlockSidewalk's size had begun to rival that of Fuck Off Google—though it was hard to measure a decentralized movement—and the organizers had pulled it off without slapping tens of thousands of stickers across the city. A hundred and fifty people showed up to their first public meeting at an east-end community centre just north of the Port Lands in April.

#BlockSidewalk got the word out not just in Toronto, but among organizers fighting Big Tech around the world. Fueser began liaising with some of the pseudonymous organizers in Berlin. Learning how different factions there had come together for the same cause, each escalating things in their own way—and winning their battle—proved galvanizing in Toronto. "They really raised expectations and pushed us to understand what's at stake," Fueser says. In May, she also went to a social-justice conference in New York called When Tech Comes to Town, where she found common ground with international groups fighting similar battles, including those who had pushed back against Amazon's hunt for a second headquarters. There was a shared sense of optimism—Amazon had been pushed out of Queens, after all—but Toronto's battle against Sidewalk was on very different terms. Quayside and the Port Lands were isolated on the lakeshore, with few neighbours to rile up, yet the consequences Sidewalk critics feared would have implications for privacy and privatization across the whole city, the country, the world. "In Toronto," the Berlin activist Sundar Panet says, "it's a completely different game."

Michael Bryant hadn't just seen governments trip over themselves to lure big-name companies to Canada. He'd done the luring himself. During his decade as an Ontario politician, he'd been both the province's youngest-ever attorney general and, as the giants of today's internet grew their power during the late-2000s financial crisis, the provincial minister of economic development. As Ontario tried to shovel itself out of the crisis, his government set up a C$2 billion fund to invest in businesses to encourage them to set up in the province, calling the move "reverse Reaganism" with "government as entrepreneur."

He was sharp, ambitious, even cocky. He harboured dreams of running the province once the Liberal premier, Dalton McGuinty, stepped down. But after a not-yet-mayor John Tory failed to remove the Liberals from power in 2007, it became clear that McGuinty wouldn't leave Ontario's helm. So, in May 2009, Bryant took a job with a municipal agency tasked with enticing big companies to set up in Toronto.

Google was one of the companies in his crosshairs. But he didn't have time to do much corporate courting. That August, on Bryant's twelfth wedding anniversary, a bike courier named Darcy Allan Sheppard grabbed on to the driver's side of Bryant's convertible, setting off an altercation that led Bryant to swerve across the road. Sheppard hit a fire hydrant, fell off the car and smashed his head. He died. Bryant was charged with dangerous driving causing death and criminal negligence causing death. He resigned from his job days later.

The charges were eventually dropped, but for years, Bryant was cast as a villain—a powerful man who'd been let off by a justice system he'd once overseen, while a cyclist was left dead. He spent nearly a decade reconsidering his understanding of justice, coming to believe that the system he'd once helmed needed to be challenged. In 2018, he found a job that let him do just that: he became the executive director of the Canadian Civil Liberties Association (CCLA).

Not unlike the American Civil Liberties Union (ACLU), the CCLA seeks to protect citizens' rights and freedoms in the face of systems that might trample on them. Bryant wanted to think bigger than the association had before and become more aggressive in court. But he also saw an opportunity to expand its work into digital rights and freedoms—to work not just like the ACLU, but like the United States' Electronic Frontier Foundation, as well.

As it happened, the CCLA had recently been working with a surveillance researcher named Brenda McPhail, and tapped her to oversee its increasingly high-profile work on privacy, surveillance and technology. Where Bryant was gregarious, she was soft-spoken. Though she had initially been intrigued by Sidewalk's vision for cities, her optimism faded with each new detail she heard. "We were always watching, so at some point, if it went off the rails, we might be looking at a legal intervention," McPhail says.

Sidewalk's brass had a sense that the national civil liberties organization might take an interest in its project, and in March 2018 sent its privacy head and general counsel, Alyssa Harvey Dawson, to meet with Bryant and McPhail. Harvey Dawson met them in the low-budget non-profit's conference room, where the decades-old ceiling was sagging and the table was surrounded by mismatched chairs scavenged from other offices. McPhail took the lead in the meeting, peppering Harvey Dawson with questions. She asked how the data collected at Quayside would be shared, and with whom. Could, say, the police have access to it? This was a standard question for a civil liberties organization to ask, but it was still the Quayside project's early days, and Harvey Dawson didn't give a particularly concrete or satisfying answer, in the CCLA staff's recollection: in essence, the police would not have access to data, but there were no rules or regulations in place yet to guarantee that, or anything around data collection.

Though the company had been distancing itself and its business model from discussions around data collection, McPhail and Bryant weren't quite ready to believe that Alphabet would stray so far from Google's core business model. So they were upfront about their feelings: Quayside didn't seem like a project they could support. When the meeting ended, Harvey Dawson invited McPhail to join Sidewalk's data governance working group. McPhail declined.

"I know from experience what you do when you try to co-opt an organization, because I used to do it when I was attorney general," Bryant says. "You line up some quotes from supportive groups, you put them on your press release, and any criticism that came, you deflect using the people that you've co-opted. That's the word for it: you're co-opting. There's a vague hope for a quid pro quo down the line, but it never gets made."

He and McPhail were soon mulling some kind of action around the Quayside project. "It seemed like the biggest corruption of our constitution from a digital rights perspective that Canada had ever faced, and it was happening right in our backyard," Bryant says. Further, what was a real estate development agency doing partnering with a subsidiary of one of the world's biggest data collectors? As far as Bryant was concerned, Waterfront had neither the mandate to do so nor the capacity to deal with the consequences. "It was just a recipe for really bad public policy," he says. "And, it turns out, the biggest digital privacy and public surveillance issue to hit Canada—and one of the biggest to hit the world—ever."

McPhail routinely attended Waterfront's public Quayside meetings, and she and Bryant met with Kristina Verner and other Waterfront executives to learn more about the project. As 2019 began, the case the CCLA wanted to make became clear. Waterfront's power derived from governments, and the choices it made had consequences under the Canadian Charter of Rights and Freedoms. Those freedoms include peaceful assembly and association; rights include "life, liberty and security of the person" and, crucially, "to be secure against unreasonable search or seizure." The July 2018 Plan Development Agreement said Waterfront and its government parents could have "potential" ownership of the data collected at Quayside, yet governments didn't have modern legislation or capacity to deal with the kind of mass collection that might happen in the neighbourhood. On top of that, publicly available documents, like Sidewalk's Quayside bid—which Waterfront had long said was not a plan but a series of ideas it would consider—described a community filled with sensors.

By signing off on the Plan Development Agreement, the CCLA believed, Waterfront and the governments of Toronto, Ontario and Canada threatened to "effect historically unprecedented, non-consensual, inappropriate mass-capture surveillance and commoditization of individuals' personal information, and give a private-sector, for-profit, corporation the right to commercially exploit it." It would be impossible to get citizens' consent for all this data capture. And even if the data was immediately de-identified, as privacy specialists like Ann Cavoukian demanded, it was possible for someone to "re-identify" that data: to combine it with other publicly available information, such as aggregated consumer data sold by data brokers, to reverse-engineer who someone was. This was the "CEO problem" that Sidewalk had helped the local start-up InnerSpace solve. Sidewalk was keenly aware of the possibilities of re-identifying data, but the company's opponents believed that the sheer volume of data that could be collected at Quayside strongly increased the chance of that happening. As Bryant, McPhail and the outside counsel they brought in for help contended, this could have tremendous implications for a person's freedom and security—especially security against "unreasonable search and seizure."

To humanize their case, they found a person whose name they could include on the court application. The group that would soon become #BlockSidewalk introduced Bryant and McPhail to Lester Brown. A retired tax auditor with a hint of gravel in his voice, Brown lived near the foot of Parliament Street, just next to Quayside, and had been active in neighbourhood groups for two decades. He was a fan of community-building, and he didn't care for the way Sidewalk was approaching his. "I would use the word 'hucksters,'" he says. "Once you started drilling down with questions, they just evaporated." Bryant's team agreed to indemnify Brown in case he faced massive legal costs.

On March 5, 2019, just weeks after #BlockSidewalk was announced, Bryant and McPhail issued an open letter to Justin Trudeau, Doug Ford and John Tory, urging them to "reset" the project and draw up modern privacy legislation before trying again. The balance of power needed rebalancing. "Ask not what your country can do for technology," they wrote. "Ask what technology can do for your country." And they dangled a threat. "The CCLA is contemplating litigation in this matter because your respective governments behave as if unaware that, constitutionally, the emperor has no clothes."

Sidewalk staff were a bit shocked, given that they'd made a point of saying they'd follow Canadian privacy laws. And yet again, Waterfront leaders felt misunderstood. They had the same concerns around privacy as the CCLA and were trying to use Quayside as a chance to encourage governments to write the same kind of legislation that the CCLA wanted. After all, this kind of data collection seemed bound to happen elsewhere, soon, and more frequently. "We had a unique opportunity to bring people together to talk about something in a focused way that would otherwise be extremely nebulous," Kristina Verner says.

A month later, the CCLA followed through on its threat, filing an application for judicial review—effectively a lawsuit—in an attempt to force Waterfront and its three government parents to shut the project down.

The legal action was not a small gamble, Bryant says, knowing how much money Alphabet could throw around to discredit the case: "It was the riskiest matter we ever brought in the history of the organization."

Waterfront staff had spent 2018 grappling with mass resignations and firings, a zigzagging government audit and a business partner that kept asking for more than the agency could offer. What they'd begun as a whole-hearted attempt to make their city better had resulted in a year and a half of lost time, energy and colleagues. And now, in just the first few months of 2019, Waterfront had been blindsided by Sidewalk's internal land grab document, the creation of #BlockSidewalk and a serious legal threat arguing that Waterfront had violated Canadians' Charter rights. "That was the moment," Verner says, "when I stopped asking 'What else could happen?'"

Something else had *already* happened.

Brenda McPhail had earned her Ph.D. at the University of Toronto under the supervision of Andrew Clement, the anti-surveillance firebrand who liked making trouble for Sidewalk on Waterfront's digital strategy panel. As the CCLA was building its case, McPhail reached out to Clement to see if he'd speak with their legal team. The digital panel had been giving him a front-row seat to watch Sidewalk's ideas play out, and he believed that both the panel and the CCLA were trying to keep Waterfront and Sidewalk accountable for their decisions. So he said yes.

At the meeting with the CCLA's lawyers, Clement made it clear he would share only insights that the panel had learned in public, and not during confidential meetings. Everyone agreed that was fine, and he walked them through his interpretation of the Quayside project. "I believed that the CCLA question, about whether Waterfront Toronto had the legal mandate to make an agreement with Sidewalk, was a fair question that deserved a public airing," he says. After the CCLA published its open letter warning Waterfront's parent governments that it was considering legal action, though, something didn't feel right to Clement. He reached out to Michael Geist, the panel's chair, letting him know that he'd taken the meeting.

Geist seemed concerned. Clement's CCLA meeting may have been a massive conflict of interest. He barred Clement from the next panel meeting, on April 1, so that their fellow panellists could discuss whether he'd crossed a line. Clement, meanwhile, wrote a long letter to the panel and Waterfront's executives, explaining himself: he'd believed he was acting in the public interest, but now recognized his lapse in judgment.

The panel kicked the public out for part of the April 1 meeting. The closed-door discussion was by all accounts rather philosophical. The grey-bearded professor had taken his frustrations public before, through the media, angering some Waterfront executives enough that Kristina Verner and her team had needed to explain to superiors that wide-ranging dialogue was actually the panel's point. Her argument for the panel more broadly—and some panellists' argument at the meeting—was that Waterfront couldn't just accept what Sidewalk was saying at face value. More discourse was better for the democratic governments they represented.

Still, some panellists felt that Clement occasionally took his role as foil a bit too far, without at least delineating that he was speaking for himself and not the whole group. He was a long-time professor who'd spent much of his life protected by academic freedom. The liberties he seemed to take when outside the panel didn't sit well with some private-sector experts in the room.

In the end, the group decided that more discourse was better. Clement stayed. He kept ravaging the project in public, but was more transparent about doing so. The months of turmoil finally seemed to be ending for Waterfront Toronto, and at long last, its executives hoped to start focusing on Quayside's draft master plan.

After John Tory had talked Stephen Diamond off the ledge back in December and convinced him to stay on Waterfront's board, the mayor began reaching out to contacts in high offices around Toronto and Ottawa. In spite of their spats, even Ontario premier Doug Ford picked up the phone and agreed with Tory: the board needed stability, and Diamond was the steadiest hand at the ready. So the municipal, provincial and federal governments set to work enacting the rarely used clause in Waterfront's legislation to appoint him as a tripartite chair.

The rest of the board filled out. Tory replaced Denzil Minnan-Wong, the deeply skeptical deputy mayor, with a fresh-faced, left-wing councillor named Joe Cressy, whose ward included Quayside. Doug Ford appointed four new directors, including newspaper executive Andrew MacLeod, a long-time Research in Motion manager who was elevated to vice-president just before Jim Balsillie exited the company, and who later tapped Balsillie to help launch a *National Post* series that examined the tech sector through an economic-nationalist lens.

In mid-March, the new board members gathered for the first time with the directors who'd survived the late-2018 housecleaning.

They met in the same bland thirteenth-floor boardroom where every other important Waterfront meeting had taken place. A couple dozen reporters, opponents and members of the public packed the small gallery in the back.

It had been two years since Waterfront launched the Quayside RFP, three months since the auditor general's report, a month since Sidewalk's land grab document leaked out, weeks since the launch of #BlockSidewalk, and days since the CCLA threatened legal action. Waterfront's reputation had been scorched, and the public had no idea what it all meant for the future of the Quayside project. Diamond was a calm and cautious man. He and the reshaped board had the power to reset the project or stay the course. No one was sure which direction they'd choose.

Diamond brought the meeting to order, then let the many new faces at the board table introduce themselves. He thanked Michael Nobrega for his patience in running Waterfront as acting CEO for far longer than expected, and reminded the room that the C$1.2 billion Port Lands flood-protection initiative was, technically, Waterfront's biggest undertaking. But Quayside, Diamond acknowledged, was "clearly the project that has attracted the greatest amount of public attention." He quickly changed tack, from good cop to bad. "At the outset, it does have the potential to provide exciting opportunities for the future of our city—but at this juncture, we really don't know if that potential exists," he said.

But the bad cop didn't turn out to be that bad, after all. He wanted to see the draft master plan—the document outlining what would actually get built at Quayside. That alone would prove Sidewalk's potential to Diamond. Formally titled the "Master Innovation and Development Plan," it was originally supposed to be written and approved within a year of announcing the partnership. Sidewalk staff had been working on details behind the scenes for months, but the document had been consistently waylaid by snowballing attention and controversy. Diamond was a process-oriented guy, and he believed the only way to straighten out the process was to get the plan finished. Not only might it quell critics' concerns about data gathering and privatizing public services—maybe—but it would also clarify whether the project was even worthwhile for Waterfront.

Yet Sidewalk seemed to be jumping the gun. The company was sharing details from the plan with a select group of people and occasionally the media, but hadn't finished it. One of Diamond's first actions as chair was to reprimand Dan Doctoroff for this in a letter. Diamond believed the plan needed to be worked out between the two partners, not in public. He especially wanted to see how Sidewalk planned to finance and monetize their proposals.

Diamond also wanted to reassure the people of Toronto that Waterfront would bring skepticism to Sidewalk's ideas. And as he looked around the jam-packed boardroom, he guaranteed that the public would play a part.

"We now understand that the plan will be delivered to us within the next few weeks," he said. "And we hope that that promise will be kept."

CHAPTER 19

Shine a Light

"IT'S NOT SURPRISING there's a widespread belief that a political fix of sorts has been in for Sidewalk Labs' preferential treatment," said Peter Kent, "since Justin Trudeau met with Alphabet's former chair Eric Schmidt at the Google Go North summit in 2017." Kent was a former broadcaster now representing a suburban Toronto riding for the federal Conservative Party. It was late February, just a week after the slide deck had leaked of Sidewalk Labs' scheme to plan and partially develop far more than the 12 acres of Quayside. He was walking the crowd in an Ottawa committee room through a series of suspicious-seeming coincidences as he prepared to demand an explanation for them from Waterfront Toronto.

Kent leaned his head on his right elbow, which itself was leaning on a deep-brown bench at the front of the room. He turned his eyes toward Meg Davis and Kristina Verner, who'd been summoned before the Canadian Parliament's Standing Committee on Access to Information, Privacy and Ethics to answer questions about the Sidewalk partnership. "Waterfront Toronto chose Sidewalk Labs to develop a data-driven neighbourhood in a process that the auditor general, I believe, did in fact question as preferential and certainly rushed." It was a mouthful of a windup before Kent got to his question: With all the resignations, leaked land grab documents and speculation about Trudeau's interference with the Quayside procurement process—speculation that the prime minister himself invited—did Davis and Verner think Waterfront Toronto's agreements with Sidewalk were still valid?

"Yes, I do," Davis said. She'd prepared for this question. She explained that a retired judge had overseen the procurement process, and that no politicians were involved in evaluating bids. "This process was as independent as any we've run."

Kent wasn't satisfied. "Do you understand why people and politicians in the GTA, in Toronto and beyond, in Ottawa, see this suddenly as more of a high-profit real estate property development project than a technology partnership?"

Waterfront hadn't asked for a developer for Quayside, but it hadn't asked exclusively for technology, either. The agency wanted to maximize sustainability, too, and needed a "thinking partner" to help with both, Davis responded. It had also asked bidders to think "at scale": to bid for 12 acres but share a much bigger vision, if they wanted—to think about electrical grids, transportation systems and more. She did not mention that she and her colleagues were often frustrated by just how hard Sidewalk pushed for more than 12 acres, nor that the leaked plans had caught them by surprise.

The next politician at bat was Charlie Angus, a left-wing New Democrat who'd spent his youth playing in a punk band and working with homeless people on the streets of Toronto. This particular multiparty crew of MPs who were running the federal ethics committee had enjoyed taking on matters of digital power, privacy and data abuse, and had even hosted meetings for the International Grand Committee on Big Data, Privacy and Democracy. Angus, who had long ago moved north of Toronto, given up the band, started wearing suits and watched his hair turn white, had spearheaded the committee's inquiry into Quayside.

"We were told that RFP was for 12 acres, and now we learn from the *Toronto Star* that it's for the whole waterfront," Angus said. This characterization was a stretch, but it made for good drama. "Why was that not made clear in the RFP?"

"As I just mentioned, the RFP actually says to think about Quayside and to think about solutions at scale," Davis said. "That means transit at scale, district systems at scale and innovative financing options at scale."

Adam Vaughan stepped up. The Liberal MP didn't sit on the committee, but he represented the lake-facing downtown Toronto riding that Quayside was a part of, and had been invited to join the session. Vaughan didn't want to see Waterfront become a casualty of the Quayside project because some loud critics wanted to fight Big Tech. He believed the agency would keep Sidewalk Labs under control, and he was tired of what he felt were exaggerations from colleagues like Angus.

Vaughan was a former TV journalist, trained to extract quotes with lightning-round questions. Process-oriented Waterfront wasn't great at producing sound bites, so he decided to help. "In terms of the waterfront properties, how much land does Waterfront Toronto actually own and have title to?" Vaughan asked.

"We own just under 12 acres," said Davis.

"That is about one percent of the entire parcel."

"That's correct."

Vaughan then raised Sidewalk's leaked plan to finance infrastructure and transit through the waterfront with a model like tax increment financing, taking a cut of taxes in exchange for its investment. "The city-owned land is still governed by city law, city requirements and city approval processes."

"Absolutely."

"Is there anything that exempts Waterfront Toronto from the legal regime that governs data in this country?"

Kristina Verner took this one. "Absolutely not," she said.

"Are you allowed to write corporate contracts that exempt you or grandfather your exemption from any changes that occur in any laws?" Vaughan asked.

"No."

After a few more questions, Vaughan then attacked the king of all rumours—on the record, cameras rolling—far more pointedly than his colleagues in opposition parties had done: "Did the federal government bring you Google?"

"No," Davis said.

"Did anybody in the Prime Minister's Office bring you Google?"

"No, we've had no contact with the Prime Minister's Office."

From the moment Justin Trudeau said at the Quayside announcement that he and Eric Schmidt had been looking to work together "for a few years," speculation had been rising that Canada's most powerful politician had somehow fixed a nice deal for Alphabet.

There was no fix. But Schmidt had stoked the prime minister's excitement about Sidewalk on a secret January 2017 call, and Trudeau was the perfect audience to sell a techno-utopian dream to. He was also prone to newsworthy relationships with the private sector, eventually losing one key cabinet minister during a sole-sourced pandemic-relief contract scandal, and another after his office was alleged to have interfered in the prosecution of an engineering company facing fraud-and-corruption charges. The world's biggest tech companies, meanwhile, branded themselves as progressive, which aligned with his vision for Canada, and he turned to them as key sources of foreign direct investment. He was thrilled that Sidewalk would come to Toronto, but according to numerous public and confidential documents I obtained and interviews with many people involved in the procurement process, he stayed clear of Waterfront as it happened. If Trudeau had crossed paths with Schmidt prior to the phone call—which would explain the "few years" comment—it was likely at the 2016 World Economic Forum in Davos, Switzerland, where both were on the guest list. "There would have been nothing extraordinary about the number of times he met Eric Schmidt compared to other senior corporate executives," says Trudeau's friend and former senior staffer Gerald Butts.

But the secrecy with which the federal government guarded the January 2017 call—it took me more than three years to prove that it happened—had significant consequences. Not even Butts could recall it, and both current and former Waterfront management were surprised when I brought it up. "I have no knowledge of any of that," Will Fleissig told me. It is possible that Waterfront's former chief strategy officer Marisa Piattelli—who dealt with governments on the agency's behalf—was made aware of it, but she died of cancer in mid-2020, long before a federal investigator forced the government to share proof with me that the call happened.

If either Alphabet or Trudeau's office had been transparent about the call, they could have stopped misinformation before much of the dialogue around Quayside succumbed to it. Instead, they created a vacuum into which misinformation could flow. Even at the committee hearing, Peter Kent tried to imply that Schmidt and Trudeau meeting *after* the Quayside announcement was evidence of some kind of "fix."

The core tension of the Quayside project sprang from a misinterpretation of how much power Waterfront had: Waterfront could only give Sidewalk 12 acres, and Sidewalk had the impression that Waterfront could eventually give it more—or at least help make that happen. The tug of war between the two organizations was breeding a whole ecosystem of confusion and half-truths that made it hard even for people involved in the project to know what was really happening.

The leaked November 2018 presentation to Alphabet executives did more than just perpetuate the idea that Sidewalk could get its hands on more of the Port Lands. It alleged that Jim Balsillie had "spun" "the majority of negative press coverage," as though Sidewalk and Waterfront weren't consistently making newsworthy decisions that spoke for themselves. Yet Balsillie didn't help matters with his own misinformation, even telling some people privately that his friend John Ruffolo had lost his job running OMERS Ventures, one of Canada's most prominent tech funds, because he opposed the project. (That explanation has "zero reasoning," Ruffolo says: after a leadership change in 2018, he'd had much more significant disagreements with OMERS over strategy.) Balsillie's friendship with Ruffolo also got an innocent bystander caught in the crossfire of misinformation.

Former Toronto mayor John Sewell once wrote a column defending Sidewalk's plans, arguing that Balsillie's opposition wasn't credible because he was "close" to a rival Quayside bid. Sewell says he heard the rumour from someone close to Sidewalk, though not Sidewalk itself. But it wasn't true; Balsillie stayed out of Ruffolo's bid. The retired RIM executive sent a defamation lawyer after the publisher of Sewell's column to correct the story.

Sidewalk's own influence campaign was among the most intense outside of conventional politics that Toronto had ever seen. Bianca Wylie tried to figure out how far that influence went, but it was difficult. Contracts and payments from tech companies often come with non-disclosure agreements, which can blend together to stifle dissent. "It's time to make the stupid spreadsheet of who is getting paid by Sidewalk Labs," she tweeted in late 2018. Soon, she wasn't the only one asking how Sidewalk threw money around. "Because Google has more money than God, it's hard to find pro-Sidewalk experts who haven't been paid by them at some point," wrote *Toronto Life* editor Sarah Fulford in a special edition of the magazine dedicated to debating Quayside.

One person who spent a lot of time on Twitter fighting Quayside misinformation was Brian Kelcey, a political consultant who'd recently become the Toronto Region Board of Trade's vice-president of public affairs. In the spring of 2019, the Board of Trade hired researchers to poll six hundred Toronto-area residents about the project. The press release announcing the results was titled "Majority Still Backs Sidewalk Labs/Quayside Project." Yet the first line of the polling firm's actual findings told a much different story: "Fewer than one-half of GTA residents are aware of the Quayside project, and only about one in ten residents currently oppose it." Discounting the suburbs, a little more than half of city residents said they supported the project. But nearly the same bare majority said they were "aware" of the Quayside project at all, so it was very likely that many of the people who said they were supporters didn't know much about it.

The Board of Trade's insistence that the majority of Toronto backed Quayside was flimsy at best. When I pointed this out on

Twitter, Kelcey got defensive: "Alternatively, the main takeaway of this poll is public opinion is 'meh,' and after months of negative hits in local & int'l news, the public still doesn't actually agree w/ the narrative that this is as alarming or as big a change as some observers insist it must be." Years later, Kelcey would tell me that the point of the poll was to point out low levels of opposition to Quayside, which was true. But in the heat of the PR battle over Quayside, even the staunchest dismantlers of misinformation were prone to distorting reality, as though the "negative" stories were made up, and not about choices made by powerful organizations.

Much of the speculation around Quayside returned to the project's mysterious origin. It's why Liberal MP Adam Vaughan pressed Waterfront executives to say on the record at the ethics committee hearing that Trudeau had no involvement—to kill a persistent rumour and help the world move on. But there was another long-time Liberal whose name kept swirling around with those rumours.

After joining Sidewalk in the spring of 2018, John Brodhead became something of a fixer—the person who got the right people in the right room at the right time, like for the exclusive "advisory council" meeting in October 2018 that Waterfront had asked Sidewalk not to host. He was a hard-working, bearded guy with tattoos that paid tribute to his kids. He also had a good sense of humour, happy to point to a sign on his desk, given to him by a friend, that mocked all the data-guzzling conspiracy theories. It read, simply: "Do I look like Google to you?"

Brodhead was a creature born of politics. Since 2004, he'd jumped back and forth between the federal and Ontario Liberal governments. At the provincial legislature, he worked closely with Justin Trudeau's close friend Gerald Butts. They stayed close when Butts went on to become Prime Minister Trudeau's senior political adviser. When Trudeau's Liberals tried to oust Stephen Harper's Conservatives in 2015, Brodhead played a lead role in writing the party's infrastructure promises. And when the party won, Brodhead became chief of staff to the federal infrastructure minister.

That role meant he kept tabs on Waterfront Toronto. He was mostly distant from Quayside, though. According to public records, he got an update on flood-protection funding in February 2017 and was asked by Waterfront that May to swat away comment requests from Bloomberg News as it sought to confirm that Sidewalk was bidding for Quayside. But he also made calls to Waterfront chair Helen Burstyn in the lead-up to the October 2017 Sidewalk announcement to bridge a connection with the prime minister's schedulers. Those calls were not about the board's impending Sidewalk decision. Trudeau had been excited about the potential partnership since the January 2017 conversation with Eric Schmidt, and Brodhead, whose job meant he had a relationship with Waterfront, was helping his colleagues learn whether the agency had set a date for the announcement. But rumours about the calls seeped out in 2018 and 2019, and people who didn't like Sidewalk Labs seized on them. In some critics' cynical interpretation, the calls seemed to validate their fears about Trudeau planning Quayside with Schmidt, democratic process be damned.

It didn't help matters that Sidewalk's leaked November 2018 slide deck showed that the company's conversations with investors included the Canada Infrastructure Bank, of which Brodhead was a key architect. Yet beyond this series of coincidences, there was no public evidence that Brodhead had done anything wrong, or that a fix had been in for Sidewalk to win Quayside. But rumours around the project were exploding, with Brodhead at the centre of them, so when the federal ethics committee hauled in Dan Doctoroff and Micah Lasher to testify in April 2019, a few weeks after Davis and Verner, Brodhead was sitting right next to them.

Peter Kent was back with another long windup. "I'm sure you'll forgive those in the GTA and the city of Toronto who have expressed skepticism and concern about the gradual evolution and the gradual revelations of the changing aspects of this project," he said. "It's somewhat like the Churchillian notion of 'a riddle, wrapped in a mystery, inside an enigma.'" Finally, a question: "When did you realize that the original 12-acre Quayside project wasn't enough for your objectives?"

Doctoroff took this one. "The original RFP that Waterfront Toronto issued mentioned not just Quayside but the broader eastern waterfront at least twenty times," he said. "It was right there from the very beginning."

The RFP did say "eastern waterfront" exactly twenty times, but Doctoroff was omitting some crucial context. Waterfront eventually wanted to turn the Port Lands into a beacon for innovative city-building, but made clear in the RFP that it was only seeking a "pilot." If the winner of that pilot hit "key objectives," the agency had written that "it may be beneficial" for some of the technologies and business partners from Quayside to be extended through parts of the Port Lands—if the city even allowed it.

After Doctoroff explained that Sidewalk had never hidden its broader ambitions—the public portion of its bid for Quayside mentioned the phrase "eastern waterfront" 312 times—Micah Lasher stepped in. He quoted two RFP segments in which Waterfront asked for financial resources and ideas to help build transit and sustainable technologies across the eastern waterfront. Waterfront had left an opening in 2017, and Sidewalk was still jumping through it two years later. He didn't mention the pilot part, or all the conditions Waterfront would impose on the winning bidder. "With respect," submitted Lasher, "I think the plans that are currently contemplated are perfectly in line with what was in the RFP."

Charlie Angus then got his turn again. The whole exercise of getting Waterfront and Sidewalk to testify had been his idea, and he'd been speaking with critics throughout the process—Ann Cavoukian and Jim Balsillie, to name two, while Julie Di Lorenzo had sent the committee a letter following the Waterfront executives' testimony a month and a half earlier. It was time, he decided, to solve a mystery.

Angus turned to John Brodhead, who was sitting to Doctoroff and Lasher's left. "What was the nature of your communication with Sidewalk Labs and Alphabet while you were working for the federal government?" asked Angus.

It was the wrong question. Brodhead had spoken with Waterfront, not Sidewalk. "I worked at Infrastructure Canada and then left and went to be chief of staff at Indigenous Services," Brodhead responded. He spoke patiently, occasionally gesturing with his hands. "I had absolutely no contact with Sidewalk Labs while I was chief of staff for the infrastructure minister. My first contact with them was after the RFP was completed, when I was at Indigenous Services." Once he was later hired, he said, he'd cleared his job offer with the federal conflict-of-interest and ethics commissioner.

Angus tried to corner him about political pressure in planning the October 2017 announcement. "I see that the auditor general noted that one of the problems with this process was that the communications and consultations that should have been done at other levels were being done at a very high political level. Who was doing that behind-the-scenes consultation at a high political level?"

"It was no one that I'm aware of," Brodhead said. This was true: though Schmidt had courted Trudeau's support, it didn't affect the actual procurement process. But Brodhead *did* make at least one call to help the prime minister's staff ask Helen Burstyn about the announcement's timing, which Angus could have clarified with the right question. The member of Parliament was leaning a little too hard on the explosive-sounding, yet vague, assertions of the auditor general's report.

Dissatisfied, Angus turned to Doctoroff.

"There was no one that I'm aware of," the CEO said.

"This is the auditor general; this isn't me," Angus said. "You're telling me you don't know who was handling this?"

Lasher jumped in and said Sidewalk hadn't spoken with Brodhead until a month after the announcement, well after the procurement process ended. "I can speak only to our communications," he said, leaving the mind's eye—but not Angus's—wandering toward Waterfront.

Angus started alluding to all the rumours about the prime minister's involvement in securing Quayside for Sidewalk. The auditor general said someone was putting pressure on Waterfront to sign the agreement, Angus said. Who?

Despite Brodhead's calls to Burstyn, no one directly involved in that process who was interviewed for this book described there being political pressure to pick Sidewalk. But several directors, including Julie Di Lorenzo, did feel rushed to vote in order to accommodate the prime minister's schedule.

"I have no idea who they were referring to," Brodhead said.

If Angus had asked what *pressure* might have meant, or even whether Brodhead had made contact with Waterfront during the RFP process, he could have gotten the answer he wanted. Instead, he extracted a response that did nothing to extinguish the fires of suspicion around the project.

Sidewalk Labs had been born from Larry Page's frustration with governments, and as its controversies in Toronto spun out of control, no government seemed capable of actually reckoning with it. That would be up to Waterfront Toronto alone.

CHAPTER 20

Maybe Tomorrow

DAN DOCTOROFF HAD a story to tell. As he recounted it, it perfectly followed the narrative arc that Kurt Vonnegut dubbed Man in Hole: well-made plans, a foil, then catharsis. Sidewalk Labs invited the public to see some of its winter-beating hardware for the first time on March 2, 2019, and—"ironically," Doctoroff later said—it snowed. "We watched the weather reports with a mixture of excitement and worry," the CEO went on. His staff worried no one would want to trudge to Sidewalk's isolated showroom across the street from Lake Ontario to see the heated, ice-melting hexagonal paving tiles and transparent, snow-blocking "raincoats" that jutted out from 307 Lake Shore Boulevard East. "The weather was striking back," he continued. "We were ready for the showdown. But we wondered if anyone would be able to see it."

By 3 p.m., Sidewalkers' fears were quelled. People did show up—hundreds, Doctoroff said, then hundreds more. He declined to mention whether they liked the paving tiles and plastic-film weather protectors. He was telling a bigger story than that. The visitors, he said, "were ready to be part of the solution and willing to give us a chance to prove we were worthy of being their partner." The humbling flourish didn't quite fit Vonnegut's framework, but it tied a nice bow around the story.

Doctoroff told this anecdote in his introduction to Sidewalk's long-awaited draft master plan, tugging at heartstrings with a humbling tale of overcoming adversity. But the people of Toronto were Sidewalk's partners only in theory. Waterfront Toronto was the steward of that partnership, and Waterfront was barely mentioned in the story.

Waterfront's chief design officer, Christopher Glaisek—the man who'd worked under Doctoroff for New York's Olympic bid, and who'd connected him with Waterfront in 2016—had warned Sidewalk that its heated paving tiles wouldn't work in Toronto. The tiles were intended to be removable, giving city workers easy access to cables, pipes and other infrastructure beneath roads and sidewalks without needing to jackhammer and messily replace a chunk of asphalt or concrete. But Glaisek told Sidewalk staff that the city always required paving tiles to be set on top of a concrete base because of the risk that they might break during frigid weather. The tiles, though nice, were a promise too far. These technologies were supposed to sell Torontonians on Sidewalk's ambitions, but they were a bad idea for, well, sidewalks. "They said they wanted to rely on Toronto for local knowledge," Glaisek says. "Unfortunately, local knowledge comes with local rules." Months later, he would tell Sidewalk to drop the plastic building raincoats from their plan, too, warning that they'd block pedestrians and be hazardous to the visually impaired.

By the time Sidewalk handed its draft master plan to Waterfront in June 2019, the C$1.3 billion development it wanted to sell Toronto was so ambitious it would ask to bypass taxpayer-protecting procurement processes and press multiple governments to rewrite laws in Sidewalk's image. "He always wants to do the best possible," Glaisek says of his old boss. But, Glaisek adds, Doctoroff "doesn't like to go small. He likes to go really, really big. And when he finds the big thing, he doesn't want to let it go."

Vonnegut had a name for this story arc, too. It was the inversion of Man in Hole. He called it Icarus—for the mythological Greek character who flew too close to the sun.

—

Sidewalk was putting Waterfront in a deeply uncomfortable position in the weeks after the agency's newly repopulated board met for the first time. Doctoroff was working the hotel ballroom circuit once again, promising in April that the plan would be made public sometime in the spring. This came as a surprise to Waterfront staff. They had agreed to make the plan public within a week of receiving it, yet as they reviewed its draft pieces, they worried that what they saw was nowhere near ready for public consumption. Innovation vice-president Kristina Verner flooded the margins of the draft pages she reviewed with warnings about what wouldn't work in Toronto, trying to convince Sidewalk to tone down what she saw as saviour complex rhetoric and asking for greater clarity. As Toronto thawed out, it became increasingly clear that Waterfront and Sidewalk wouldn't be able to agree on a plan in time for Doctoroff's solstice deadline.

Much of the delay had to do with the sheer volume of work that dozens of Sidewalk staff and consultants were putting into research. They'd been studying Toronto in great depth. That included an analysis of the energy performance of close to a hundred buildings in Toronto, which found that those buildings used 13 percent more energy than their builders had estimated. This lent support to Sidewalk's proposal for real-time energy metering—and, in turn, for using that kind of data to help with building design.

Staff dove deep into the benefits of building with wood, too, with sections prebuilt offsite. If all 2.6 million square feet of Quayside were built this way, trading away the massive carbon footprints of concrete and steel, Sidewalk staff calculated it would have the same impact as pulling twenty thousand cars off the road each year—with an added bonus of sped-up construction time.

Sustainability ideas were front and centre in the plan. If Toronto decided to buy into Sidewalk's efforts, which ranged from micromanaging building energy usage to convincing people to use shared bike services, Sidewalk said it could limit greenhouse gas emissions per capita at Quayside to just 15 percent of other city neighbourhoods. Living units would be designed for "ultra" efficiency, making more space for affordable housing, in buildings made from a "library" of prebuilt parts. A portion of the community's building heating and cooling would be generated geothermally, by loops of hot or cold water flowing through the buildings, with help from the bedrock underneath Quayside as a source of summer coolness and winter heat. Some energy might even come from local wastewater. Whole blocks might be dreamed up by Sidewalk software that algorithmically sketched out neighbourhoods to maximize traits like open space, density and natural light.

Sidewalk's draft master plan was designed to strike awe. It was Googley. If Toronto gave Sidewalk everything it asked for, in fact, Sidewalk said it could build the "largest climate-positive district in North America," making good on former Waterfront CEO Will Fleissig's ambitions for Quayside. This was a massive promise, and it would require an improbable number of things to work in Sidewalk's favour. Moreover, the plan pushed Torontonians into a corner over the question of the Port Lands: if they wanted to reach this climate-positive goal—or host a Sidewalk-built neighbourhood at all—Quayside wouldn't be large enough.

As foretold many times, Sidewalk wanted more. Now the city would learn what more would look like. Geographically, the draft master plan appeared to present a scaled-down version of the plan in the leaked slide deck. Executing it would require Waterfront and the city to hand over more than double Quayside's acreage, extending Sidewalk's reach onto Villiers, the new island planned for the Port Lands during the flood-protection project. Waterfront and the city would also have to grant Sidewalk significant privileges to plan details, finance infrastructure and deploy technology across even more land—190 acres, including through the Port Lands. "The scale of Quayside alone is not sufficient, in and of itself" to achieve Waterfront's goals in the RFP, the authors wrote, even though Waterfront had long made it clear that accessing the rest of the Port Lands hinged on success at Quayside. But the plan also brought more clarity to Sidewalk's thinking in ways not seen since its RFP response nearly two years earlier.

The push for more land was not just about unbridled, Google-sized ambitions; there were genuine questions about the feasibility of building out the company's ideas on just 12 acres. "We wanted all the innovation teams to be thinking at two different levels," says Eric Jaffe, Sidewalk's editorial director and the draft master plan's lead author. ". . . Is the Quayside achievement financially viable just on its own? Sometimes it was yes, sometimes it was no."

Sidewalk staff had spent many months conducting formal consultations, and routinely pointed out how many people they'd consulted: at least twenty-one thousand. Doctoroff and his team often cited this number as evidence that they were designing the plan *with* Torontonians. Yet some of the proposals that the company explicitly said responded to local concerns—streets that focused on pedestrians, dedicated biking infrastructure, public space designs that shielded people from the elements, modular building components—had been part of Sidewalk's plans since the Yellow Book, and some were even part of Javelin's plans before that. (One person who contributed to the draft master plan said that Sidewalk wanted confirmation that the public would embrace these ideas, and hoped to refine them.)

Much of the draft master plan seemed to directly respond to critics. Anyone who read all four volumes would find that Sidewalk staff took pains to remind readers that Sidewalk was "not trying to develop the Port Lands." It would not sell personal data or use it for ads. It would share data and use open standards. It was not "motivated by a desire to export Canadian talent or intellectual output to the United States." It would make room for local entrepreneurs. Yet many of these promises skirted the substance of criticisms levelled against Sidewalk. Sidewalk didn't want the whole of the Port Lands, but it did ask for powers through nearly a quarter of them. Similarly, the data Sidewalk could collect was never meant to be sold against ads, but it could still be used to develop new products that would indeed benefit its parent company in the United States. And Sidewalk wanted to share only 10 percent of profits from products developed at Quayside with Waterfront, despite intellectual property experts' push for a more even split.

At some points, the plan padded its claims of technological novelty, going so far as to call an elementary school an "innovation" and planting trees a "technical system." At others, concepts Waterfront had thought were dead suddenly appeared again. As she went through the drafts, every time Kristina Verner saw "urban data," the makeshift term that had been discredited by some of Canada's top privacy experts when Sidewalk introduced it in 2018, she flagged it. Yet the term still appeared in the final version, because Sidewalk felt it needed to address how data from the physical world would be governed, whether at Quayside or in a future project. In that sense, the plan was "almost like a catalogue of how they approach digital technology" for "anyone, anywhere," Verner says.

Sidewalk doesn't disagree with this characterization. But in the lead-up to the plan's release, Waterfront Toronto tried to warn the company that proposals like urban data simply wouldn't fly in the city it was trying to build in. "There was never a version of the content that I ever saw, including what was published, that reflected the concerns we had expressed through the draft development process," says Verner.

By the beginning of June, Waterfront still hadn't seen the final details. Stephen Diamond says he asked Sidewalk to defer the plan's release so the partners could align themselves on whatever financial proposals Sidewalk were planning to include. But Sidewalk stayed the course.

Even inside Sidewalk, some people were worrying that they should have been finessing a plan for just 12 acres. Some of the people Sidewalk was sharing drafts with also tried to get the company to slow down or reconsider elements of the plan. One was Suzanne Kavanagh, a community member who lived near Quayside. When staff shared the massive, 190-acre plan with her and a small group at 307 Lake Shore, she says she warned that such a sweeping proposal would only make Sidewalk's opponents more angry. "Stick to your knitting," Kavanagh told them.

The company didn't listen.

—

Waterfront had at least one reason to be happy in June. A year after Will Fleissig's resignation, the agency finally hired a new CEO. His name was George Zegarac. He was, in many ways, Fleissig's opposite: a detail-oriented lifelong bureaucrat who didn't just understand the nuances of Canadian development but knew Waterfront intimately. He'd spent two decades as a deputy and assistant deputy minister, handling some of the biggest files Ontario could throw at him: health, education, universities, agriculture and, for two years, infrastructure. There, he got a deep understanding of how Waterfront worked, helping secure flood-protection money from the province while Fleissig was still CEO. If Fleissig was all vision, and if interim CEO Michael Nobrega was all process, Zegarac tried to present himself as all execution. In him, Diamond found a kindred spirit—someone who could see past drama to focus on facts, and maybe even get a few things done.

One of Waterfront's government parents was finally trying to get things done, too. The City of Toronto began wading more deeply into the Quayside project in early June. That month, a deputy city manager revealed that Toronto wanted to dedicate C$800,000 in 2019 to pay for staff and consulting work to review project details, confer with the public and begin developing a framework for data and privacy. The smallest of the three governments that oversaw the Sidewalk project was the most willing to take on its biggest burden.

But in spite of his new CEO and the newfound help from the city, Diamond still had to spend most of June doing damage control. When the draft master plan's business terms finally arrived, they were full of ambiguity. The plan listed eleven "potential" sources of revenue, some of which would be returns on Sidewalk's investments: in real estate, in new start-ups, in infrastructure, in transit. It wanted "performance payments" if its next-gen community was a success. The idea of using tax increment financing re-emerged—to build out light-rail transit—despite having been pilloried after the slide deck leak. And Sidewalk wanted to invest in a "limited set" of new technologies in the neighbourhood, appearing to respond to public outcry over how much data collection might happen in a Google-affiliated neighbourhood by repeatedly emphasizing how few digital technologies the company would deploy.

The total cost of Sidewalk's ideas was hard to pinpoint. Despite the plan's sheer volume—1,524 pages—little in it was concrete, and some of its numbers were fluid. The company and its potential partners would lay out C$900 million, with an additional C$400 million available to help finance light rail and other infrastructure through the neighbourhood if the city let it. An additional billion or more was potentially available to build out tech and tech-laced infrastructure.

Then there was the matter of land. Quayside was worth about C$590 million. The western portion of Villiers Island, where Sidewalk hoped to build a Google office, would be worth nearly C$400 million. Sidewalk expected a discount on the land because of the value it hoped to add to the area. Waterfront often granted these kinds of discounts, but the architecture critic Alex Bozikovic consulted industry sources and estimated that Sidewalk's asks would give it a discount of almost half a billion dollars.

The total cost to the public was incalculable. In order for Sidewalk's plan to work, governments would need to set up a public administrator for the neighbourhood to handle planning, transportation and construction issues, and a set of non-profits and trusts, including a data trust. (The Toronto Region Board of Trade had earlier pitched the idea of having the city's library host the trust, since it was designed to make information available to the public, which Sidewalk said in the plan that it would consider.) Sidewalk's plans hinged on asking the city and provincial governments to change laws and regulations, too. Traffic proposals, like charging vehicles for waiting at a curb and making streetlights adaptive to different situations, would require amending the City of Toronto Act, the city's municipal code, and the Ontario Highway Traffic Act. Constructing thirty-storey towers out of wood would require changes to Ontario's Building Code. Noise and zoning bylaws would need to be adjusted to allow the blend of residential and commercial uses Sidewalk hoped to build.

When Waterfront staff saw the final version of the draft Master Innovation and Development Plan, or MIDP, in June, they found it jumbled and repetitive. The plan disregarded many of Waterfront's suggestions, leaving the agency's negotiating team in the very awkward position of having to push back against a document that was supposed to represent the power of a first-of-its-kind partnership. They were also trapped, having promised to release it to the public within a week of receiving it.

So, backed into a corner, Stephen Diamond wrote an open letter.

"It is important to know that Waterfront Toronto did not co-create the MIDP," Diamond wrote. Waterfront found some of Sidewalk's ideas "exciting," but the organizations diverged significantly on many ideas. Sidewalk wanted powers over more than 100 acres that Waterfront couldn't promise it, was asking for commitments and regulatory changes that Waterfront was powerless to make, and was re-pitching digital governance ideas that Waterfront had already rejected.

If the draft master plan read like a threat, Diamond ended his letter with an implicit threat of his own: he was willing to walk away. "Whether the Quayside project proceeds or not, the conversation we are having is important for all of Toronto," he concluded.

I got Diamond's letter by e-mail while sitting in a cab on the way to Waterfront's office on June 24. I hadn't seen the plan yet—thanks to some strange scheduling, many media had to speak to Diamond before Sidewalk would show them details at a technical briefing. I met Diamond in the same aging boardroom where everything important related to Quayside seemed to happen, with its expired-candy-laden scale model and blocked view of the lakeshore. He was as pissed off as his stoic demeanour would show. "I don't want to be perceived as being unduly negative towards them," Diamond said. "But, you know, from a professional and personal perspective, I would have preferred if they would have not made requirements for a much larger area." It was aggressive, he added, but "we're just saying, 'Hey, slow down a bit.'"

Soon after, media descended on an event space just north of Quayside, where we got our first glimpse of the draft master plan. The four-volume set, called *Toronto Tomorrow*, weighed 18 pounds. Dan Doctoroff gave a brief intro, Josh Sirefman presented reporters with the big-picture details, and then media tried to digest it all, rushing around the room to talk with Sidewalk's subject-matter experts. Strangely, the company didn't have a figure ready for the total cost of its proposal. After I asked, a PR director took nearly twenty minutes to deliberate with colleagues and settle on C$1.3 billion—one bare-bones permutation of the many potential investments Sidewalk said it might make in its choose-your-own-adventure of a plan.

The document carried the grandeur Sidewalk hoped to attain after it first conferred with Will Fleissig in 2016, and seemed to ignore the many warnings the company had received about overreach in the years since. And so the headlines that appeared online after the media briefing weren't particularly kind. *Vice* called it a "democracy grenade," while John Lorinc, one of Toronto's best-known urban-affairs writers, wrote that the draft master plan "seems to be about pushing back the City's authority and replacing it with semi-privatized bureaucratic SWAT teams that owe their existence to [Sidewalk] and will likely never forget or challenge that connection."

Bianca Wylie saw the plan as an overwhelming document, something that people might be afraid to comment on because it was almost impossible to find time to read the whole thing, let alone get comfortable with it. "That puts a chill on speech," she says. A political science professor from Brock University named Blayne Haggart spent the summer blogging about each section of the plan. "I'm not sure that Sidewalk Labs has fully grasped how badly it is screwing up, not just this job, but its future business prospects," he wrote near the end of his 1,524-page adventure. "It really is that bad a plan. Also, it's not a plan." Some of the plan's slick renderings showed a neighbourhood with so many exposed-wood features that longtime

Quayside watchers wondered whether the company and its designers were aware of Toronto winters. The provincial and federal governments hesitated to weigh in yet again, but progressive city councillors couldn't believe the details. Downtown councillor Joe Cressy, the Waterfront director, insisted that any land beyond Quayside was "not up for sale."

Sidewalk Labs had managed to embolden and grow the ranks of its enemies. The draft master plan rankled Sidewalk supporters who had warned of its overreach; it sowed suspicion among elected politicians who had previously tried to stay neutral; and it gave the federal government even more reason to keep a distance from the country's highest-profile city-building project.

But none of that mattered when it came to building Sidewalk's neighbourhood of the future. What mattered was satisfying the man on the other side of the negotiating table: Stephen Diamond.

CHAPTER 21

Underwhelmed

THE WEEK AFTER Sidewalk Labs released its draft master plan, a municipal affairs columnist named Matt Elliott took a walk along Lake Ontario. He liked watching how Toronto changed over time, especially all the "raw and in-the-works parts." His walks routinely took him through the Port Lands and past Quayside, which inevitably took him to the Sidewalk Labs office. On this July day, he stopped to look at the technologies the company had on display outside—the ones Dan Doctoroff waxed poetic about in the draft master plan. He found it confusing. The building "raincoats," in particular, didn't make sense to him. They were basically just awnings.

Then Elliott noticed something new. It was a billboard, in black and white, that soared directly above the bright-blue Sidewalk Labs office. The billboard was from Apple: a photo of its latest iPhone juxtaposed with a block of sans serif text. *We're in the business of staying out of yours*, it read. *Privacy. That's iPhone.* Elliott took a photo of it.

The computing company had been putting up location-specific billboards and other physical ads throughout the year, including a massive building-side advertisement during the Consumer Electronics Show in January, playing on a slogan for Las Vegas: *What happens on your iPhone, stays on your iPhone.* Billboards began popping up in Toronto that summer. On King Street, one said *Privacy is King.* Especially since the Cambridge Analytica scandal broke a year earlier,

Apple had been using publicity campaigns to distance itself from its data-hoovering biggest-companies-in-the-world brethren, Facebook and Google. And just below its Lake Shore Boulevard billboard sat Sidewalk, Google's little sister.

Elliott was a long-time chronicler of Toronto's follies, and though he'd at first supported the project, he also knew big, splashy plans usually didn't turn out as promised. He'd been following Sidewalk's "phalanx" of lobbyists who had marched through City Hall, and felt that the company's failure to explain its ideas for Toronto had left the fate of this lakeside gem in the throes of unnecessary drama. Apple's billboard did not look like a coincidence. "I was standing on the undeveloped waterfront, a place packed with potential for public spaces and affordable housing, and the photo was one tech giant trolling another tech giant," Elliott says. "It kind of just cemented how far away this process had gotten from what neighbourhood planning and city-building should be all about."

So he tweeted the picture.

It shot around the web, with Forbes, Insider and even Breitbart applauding the trolling. With compounding accusations of a massive overreach in Sidewalk's draft master plan, the world wasn't taking Google's smart-city sister company nearly as seriously as it had two years earlier.

Toronto itself was still split on the matter. The Board of Trade issued an open letter that same week with thirty signatories—including three university presidents, two former mayors, a future ambassador to China and Richard Florida—encouraging governments to stay the course and give Quayside a clear-eyed review. #BlockSidewalk spent the next few months gathering nearly three times as many signatories, many from the worlds of academia, activism and organized labour, for an open letter of their own. They argued that Waterfront and Sidewalk's negotiations needed to happen in the public eye to avoid any more 190-acre surprises.

Stephen Diamond—the great compromiser, the man installed by three levels of government to bring calm and order to Quayside—was torn. Sidewalk had ignored many of his and Waterfront's pleas to make sure they were aligned on the draft master plan, yes. But he was not a catastrophizer. He believed that if Waterfront could bring Sidewalk to its terms, they could still build something great.

He spent the summer trying to wrestle their relationship back under control. He and Waterfront's core Quayside team began to outline a set of key "threshold issues": sticking points on which they refused to budge. The biggest was the land, of course. But there were other issues that Waterfront decided could make or break their partnership. For one, they'd have to hold a fair procurement process to develop or plan any land beyond Quayside, not just hand it over to Sidewalk. Similarly, though Sidewalk needed a commitment that light-rail transit would be built through the area, Waterfront needed the company to recognize that the agency had no power to make that happen. The agency also wanted Canadians to get fairer access to Sidewalk's patents, and Waterfront's government parents to get a share of revenues from new technologies that Sidewalk piloted at Quayside—as opposed to profits, which can be manipulated with clever accounting. And Waterfront wouldn't budge on the fair market value of the Quayside lands. It had recently been appraised for C$590 million. That's what Waterfront wanted, no matter what Sidewalk claimed about needing a discount because of the costs of its sustainability and innovation plans.

In early August, Sidewalk and Waterfront set a deadline for coming to terms: October 31. Meg Davis called it "the Halloween Accord." If Sidewalk couldn't rewrite the draft master plan to accommodate Waterfront's threshold issues by then, Waterfront would walk away.

Dan Doctoroff's negotiating tactics were, by the summer of 2019, well known to Diamond, Davis and Verner. When he was deputy mayor of New York a decade and a half earlier, he would deliberately lowball the amount of affordable housing the city would commit to include in massive rezonings. His team might offer to make 25 to 28 percent of units in an area affordable, knowing local politicians and residents would want a figure closer to 30 to 35 percent. Toward the end of negotiations, Doctoroff's team would appear to acquiesce and meet in the middle. His New York planning colleague Amanda Burden called this tactic "leaving a little juice in the lemon."

If Sidewalk couldn't build out infrastructure and deploy technology across 190 acres of the lakeshore, the company still had some juice to squeeze out of the Port Lands. The draft master plan proposed that the company should develop 20 acres on a patch of land southeast of Quayside called Villiers West—a sizeable chunk of the new island that would be formed by the rerouting of the Don River. That was where Sidewalk wanted to build the new Canadian Google headquarters it had been dangling in front of Toronto since it first bid for Quayside. Tacking Villiers West onto Quayside would make investing in Toronto a lot more financially feasible than betting on the future of cities with just a 12-acre lot. As Sidewalk and Waterfront met throughout the summer and early fall, Villiers West became a sticking point for Sidewalk.

At the negotiating table, though, the dynamic between Waterfront and Sidewalk had begun to shift. Diamond took a much different tack than Will Fleissig or Michael Nobrega had. He was a compromiser, yes, but he didn't care much for theatrics; he simply wanted Sidewalk on board with Waterfront's terms.

Doctoroff once again threw a fit, angrily raising his voice in one meeting in a bid to get his way. This time, something different happened. Diamond stopped the CEO.

"Look," he said, raising his own voice from across the table. "If you want me to yell back at you"—he was yelling himself now—"I'll yell back."

People chuckled. Doctoroff didn't seem used to this. The Sidewalk chief stood down. The talks paused, awkwardly, and the parties took a break. When Waterfront and Sidewalk came back, the negotiations resumed as though nothing untoward had happened.

In rejecting the draft master plan, Waterfront had cast aside Dan the Storyteller. Now it was putting a dent in Dan the Dealmaker.

Even some of Sidewalk's own staff had come to realize that their CEO was hurting their cause. They wanted a friendlier touch in negotiations. But they felt that Doctoroff tended to take advice better if it came from members of the community, as though he could tell himself that he was gathering important intel from the field. So a Sidewalk manager called up one of the company's favourite local community leaders, Suzanne Kavanagh, and asked her to be candid with Doctoroff about the efficacy of his approach. She agreed and met the CEO for coffee.

Despite Doctoroff's hectic schedule, they spoke for an hour and a half. Kavanagh was happy to be blunt. "I told him to get on his bike in Italy and just let his staff do their jobs," she says.

Soon after that, Waterfront staff noticed that Josh Sirefman was taking the lead on negotiations.

Sidewalk had made a powerful connection in Toronto—one whose money and influence stretched far outside the city's boundaries. In August, Sidewalk announced that it had partnered with the Ontario Teachers' Pension Plan, which managed more than C$200 billion in assets, to launch a new company called Sidewalk Infrastructure Partners. The draft master plan had briefly described its goals. It would invest in "technology-enabled infrastructure"—the backbones of next-generation cities. Together, Alphabet and Teachers invested $400 million into the partnership to finance projects across North America. Sidewalk Infrastructure Partners would go on to commit $100 million to a start-up that would try to conserve energy and minimize blackouts with a mix of smart-home technologies and electrical grids that incorporated consumers' own solar energy. And a year after Sidewalk and Teachers announced the fund, it revealed plans to build roads designed for self-driving cars, starting with a 40-mile stretch in Michigan between Ann Arbor and Detroit, cities deeply familiar to Dan Doctoroff and Larry Page.

The fund was a return to form for Doctoroff, who'd spent the first half of his career as a private-equity mogul. He sat as a director of Sidewalk Infrastructure Partners. While his reputation as a deal-maker had been chipped away by successive controversies in Toronto, if Waterfront made good on its threat to walk away from the Quayside deal, at least he had something to show for his time in the city.

At the same time, Sidewalk was trying to turn some of its urban-technology ideas into reality. Its staff had been filing for patents since at least 2017, and the company made initial applications to the U.S. patent office that summer for ideas connected to its robotic underground freight delivery and garbage-hauling system and a user-authentication system that would use a person's physical characteristics, like facial recognition, with greater security than existing systems. When tech news website the Logic reported on Sidewalk's patent filings after they were finally made public, Waterfront's digital strategy skeptic Andrew Clement called the project a kind of "IP grab"—even if Sidewalk left Toronto, the ideas it had generated while working in Toronto could make the company a mint elsewhere.

Fresh blows kept landing for Sidewalk. After months of scouring the draft master plan, Waterfront's panel of digital strategy advisers released a nearly hundred-page response in September 2019. This was the kind of substantive work that Saadia Muzaffar had criticized Waterfront for not undertaking a year earlier. The remaining panellists made their qualms clear. Quayside was supposed to be a place where citizens were "at the centre," yet the panellists questioned whether Sidewalk was truly responding to the public's needs. They felt Sidewalk was struggling to prove its own innovativeness—that "the unique value proposition put forward is not always convincing." And the panellists were surprised, too, that Sidewalk had sidestepped addressing the inevitability of a data breach or other major privacy violation, warning the company to outline clear plans to deal with one. The panel rarely shared a unified point of view, with economic boosters like Mark Wilson sometimes sparring with more aggressive critics like Andrew Clement. Yet they found common ground when analyzing the plan. Despite Sidewalk having had nearly two years and $50 million to complete the plan, the panel largely agreed that Sidewalk's ideas were "frustratingly abstract."

Two weeks later, Ontario's privacy commissioner, Brian Beamish, sent a letter to Stephen Diamond. His team had been concerned about Sidewalk since the company first floated the concepts of urban data and a data trust. Like Waterfront's digital panel, Beamish's team had needed a couple of months to sift through the draft master plan, and they were finally ready to share their thoughts. They found it full of holes, to the point where one version of the proposed data trust appeared to lack independent oversight and might circumvent Ontario's privacy laws. Even if someone consented to data collection at Quayside, the commissioner warned that it wasn't *really* consent if a person needed the services tied to that data collection. The separation of urban data from information collected by apps or phones created further privacy headaches.

There was an opportunity here, Beamish said. In the absence of updates to federal privacy laws, Ontario could take the lead in protecting its own residents. It could mandate "data minimization" to prevent corporate overreach, implement "ethical safeguards" and give privacy commissioners more power to audit and investigate potentially bad actors.

Doug Ford's government did not leap at the opportunity until 2021—not even to one-up Justin Trudeau's federal Liberals. Two years after Sidewalk came to town, governments were still passing the buck when it came to leadership in safeguarding the cities of the future.

Waterfront Toronto already had a relationship with the Mississaugas of the Credit, the Indigenous Nation that held the treaty with Canada for the land on which Quayside might be built. But members of the First Nation were piqued enough by the project that its consultation director, Mark LaForme, reached out to Dan Doctoroff. The Sidewalk CEO met with LaForme, gave a presentation to the chief and council, and kept in touch with the Mississaugas "not simply as a stakeholder, but as a Nation that had to be recognized as part of the process," the director says. The First Nation wasn't ready to endorse Quayside, but its council was happy to be part of the conversation.

In November 2018, Sidewalk had also hosted a workshop with Brook McIlroy's Indigenous Design Studio, whose work was informed by Indigenous cultures and sustainable practices. Duke Redbird, an Ojibway artist, author, filmmaker and poet, was the elder in attendance. Sidewalk spent the workshop seeking input from Indigenous people from the design, architecture, art and education sectors to hear how Quayside could address Indigenous people's needs. Fourteen recommendations emerged from the workshop, including to establish Indigenous education programs at Quayside and to host dedicated affordable housing for Indigenous people.

Nearly a year passed. Then Dr. Redbird sat down in October 2019 with participants from the original workshop and other Indigenous community members to read the draft master plan. As far as they could tell, Sidewalk hadn't included their recommendations, despite name-checking the group's involvement. So they issued an open letter. "This resulted in a grossly misleading implication of endorsement by the Indigenous community of Toronto," Dr. Redbird wrote alongside Calvin Brook, Brook McIlroy's co-founder. "Indigenous Peoples are accustomed to tokenism when it comes to consultation—'check the box' gestures—that are labelled as inclusive because a meeting was held, but the scope of Sidewalk Labs' insincerity reflected in its [draft master plan] and media campaign is truly shocking," they continued. "It should be emphasized that the Indigenous community did not ask to be included in the consultation process but were rather invited assuming that a valuable contribution could be made. It appears that instead the process was used to manufacture a politically correct endorsement."

When I asked Sidewalk for comment, the company directed me to the Mississaugas of the Credit. Consultation director Mark LaForme didn't pick up the phone that day, but when we spoke a couple of years later for this book, he was dismayed about the Brook McIlroy consultation for a different reason. He was frustrated by the idea of consulting "Indigenous communities," which had sidestepped the Mississaugas as treaty holders. "It diminishes our status as a First Nation, who has the legal obligation to be consulted," he said. "When it comes to land development, these are our treaty lands." The Nation's chief, R. Stacey LaForme, echoed his concern. "We want to do the engagement when it comes to reaching out to Indigenous populations," said Chief LaForme. "It's our treaty lands. We understand more of the Indigenous mindset."

Even the First Nation that Sidewalk had tried to use to deflect criticism about its relationship with First Nations was frustrated by how little the company seemed to understand how to work with stakeholders. The Mississaugas of the Credit didn't know it, but some of Sidewalk's own employees had been privately describing the company's work as "colonial."

George Zegarac had formally started as Waterfront's CEO in August, arriving on the thirteenth floor of 20 Bay Street eager to steer things in a better direction. Yet by late September, he worried that the agency couldn't reconcile its differences with Google's urban planning sister company. Following the release of the draft master plan, Waterfront had drawn clear lines around their "threshold issues," but Sidewalk was concerned that if they met those terms, the project wouldn't be profitable enough.

So Zegarac, Stephen Diamond and Meg Davis walked eastward along Lake Shore Boulevard from Bay Street to Sidewalk's office to meet with Dan Doctoroff and Josh Sirefman. Zegarac brought a matter-of-factness with him forged by decades of navigating Ontario's bureaucracies. Like Diamond, he knew when it wasn't worth wasting time on bullshit. "I've been around long enough to know that if you're going to be in a tense negotiation, you want to know if there's actually a chance of agreement," he says.

Sidewalk staff gave Zegarac the grand tour of the blue-splashed building before they sat down in a meeting room. The company still wanted more than 12 acres to make the Toronto investment worthwhile. Waterfront was willing to try to help them find a compromise, and even reached out to City Hall about including the western part of Villiers Island in a Sidewalk deal that could include a new Google headquarters and a hub for tech companies. This was politically appetizing: a shiny new campus beckoning people to the Port Lands, bringing thousands of new jobs to an underdeveloped part of Toronto that could serve as a gateway to a whole new extension of downtown. They discussed a handful of ways to make the Villiers plan work, including a land swap with another city agency for a small piece of Quayside. But as usual, the agency could make no guarantees beyond the original 12 acres.

Doctoroff told them that Sidewalk could plan and build something significantly better if the company could just access more land.

But Zegarac said that was out of Waterfront's hands. "You need to be frank with me as to whether you can agree to these terms," he told Doctoroff. "I don't want to spend a lot of time negotiating if we're not going to agree."

Doctoroff turned to him and said, "I think we can find a way."

It started with an e-mail campaign. The Canadian branch of the Association of Community Organizations for Reform Now (ACORN) set up a form on its website that let Torontonians send a note to Waterfront's directors and a handful of local politicians. The automated message warned that if they approved Sidewalk's draft master plan, "the city will set an extremely disturbing precedent for the country as a whole." People like Stephen Diamond suddenly saw their inboxes flooded with demands for a greater focus on human rights at Quayside—for better affordable housing, privacy provisions and public services.

In two years of fighting the Quayside project, its critics had not used particularly aggressive tactics. There had been no massive rallies outside 307 Lake Shore like the ones Amazon opponents had held in Queens, no blockade of Big Tech–branded buses, as in San Francisco. Opposition in Toronto mostly took the form of meetings, academic papers, op-eds and public remarks. But as the Halloween vote loomed in the fall of 2019, ACORN and a handful of other Sidewalk opponents decided to one-up their e-mail campaign and take a cue from their counterparts in Berlin.

As Waterfront's board met in its Bay Street boardroom on October 24, about twenty activists gathered in the downstairs atrium, most of them clad in bright-red ACORN T-shirts. They took the elevator to the thirteenth floor and began chanting as they moved toward the glass doors of the Waterfront office.

"Who are we?"

"ACORN!"

"What do we want?"

"Affordable housing!"

"When do we want it?"

"Now!"

As they filled the agency's lobby, they were one set of doors away from the boardroom, but a Waterfront employee stood to block it. The staffers who greeted the protesters tried to calm them as quickly as possible, and within a couple of minutes offered to take the group to a different meeting room, where they assured the protesters that Stephen Diamond would come and speak with them. "They received us like we were VIPs—'Come here, sit down in this nice room,'" says Alejandra Ruiz Vargas, who chaired a local ACORN branch.

Ruiz Vargas had been one of the first Torontonians to stand up to Sidewalk's ideas, chastising Dan Doctoroff about affordable housing at the first public Quayside meeting in 2017. Here she was again, on the eve of the most fateful moment in the project's history, attempting to occupy Waterfront's office and being received like a distinguished guest. It was almost funny, she thought.

Eventually, Diamond came to the meeting room, as did several Waterfront staff. Wearing her bright-red ACORN shirt, Ruiz Vargas stood up and read from a letter. While Sidewalk's draft master plan, she said, is "a good opportunity to develop the waterfront and provide affordable housing and decent jobs to people, the company actually proposes huge profits for itself and deeply [compromises] the human rights of people." ACORN wanted more than half of the housing at Quayside to be rented for far below market rate, and for the public sector to have greater control over the project.

"On October 31st, stand with people," Ruiz Vargas concluded. "Stand with us."

Diamond thanked her for the letter and told her the agency hadn't decided yet on the project. But he appreciated the group stopping by, he said, because they wanted to hear from the whole community.

This was a far different experience than the mass occupation of Berlin's Google campus. The result was, in its own way, the most Canadian one possible.

After chatting for a bit, Diamond said they should reach out to him anytime—but issued a small plea of his own. "Next time," he said, "please don't send all those e-mails."

He was tired. There was a lot going on that month.

Ten days before Waterfront Toronto's board would vote on the future of the Sidewalk partnership, a deputy city manager sent CEO George Zegarac a letter with bad news. The city wasn't willing to just hand Waterfront a chunk of Villiers Island. "Any disposition of land in Villiers Island would require an open and competitive procurement process," the manager wrote. To go forward, the city would need council approval to spend at least a year reviewing and updating its plans for Villiers, *then* put the land out to open competition. Nothing could happen soon, nor was Sidewalk guaranteed any land when the process was done.

Waterfront's leadership had expected this might happen. The agency had done as much as it could to reach a compromise with Sidewalk, but the decision was out of Waterfront's hands. Sidewalk, however, was not exactly prepared for this scenario. "We were quite confident, because Waterfront Toronto seemed quite confident, that we would be able to include Villiers West in the proposal," Doctoroff says. The city's refusal prompted an existential question—one that ignored the point of fair procurement processes in the first place yet weighed heavily on the minds of some of the company's staff, especially after the deputy city manager's letter. If the City of Toronto was the authority that would grant land Waterfront couldn't, but couldn't compromise the way Waterfront was willing to, did Sidewalk even want to build in Toronto?

Take the company's vision for tall timber buildings: if the city didn't approve the regulatory changes they'd asked for in the draft master plan to allow for skyscrapers to be built from wood, Sidewalk would need to use materials that would make its buildings look a lot like the regular towers of today. Where was the innovation in that? Sidewalkers had already been scaling back aspirations for months, trying to align their plans with the new terms Waterfront had outlined that summer. With the city blocking Villiers West, they would have to scale their plans back even further. Some staff began asking themselves: how much of a dream can you shave away before it isn't really your dream anymore?

Halloween was Thursday, and by the Monday before, no one knew if the Quayside project would survive the week. On Tuesday, George Zegarac sent Sidewalk president Josh Sirefman a letter asking Sidewalk to confirm it would agree to Waterfront's terms. Then all he could do was wait.

Bianca Wylie had spent the past few months working on something new. Though she had gradually stepped back from the fracas once #BlockSidewalk had been created, she was still often upheld as the de facto figurehead of Sidewalk opposition—the person to whom the world turned to understand that Quayside's greatest consequences wouldn't be how it shaped the future of technology in cities, but how it shaped the future of power and democracy. She had been christened "the Jane Jacobs of the Smart Cities Age," a clear counterpoint to Doctoroff, who was once anointed "a modern-day Robert Moses." Wylie injected nuance into a conversation that could easily have been written off as anti-tech voices yelling at pro-tech voices. Now she was going to show the world just how deep that nuance ran.

After nearly two years spent writing essays from a collapsible table in her attic, she began to solicit essays from others. Wylie, Digital Justice Lab director Nasma Ahmed and a few others had been quietly asking big-picture thinkers who cared about cities to write about the wide-ranging consequences of the Quayside project and smart cities in general. In late October, they gathered all of the essays—nearly a hundred in total—and launched a website called Some Thoughts.

The organizers envisioned the collection as a series of conversations around a table. At this imaginary table, you had the pro-Sidewalk urban designer Ken Greenberg in one corner, arguing that the company would give Toronto a chance to explore new ways to solve urban problems, and Jim Balsillie across from him, arguing that governments should be making the kinds of decisions about data that Sidewalk had been putting forward on its own. Dan Hill, the level-headed, bearded British smart-cities consultant who stopped showing up to Sidewalk's offices in 2016, wrote that "it's easier than ever to see that the private sector cannot do the public sector's job, particularly in cities."

There were ruminations on data trusts, equity in technology, and colonialism. Some Thoughts came across as a grand summation of two years of strain. No matter what Waterfront's board decided on Halloween, here were more than ninety essays to help cities navigate the tensions that arose when tech came to town.

As Wylie and the Some Thoughts team were steering the narrative around Quayside further into critical territory, Stephen Diamond and George Zegarac still weren't sure if Sidewalk would even stick around in Toronto. They'd come close to a verbal deal to keep working together, but the company hadn't signed off in writing. Tuesday turned to Wednesday. The hours began ticking away on Wednesday morning. The vote was less than twenty-four hours away, and it was entirely possible that the Quayside project was about to die. Diamond was so unsure of Sidewalk's decision that he prepared two speeches for the next day's board meeting: one about pushing forward with the Sidewalk partnership, and one about ending it altogether.

At around noon, he received a letter from Josh Sirefman.

As frustrated as the Sidewalk brass were that Diamond had forced them to scale back the June draft master plan, they also felt he and Zegarac were trying to make the partnership work in good faith, and could help open doors to more land later. So the company, Sirefman wrote, would agree to Waterfront's terms. But only conditionally. After all of the allegations of overreach over the past two years, Sidewalk was forced to reconsider its entire plan over an ordinary government decision. With City Hall cautioning Sidewalk that it wouldn't be able to guarantee a piece of Villiers Island right away, the company needed to keep running numbers: "That timing change has an impact on the innovation agenda and economic underpinnings of our proposal."

It was time for the Halloween Accord. On Thursday, the public once again packed Waterfront's aging thirteenth-floor boardroom. Sitting at the head of the board table, as far from the public gallery as possible, Stephen Diamond called the meeting to order just after 9 a.m.

"Our board felt that there would be no point in wasting the public's time or money if we could not reach an understanding on some of these basic threshold issues," Diamond said. Even from afar, he looked tired, but not despondent. "This view was, in fact, confirmed by the public consultation process that we carried out over the summer." He was dragging the announcement out. "As a result, the board, with the consent of Sidewalk Labs, agreed to work until today, October 31, to determine if we could resolve our differences and realign on these issues."

He clarified that the Halloween vote would not be to approve the draft master plan, which was well beyond the scope of what Waterfront wanted, but to determine whether Waterfront should start reviewing a reworked version of the plan.

And then, because so many members of the public had turned up to the meeting, Diamond asked the directors to follow him to another meeting room to have an in-camera discussion about Josh Sirefman's letter. Dozens of people sat waiting in the boardroom. They included members of #BlockSidewalk, who were waiting to dramatically unfurl some banners, one of which read, *No more private, unsolicited deals for public assets behind closed doors.* The local federal politician Adam Vaughan was there, too, and a whole bunch of Sidewalk staffers, all waiting for Waterfront's decision. It still seemed like the decision could go either way: On top of everything else, just hours earlier my colleague Tom Cardoso and I revealed the contents of the Yellow Book to the public for the first time, prompting at least one director to waver in their willingness to support the project.

Twenty minutes later, the board came back.

Diamond sat down at the head of the table. "In fairness to Sidewalk Labs, from the outset, they have indicated a desire to expand beyond the initial scope of Quayside," Diamond said. But now, he continued, "the project has clearly been defined as an initial phase as being the 12 acres of Quayside only." If Sidewalk wanted more, he said, the company would need to compete for it when the time was right. But beyond land, he said, the board was happy to announce that Sidewalk had agreed with all of Waterfront's other "threshold" terms, including, more than a year after they'd introduced it, to ditch the legally confounding concept of urban data. "Our objective is to require that the control and collection of data would be democratically accountable," the chair said.

Diamond asked the board to vote on whether they should forge ahead, consulting the public along the way, and hold *another* vote over sticking with Sidewalk the following spring. Everyone around the table raised their hands at once. Waterfront Toronto would keep working with Sidewalk to try to design the city of the future.

But before the meeting ended, thirty-eight minutes after it

started, Diamond left the room with a warning. "This is one step forward," he said. "But this is not the final step."

A few hours later, the *Globe and Mail*'s architecture critic, Alex Bozikovic, and I walked to 307 Lake Shore to interview Dan Doctoroff. We were in a fishbowl-style meeting room, with Doctoroff facing Lake Ontario and us facing the elevated expressway that Waterfront Toronto's founding chair had hoped to tear down two decades earlier.

Bozikovic started: "With just Quayside to focus on, what will the site look like in a few years, and how will your business be focused?"

"The vast majority of the planning work was actually focused on Quayside, right? That's where we went into great detail," the CEO said.

His answer ignored that the draft master plan said Quayside "becomes possible only when considered in combination" with more land—that just months earlier, he was using it as a bargaining chip for much more. So I asked how he came to accept a 12-acre deal.

"We've thought some about how we would arrange things if it was only Quayside," Doctoroff told us, without going into details. "At the end of the day, I spent a lot of time in government. And even though people may not necessarily have seen this, I have significant empathy for the pressures that government has placed on it." (Waterfront would not characterize his approach as empathetic.) "One of the things that we just kept hearing, over and over and over again, is that it's really important that before we hand over more to you, that you prove that what you say is doable." (This was true, but he omitted how often his company had tried to ask for more without proving itself.) "To be honest, I found it kind of hard, putting on my government hat, to disagree with that." (Did he just find this hat?) "And I have enough confidence in what we have proposed to believe that what we can do in this, hopefully, first phase could be pretty amazing."

This was a significant about-face from the expansive dreams of the draft master plan. But only after his soliloquy did Doctoroff admit that part of his turnaround had come from acknowledging his company's strategic faults: "It's a combination of growing confidence in our ability and a recognition that asking us to prove ourselves just is not unreasonable. In fact, it's fair. And it took time to understand that, in a way."

Bozikovic jumped in and asked about Villiers West. "As you say, you've worked in government, and real estate, and you can't have been surprised that attempting to short-circuit a procurement process—"

Doctoroff interrupted him. "The procurement issue for Villiers was always complicated."

It wasn't. There *hadn't been* a public procurement process for Villiers West, but the city had already developed extensive plans for the island years earlier, with help from Waterfront. The city's procurement processes appeared to be complicated only from the perspective of a company that wanted more than a particular procurement process had put on offer—though the Sidewalk CEO pointed out that Toronto had occasionally sidestepped similar procurement processes in the name of "economic development." Then Doctoroff said that Sidewalk had always been transparent about wanting more land, and that the company had mentioned the "eastern waterfront" hundreds of times in its bid for Quayside. This was the same talking point he'd raised at the federal ethics committee hearing months earlier. "That possibility was out there."

The CEO acknowledged that Sidewalk had needed to cut some of its plans to meet Waterfront's terms, but hoped to hold on to "the substantial majority" of innovations in its plan to prove it could make cities more efficient, affordable and sustainable. Wooden towers, underground freight tunnels and efficient neighbourhood-wide energy systems were all still on the table. As the interview wound to a close, I asked, "What does a deal-breaker for you look like?"

"I'm not even going to speculate on that," Doctoroff said, before immediately speculating. For the first time in two years of conversations, he shed his usual confidence. "At the end of the day, if we don't believe we can achieve the innovation standard that I articulated, or the financial objectives that I did not articulate—a reasonable return, given the risk—then I would say that's just a reality. We have to achieve what we're trying to achieve."

CHAPTER 22

Make and Break Harbour

BY THE END OF THE 2010S, Larry Page was a ghost. Ever-growing Alphabet had ballooned into a crisis factory: it faced antitrust crises, regulation crises, data crises, and a widely publicized sexual-harassment crisis. Yet Alphabet's CEO was nowhere to be found, neither in the public eye nor, increasingly, inside his own company. Faced with growing scrutiny, Page and his co-founder Sergey Brin had lost the verve they'd once brought to Google. When reporters pressed Alphabet about Page's absence, the company would say he had been focusing on bigger-picture ideas and pet projects, such as Sidewalk Labs. Yet even within Sidewalk, few people saw him involved in day-to-day decisions. Page would check in with Dan Doctoroff every week or so, and he paid another visit to Toronto in July of 2018, but otherwise seemed scarce. He did, however, appear to be spending a lot of time on his private Caribbean island.

In December 2019, Page and Brin shocked the tech sector when they announced they were stepping down as Alphabet's CEO and president, respectively. Sundar Pichai would rise from Google's CEO to Alphabet's. The company was twenty-one years old—old enough, the co-founders joked, "to leave the roost." Their decision marked a shift in the dynamics of the digital world: though the company's share structure ensured the co-founders still had controlling votes on the board, two of the biggest power brokers of the tech sector's pre-social-media era had ceded much of their influence.

The culture change was swift. Pichai made it clear, just weeks into the job, that he wanted to bring more financial rigour to Google's side businesses like Sidewalk Labs. "While we take a long-term view, we also want to marry that with the discipline of making sure they are doing well," Pichai said. He later expanded on this: "How do you assess the value of the entity you're creating and how do your other partners and other stakeholders? It's a direction we had already gotten underway. But you will see me focused on that more and emphasize that more."

This was a fundamentally different approach than that of Page, who preferred to approach value with a much different question: "Why can't this be bigger?"

The strategic shift was fantastic news for Ruth Porat, Alphabet's long-time chief financial officer, who had never been a fan of the spending habits of Page's experimental playthings. Moonshots usually turned out to be money pits: Alphabet's revenue for what it called these "other bets" in its financial statements was $659 million that year, 0.4 percent of its total income, yet they lost $4.8 billion. After restructuring to make more room for moonshots, shedding the name of one of the most popular new verbs of the twenty-first century, Alphabet was still mostly Google. Close observers began to wonder whether Alphabet would survive as a holding company—whether Google might just become Google again, leaving some of its costly experiments behind.

In the lead-up to the Quayside-saving Halloween vote, months before Page and Brin would announce their retirements, Waterfront Toronto's negotiating team began to notice that Sidewalk Labs staff were mentioning their parent company much more often. "They would go so far as to say, 'Good idea, but we have to check with Alphabet,'" says Kristina Verner, who was brokering Waterfront's terms on privacy, data governance and intellectual property.

This was happening regularly. As Sidewalk raced to make Waterfront's new terms work ahead of the Halloween deadline, "they clearly had to get a blessing from Alphabet," says George Zegarac. "And they got that blessing the day before the deadline. That became a trigger point: it's not a two-party negotiation, it's a three-party negotiation."

When Waterfront forced Sidewalk to meet its "threshold issues" on Halloween, that meant casting aside the urban data concept and guaranteeing that Sidewalk would follow all current and future privacy laws. This let Verner and her digital issues negotiators focus on developing a plan for intellectual property that would give the public sector a share of any patents that arose—a potentially world-leading prosperity guarantee from what was designed to be a world-leading public-private partnership. The new October terms were the most beneficial for the public sector yet: Sidewalk would share a to-be-determined portion of revenues from technologies it would pilot in Toronto, versus the 10 percent of profits it had proposed in June. And they included an expanded "patent pledge" that would let Canadians access Sidewalk's global cache of patents, giving them the chance to build upon Sidewalk innovations without fear of being called out for infringement.

In the late fall and early winter, Verner's team began to negotiate what the actual contracts around these promises would look like. Waterfront even retained the services of Natalie Raffoul and Jim Hinton, two intellectual property lawyers who had attacked Quayside's IP provisions as they were being negotiated earlier in 2018 and 2019. With their help, Waterfront was trying to expand the terms of Sidewalk's proposed patent pledge to any company Sidewalk might bring in. One of the world's fiercest patent filers, after all, was Google. Sidewalk "thought they were going to give up nothing and get something," Hinton says. "They can't hold back all the good stuff for themselves and not share."

Still, Waterfront struggled to get assurances from Sidewalk. The Canadian agency was trying to maintain the advantage it had gained on Halloween, only to learn that Alphabet wasn't happy about being pulled into Sidewalk's mess of compromises. As fall turned to winter,

Verner's team kept hearing the same phrase: "We'll have to check in with Mountain View."

On the rooftop patio of an east-end Toronto hotel that past summer, Dan Doctoroff had tried to woo fifteen luminaries from Toronto's venture capital community, including former *Dragon's Den* TV-celebrity investor Bruce Croxon and Kim Furlong, the head of the Canadian Venture Capital and Private Equity Association. Sidewalk planned to seed C$10 million into a venture fund for early-stage companies whose work focused on cities, and it was scouting for potential co-investors. By November, it had found its partner: Plaza Ventures, the start-up wing of a Toronto real estate developer that also happened to own the land in the elbow of Quayside on which Sidewalk had set up its big blue showroom. Quayside Venture Partners would start with between C$20 million and C$30 million in funding—not a massive amount of money, but enough to help a dozen or two early-stage companies refine their products. It was an invitation to be part of Quayside, a real chance for local start-ups and investors to be involved in a project that many critics had positioned as a net negative for Toronto tech.

Sidewalk also released an appendix to its draft master plan that shared greater details about its technologies while acknowledging both the criticisms that Waterfront's digital panel had launched in September and the new terms of its relationship with Waterfront. It could seem, more and more, like Sidewalk wanted to be a team player, rather than just talking about it. The company pledged to prioritize the work of Canadian companies at Quayside "wherever possible."

But the appendix still managed to avoid answering important questions, especially regarding how much the public would pony up for the public-private partnership. As an example, the company laid out how the business case might work for standardized mounted sockets that it would string through Quayside like electrical outlets, but for plug-in sensors or other technology—perhaps products designed by local companies that Sidewalk would invest in. The sockets were called Koala mounts, and they would let whichever company plugged into them connect to the local network, simplifying sensor installation and making data collection easy. In the appendix, Sidewalk described different financial scenarios depending on whether a network of Koalas was owned and operated by the public sector or the private sector. But despite the appendix's 482 pages, it offered few cost estimates for such technologies. Sidewalk said that if the public owned the mounted sensor adapters, taxpayers would need to cover yet-to-be-defined upfront costs, maintenance contracts and annual software payments. Those were crucial details. It was still impossible to tell, this far into the project, if partnering with Sidewalk would put a burden on the city.

And despite strong promises to protect Torontonians' privacy, the original version of the appendix contained a line proposing that Sidewalk would monitor residents' electricity use. When *Financial Post* reporter James McLeod asked the company about this, its staff scrambled to contain what appeared to be a sloppy error. Four different officials tried to explain what happened. "I don't know," Sidewalk's director of sustainability told the reporter. "Maybe I would take it out if I were to do it again." A few hours later, the company removed the phrase and re-uploaded the document to its website. Sidewalk was finally playing on Toronto's terms, within the very real, messy constraints of an actual city, and it was stumbling.

"When I walked in, a couple people did ask me whether Dan and I were going to arm-wrestle today."

It might have been the closest thing to a joke about Quayside that Stephen Diamond had ever made. He was sitting across from Dan Doctoroff onstage at a Toronto Region Board of Trade lunch in the country's tallest office tower. With the upper hand in Quayside negotiations since October, Diamond was revealing a tinge of light-heartedness on top of his usual sense of caution. "Where we are today is in a great position," he said.

It was January 2020. Board of Trade CEO Jan De Silva sat between them. She'd worked closely with both men over the years, and saw this lunch meeting as a chance to show the public that Waterfront and Sidewalk's relationship was working well under their new terms. She turned to Doctoroff. "Originally, Sidewalk had a big vision for what you wanted to achieve in Toronto. What was behind the decision to get on board with the realignment, and working initially with the 12 acres?"

"The simple answer is, we listened," Doctoroff said. This response would have been plainly revisionist if he hadn't followed up with a caveat: "Sometimes it takes us longer than maybe it should to hear what people are saying." He praised Diamond for working to salvage their partnership, rather than cancel it altogether. "We feel like we've been, despite the controversy, just so well received here."

But Doctoroff was also unusually careful. He managed to couch his caution in confidence, but it was caution nonetheless. "This has to be viewed as the most innovative district in the world," he said. "We've never been in this, primarily, for this project to make a killing. But we are a part of a public company. We do have real financial responsibility that we are held accountable to."

For the first time, Alphabet's investment in Sidewalk seemed to matter to Doctoroff. For years, he'd usually brought Alphabet into the conversation only to point out its "patient capital." Now that money was looking impatient.

But Waterfront wasn't going to give Sidewalk any special deals to please Alphabet. Diamond threw back to his earlier joke. "I'm sure there will still be some arm-wrestling on some of the financial issues."

In early February, Alphabet CFO Ruth Porat came to town. In a packed room at Google's downtown Toronto office, she and her colleagues announced that the company would triple its Canadian workforce to five thousand people over the next three years in Toronto, Montreal and the tech hub of Kitchener-Waterloo. There was something off about the announcement, though. The company said it would house all the new Toronto employees in a new office just around the corner—not at Quayside, about a kilometre to the southeast. Just weeks earlier, Dan Doctoroff had told the Board of Trade audience that Sidewalk was working with Google on a plan to build a Canadian headquarters for the company at Quayside, since the original site it wanted to build at, Villiers Island, wasn't guaranteed to the company.

I asked Porat about this.

"We're still in the midst of discussions with Quayside," she told me in a makeshift interview room down the hallway from the press conference. "That's certainly an opportunity and an option to make that the headquarters, but that's still in the process."

She'd said "we." Was she involved in the negotiations?

"Sidewalk is a separate 'other-bet' company. That's being run by Dan Doctoroff and his team."

But she was the CFO, I said. She couldn't ignore side businesses like Sidewalk.

"But they run their negotiations."

Doctoroff was worried that 12 acres might not make Alphabet's investment worthwhile. Did she share his concern?

"What we're looking at with Sidewalk is the opportunity to re-imagine what the city can be. It's better for the community that it really leans into how technology can create a better experience in the city, and they're still working through those specifics."

If those specifics cost a lot—say, half a billion dollars—would the company be willing to eat that expense?

"I've already addressed it."

I pointed out that she hadn't actually answered the question.

Porat, who commanded one of the most influential corporate budgets in the world, responded with silence.

Yet that same ice-cold day, she was shuttled over to 307 Lake Shore, where she and a finance director for Alphabet's side businesses met in a boardroom with Waterfront's Steve Diamond and George Zegarac. She was keeping an eye on Quayside talks, after all.

It was Zegarac's first time meeting Porat and her team, and it was cordial. But it was clear that each side of the table needed something from the other. Zegarac and Diamond needed confirmation from Alphabet that it would provide financial guarantees to Waterfront if their project went south or if Sidewalk disappeared. Sidewalk wasn't bringing in much, if any, revenue yet. At minimum, someone needed to pay for the land. "What would happen if there's a big hole in the ground in two years?" Diamond explains, looking back at the meeting. "You need guarantees."

Porat acknowledged Waterfront's concerns, but the Alphabet envoys had problems of their own to discuss: with Larry Page's influence diminished, Sidewalk was under scrutiny. Since Sidewalk couldn't take advantage of the economies of scale that would come from working across the Port Lands, Alphabet needed Quayside alone to make money. It didn't need to make a *lot* of money, but it couldn't lose money.

This was the arm-wrestling match Diamond had warned of at the Board of Trade lunch. Sidewalk, Zegarac says, "needed a rate of return they could justify—and had to justify it to Alphabet."

Sidewalk was willing to take a lower return on its investment in Quayside for a chance to use the neighbourhood as a test bed for new innovations and approaches to infrastructure. But Waterfront's terms and restrictions from City Hall constrained even the chance for a low return. Without access to some of Villiers Island, the 12 acres of Quayside alone didn't give Sidewalk much space to scale its new technologies. And without changes to regulations, like those that still forbade tall wooden buildings, it was hard for Sidewalk to justify building at Quayside. Shorter wood buildings already existed, which would make them harder to sell around the world as a cutting-edge innovation.

It was a strange turnaround for Sidewalk Labs. Two people at Waterfront, including Kristina Verner, recalled a Sidewalk senior staffer telling them in a meeting earlier in the project that the company was willing to lose a significant amount of money in Toronto, maybe a billion dollars or more, to test out its ideas. Now the company was scrounging for pennies. And complicating things further was the fact that the money would need to come through Alphabet.

Zegarac's earlier hunch was proving correct: Quayside was a three-way negotiation with both Sidewalk and Alphabet. This mirrored the unease that had been forming among some Sidewalk leaders since they'd arrived in Toronto: that they had bid for Quayside thinking Waterfront was a one-stop shop for dealing with governments, only to find out it was an agency with few powers, beholden to three governments with priorities of their own. As close as Waterfront and Sidewalk were to striking a deal, the battle for the future of cities had become a proxy war that no one was winning.

Zegarac and Diamond left the meeting with no new answers. As they walked out of 307 Lake Shore, there was a growing sense among the people still inside the building that all this compromise might not be worth it.

If Quayside really was a battle, Jim Balsillie had stopped fighting. He'd come something close to winning. By early 2020, he was having dinners and meetings with both Zegarac and Diamond, trying to impress upon them the importance of good data governance and the economic imperative of intellectual property. Waterfront had even hired two of his favourite IP experts, Natalie Raffoul and Jim Hinton, as advisers.

The agency's panel of digital experts was also pleased with how Sidewalk was coming around to their concerns. They soon released a report on the digital innovation appendix to Sidewalk's draft master plan that had been published in November. Most of the panellists found it much better than the first plan, but they still wondered whether some ideas "create sufficient public benefits."

Waterfront staff, meanwhile, had been slowly evaluating the ideas Sidewalk had published in the draft master plan. It contained

160 proposals that Waterfront was willing to entertain after the October vote. In February, they announced they'd endorsed all but 16. They rejected sourcing heat for the neighbourhood from sewage waste at a facility four kilometres away—it was too far—and declined to accept Sidewalk's ultra-small "efficient" apartment units, declaring them so tiny as to be effectively uninhabitable.

But people who attended the latest of Waterfront's semi-regular public consultations might not have known that Waterfront had drawn these conclusions. The meeting was jammed with union members, dozens of them, wearing hats and shirts proudly declaring their locals. Many knew little about the Quayside project; they wanted to talk about jobs. It turned out that Sidewalk fixer John Brodhead and a few others had asked their labour contacts to send out potential supporters in an effort to counteract #BlockSidewalk's presence at consultations. The plan worked *too* well. The union members so dominated the meeting that #BlockSidewalk organizer Thorben Wieditz found Waterfront's long-time event facilitator Nicole Swerhun, normally a bastion of calm in a rowdy room, fuming in the hallway, struggling to manage the crowd.

In general, though, Waterfront was still relishing its newfound confidence following the Halloween vote. It even tried to stop the Canadian Civil Liberties Association's legal efforts to shut down the Sidewalk partnership. The agency filed documents in court that called the CCLA's efforts to cancel the Quayside project on grounds of non-consensual surveillance "speculative" and "premature." The project hadn't broken ground yet, Waterfront pointed out, and when and if it did, it would be subject to any and all privacy laws and regulations.

But the CCLA was confident, too. It had lined up affidavits from experts who shared its views, including *The Smart Enough City* author Ben Green, the technology and society researcher Zeynep Tufekci and data-trust expert Sean McDonald. Julie Di Lorenzo had even joined the association's board. Even if the CCLA couldn't get the project cancelled, it still wanted to strong-arm Waterfront into squealing on Sidewalk. Michael Bryant and Brenda McPhail hoped to get their hands on internal Sidewalk documents, and the road to Sidewalk led straight through the public agency. With its Charter challenge, the CCLA thought it could force Waterfront to disclose Quayside documents in a discovery process, and could cross-examine Waterfront executives to get as much as possible about Sidewalk's plans on the public record. The Divisional Court of the Ontario Superior Court of Justice gave CCLA a hearing date of March 25.

The hearing never happened.

On February 27, 2020, I wandered through Hudson Yards, admiring all the stores I couldn't afford to shop at as well as the Vessel—the stairway-laden, sixteen-storey piece of public art designed by Heatherwick Studio that looked like a rotisserie of shawarma meat. The public art piece had another defining feature: its railings were only slightly above waist height. This, unfortunately, attracted a particular kind of visitor. Just weeks before I stared up at the structure, a person had leapt from the Vessel to their death. At least three others would go on to do the same, prompting the companies that oversaw Hudson Yards to close the massive art installation in 2021. In prioritizing style over substance, Heatherwick's design for the Vessel—the centrepiece of Hudson Yards, Dan Doctoroff's greatest legacy—had lethal consequences.

Across the way, ten storeys above the Vessel's summit, staff from a company run by Doctoroff that had previously commissioned renderings from Heatherwick were hard at work trying to design something much different: a neighbourhood that would actively make life better for people. For years, Sidewalk Labs, too, had been accused of putting style before substance, talking constantly of innovation but largely on terms that wouldn't work, seeming to change tack only after Waterfront Toronto threatened to end their partnership. Now its staff were trying to make the math work for a much smaller neighbourhood with fewer innovative building ideas than they'd hoped. It wasn't an easy task.

Then it got even harder. Three days later, a Manhattan health-care worker who had recently travelled abroad was diagnosed with COVID-19. This novel coronavirus, a mysterious flu-like disease that had seemed like a distant threat over the past couple of months, had officially landed in New York City. Over the next three weeks, most of the Western world went into lockdown. The gleaming towers that Doctoroff had willed into existence emptied out. Sidewalkers were sent home. By the end of March, New York had become one of the epicentres of a global pandemic.

Across the country in Mountain View, Alphabet was staring at a real estate crisis. The possibility of a future where many people worked from home seemed increasingly realistic, especially since no one could tell how long the pandemic would last. By late April, Alphabet had bailed on negotiations for more than 2 million square feet of commercial real estate across the Bay Area. In one deal, in a development at a former shipyard in San Francisco, the company walked away from leasing square footage on a scale equivalent to the city's Salesforce Tower.

One of the world's digital titans was picking its battles. Sparring for a bigger stake of the physical world wouldn't be one of them.

The situation in Canada was no better. Home sales collapsed in April, and the country's national housing agency soon warned that prices could plummet by 18 percent. Sidewalk was in a double bind. It was facing a parent that was actively spurning commercial real estate investments yet expected fiscal rigour for a project that depended on extracting value from both commercial *and* residential real estate, neither of which was doing well.

In mid-April, Dan Doctoroff wrote an op-ed for the *New York Times* headlined "I Helped New York Rebound from 9/11. Here's How to Recover After the Pandemic." The city, he said, could revitalize itself by investing in medical research and by focusing on "inclusive, sustainable and resilient growth," and it could start right now. Doctoroff snuck in a little advertisement for his recent work—"It's now possible to make tall buildings out of wood and produce core building components in factories"—and seemed to acknowledge some of the lessons he'd learned from the Quayside partnership. "Communities want more of a voice in decision-making," he wrote. It was a beautiful rallying cry for New Yorkers. But he didn't mention Toronto once.

Two months had passed since Stephen Diamond and George Zegarac had met with Ruth Porat. The pandemic had slowed business dramatically in Toronto. Waterfront's staff had managed to finish their evaluation of Sidewalk's tech and building innovations a few weeks before the city shut down, but the board still needed to have a make-or-break vote on partnering with Sidewalk Labs. And to do that, they needed monetary guarantees from Alphabet.

So Waterfront delayed the vote from May to June, and its chair and CEO sent Alphabet a letter asking for details in writing to confirm the parent company's financial support for the project. Then they waited.

From kitchen tables and spare bedrooms across New York and Toronto, Sidewalk staff were still trying to make Quayside's new math work. The company had been looking at razor-thin returns before the pandemic, but the sheer uncertainty of COVID-19's early months tipped the balance farther from Sidewalk's favour. Its negotiators sought new compromises from Waterfront on Quayside's price. The RFP was only for a plan for the community, but Sidewalk had made it clear in the draft master plan that it wanted to buy the site. Waterfront was comfortable with the idea as long as Sidewalk was just a financial partner and a traditional developer was brought in. The agency had appraised the site at about C$590 million, which already included a discount related to land for affordable housing. But the exact price that Sidewalk would pay—or, rather, that Alphabet and a group of developers might pay—became a source of tension. Doctoroff says discussions were "never about getting a discounted price," but Zegarac remembers the talks differently: "I do think that they wanted us to eat a bit of the value of land."

Sidewalk staff talked about stretching land payments out differently, reappraising the land or giving Waterfront a greater share of future income—anything to make Quayside's upfront costs more feasible in an unstable real estate market. Waterfront didn't want to budge. The agency needed the proceeds from Quayside's sale to pay for infrastructure, for public spaces and to keep its lights on.

Doctoroff says he understood Waterfront's position, though he calls the agency's stance "largely a matter of bureaucratic self-preservation." Waterfront was not willing, he said, to be "flexible at a time of incredible uncertainty. And as we looked at it, just given the uncertainty, we said very simply that in order to justify doing this, we're going to have to cut back on the innovations substantially in order to achieve adequate returns."

For weeks, Josh Sirefman had been marshalling Sidewalk staff to run numbers to see if that was even possible. Soon, he called a meeting of Sidewalk's top leadership. The company's chief decision-makers logged on to a video conference, and Sirefman laid out the risks. His staff had to confront an existential question: Was Quayside still worth it? Could they get a better deal on land somewhere else, without all these headaches? The grand plans they'd laid out when they bid on Quayside in 2017 had been gradually chipped away, with their requests for more land and regulatory changes being greeted by, at best, indefinite delays. How many of their plans for technology, sustainability and quality of life would they need to give up just for the privilege of building on a small slice of the Lake Ontario shoreline?

One by one, Sidewalk's leaders offered their thoughts. Each time a new face dominated the screen, they offered a version of the same difficult answer: this was no longer the project they'd signed up for. Quayside wasn't worth it anymore.

It was time to leave Toronto. Even Doctoroff agreed.

At least one person in the meeting noticed that Sirefman seemed to come the closest to wavering. He had been a quiet but constant force within Sidewalk: the calm hand at the wheel, a diligent counterweight to Doctoroff's aggression, and the person some staff went to when they feared Doctoroff's reaction to bad news—or even just ideas that didn't look like his. Sirefman had thrown his whole life into

Quayside, taking blows from Waterfront, critics and sometimes even his own boss as he tried to steer the project into sanity. But he was also the person who'd called the meeting. He'd done the math, over and over again. It was time to go.

The decision was unanimous.

Cancelling the Quayside project was a momentous decision, and Doctoroff and Sirefman needed to explain themselves to Alphabet. But they had a surprisingly soft landing. The company had spent so much money and achieved so little after two and a half years in Toronto that CFO Ruth Porat had already been agitating for some kind of change. The 12 acres of Quayside, meanwhile, was far smaller than the dream Larry Page first envisioned for a Sidewalk community. Getting their blessing, in early May, took surprisingly little effort. Then the Sidewalk executives had one last leak to prevent.

Waterfront Toronto managers were in the middle of an all-hands video conference about performance reviews on May 6, a Wednesday, when George Zegarac received a message from Josh Sirefman. He said he needed to talk—urgently. The Waterfront CEO told his colleagues only that he needed to step away from the call. He was gone for a long time. Then Meg Davis left the screen, too. Still logged into the Microsoft Teams meeting, Kristina Verner began to wonder what was going on.

Sirefman, meanwhile, was sharing the news with Zegarac. The economics of Quayside no longer worked, he said. Sirefman spoke matter-of-factly, the way people do when their minds are firmly made up. Zegarac would soon receive a letter, he explained, outlining the details.

The Sidewalk president called Stephen Diamond, too. Diamond could hear some disappointment in Sirefman's voice. Sirefman was apologetic: both organizations had spent months trying to make the deal work, and it was a jarring end. Doctoroff soon followed up with Diamond, but by then the Waterfront chair was distracted. He had to figure out how to tell the public.

Diamond, Zegarac, Davis and a small crew of Waterfront staffers began working on an announcement strategy. They made a flurry of calls to governments, too—the ones who'd put their faith in the agency to try something innovative with Quayside, but who'd rarely taken leadership on the governance issues the project presented.

The news was leaking out of Ontario government circles before midnight. Toronto's architecture community was the next to start chatting about it. I spent hours trying to confirm the news with a primary source, but no one who could verify facts first-hand was picking up the phone. By 9:30 a.m. Thursday, I'd reached enough people that government officials had begun e-mailing each other that I had the story, and they were racing to get the news out before the *Globe and Mail* did. Then, just before 10 a.m., a source texted me a skull emoji. Quayside really was cancelled. Sidewalk made the announcement less than twenty minutes later.

That same day, *Fortune* magazine reported that Sidewalk Infrastructure Partners had raised $400 million for its tech-infused infrastructure fund. It was great news and great timing for the Sidewalk spinoff firm—especially since the "exclusive" story was based on an eight-month-old U.S. Securities and Exchange Commission filing. Sidewalk might have lost Quayside that day, but the company still found a way to look like it had won something, just as Doctoroff had when his bid to host to the Olympics faltered. "Failure is only failure," he once said of that loss, "if you define it that way."

Sidewalk's failure to win a piece of Toronto was inseparable from the end of a more significant era at Alphabet. One person who worked with Larry Page around the time of Sidewalk's creation told me that what happened in Toronto was the final paragraph of an "obituary" for Page and Sergey Brin: they'd founded Google, turned it into the world's most efficient library and biggest ad company, then gradually lost their sense of innovation over the course of the 2010s. They were brilliant kids with a brilliant idea and brilliant timing, launching Google just as the internet was rising, but struggled to reinvent themselves as the internet's potential spilled into the physical world.

It was Page and Brin's timing, the person told me, that was perhaps the most brilliant. Maybe they were lucky to be born in the mid-seventies, and to come up with the idea for website-ranking search technology before anyone else. Maybe they were lucky to meet Eric Schmidt after they had their good idea, and lucky to have good acquisitions lawyers. Though Page no longer had to worry about dying in obscurity like his hero Nikola Tesla, and though Google and Alphabet continued to buy up the world's greatest ideas and companies, he and Brin were no longer coming up with new ideas to change the world.

Canada changed, at least a little, because of Quayside. Six months after Sidewalk bailed on the country, Justin Trudeau's Liberals finally introduced a new privacy law governing the private sector that actually had some teeth. Canada's privacy commissioner would get more power to force companies to comply with citizen-focused rules that were closely aligned with Europe's General Data Protection Regulation. The law threatened multi-million-dollar fines for violators. It even seemed to lay the groundwork for regulating data trusts. The bill wasn't perfect, but it could have been shaped into something even better through Parliamentary debate. Except Parliament didn't do anything with it. By the time Trudeau called an election the following August, the proposed law had been stalled for months. After he won another minority government, his new innovation minister, François-Philippe Champagne—who had earlier taken little ownership of the Sidewalk file as infrastructure minister—didn't reintroduce the law until June 2022. Trudeau, meanwhile, had stopped showing up to Big Tech announcements well before the pandemic. Eventually, so did his deputies. If you talked to enough Liberals, you'd eventually learn why. On the heels of Cambridge Analytica, Sidewalk Labs was one controversy too many. The progressive optics just weren't there anymore. And for

John Brodhead, life came full circle: after years battling unfounded allegations that he might have helped fix Sidewalk as the Quayside winner for Justin Trudeau, he took a job in the Prime Minister's Office, brokering its relationship with Canada's provinces.

Doug Ford's Ontario government turned to the private sector to figure out how tech would fit into the province's future, asking Jim Balsillie and a bunch of his close advisers to help the province determine whether its universities and accelerators were doing enough to bring intellectual property to market. The province asked the head of another private-sector group to help with its data strategy. Miovision's Kurtis McBride, who'd spent years with Waterfront trying to work some sense into Sidewalk's digital governance proposals, helped launch a non-profit called Civic Digital Network, which designed digital infrastructure plans for governments. In 2021, Ford tapped its CEO, Andy Best, to guide a new provincial authority that would make sense of the messy world of public data. And in Toronto's City Hall, Waterfront board member Joe Cressy kept pushing for Toronto to develop a smart-city framework to deal with future projects.

Some Torontonians came to feel that the Quayside saga was a crucial lesson in innovation: sometimes you need to fail to learn how to do things right. "You can't innovate without pushing the boundaries," says Waterfront director Mazyar Mortazavi. Others took a much different perspective. "People kind of framed this as: government isn't ready to deal with technology," says Siri Agrell, who watched the Sidewalk file for Mayor John Tory for several years and went on to run a start-up hub. "And I completely think it's the other way around—that the tech sector has no idea how to combat the actual challenges of cities."

That's what Bianca Wylie had been saying all along. When Sidewalk pulled out of Toronto, she felt relief, but not joy. Like the Berlin activists who'd helped kick Google out of Kreuzberg, she came to see Quayside as one battle in a much broader conflict. Tech companies have only amassed more power and influence during the pandemic. "The thing that carried right to the end, that carries right to this day," Wylie says, "is that this can happen again."

The death of the Quayside project was fundamentally different from that of Google's Berlin start-up campus. The campus was smaller and less ambitious, and was surrounded by immediate neighbours from all walks of life who made it very clear that a gentrifying force like Google wasn't welcome there. Sidewalk Labs left Toronto because of math. The company relied on an equation that tried to turn 12 acres into something much bigger. But over two and a half years, the variables changed. Sometimes they changed slowly, and sometimes they changed in the time it took Stephen Diamond to write an angry letter. In Diamond, Dan Doctoroff found a mirror: another aging male executive across a boardroom table who, like him, refused to say no. But reflections move in opposite directions. For decades, Doctoroff had intertwined the public interest and profit. Diamond liked profit, sure—he ran a development company that shared his family name—but he wanted Quayside to go forward only if it actually worked in Toronto's interests.

Though the people who loudly questioned Quayside didn't cancel the project, they changed the math that did. They chipped away at Sidewalk's glossy veneer, separating niceties from intentions and truth from marketing. Toronto is big enough that most people know of it, but small enough that they don't know much about it—the kind of place sometimes overlooked by Americans when it isn't winning an NBA championship or offering the world another pop star. Sidewalk's critics made the world care about what Larry Page's plaything was trying to do to cities, starting with their own.

Many people who tried to make the Quayside project happen, and who watched the calamity unfold, wondered whether a different company with a gentler strategy would have succeeded in Toronto. Doctoroff does not believe this is the case. "The only thing that ended up making this project not happen was COVID," he says. "We were prepared to go ahead. I think that's actually incredible evidence of our flexibility." He blames shifting dynamics in Canadian governments, too: the provincial Liberals were ousted by Ford's Progressive Conservatives soon after Sidewalk arrived, and Justin Trudeau's majority was downgraded to a hamstrung minority in 2019. "We were trying to solve a fifty-sided Rubik's Cube with three orders of government, coordinated by a government agency that was responsible to all of them." He's not sure if he would have done anything differently. "That was a very messy process," he says. "I've debated a lot about whether we should have been more explicit upfront about what the plans were, rather than co-creating them in public. I've debated this. I don't know the right answer."

In the two years since Sidewalk left Toronto, tech companies have only crept deeper into the places people live. Elon Musk's SpaceX tried to buy a village in Texas, bit by bit, near its rocket launch facility. He pursued what the *Wall Street Journal* called a "war on regulators," pushing back against U.S. regulatory agencies or ignoring them altogether through both SpaceX and Tesla, as he cemented his spot as the world's richest man. The man he seized that title from, Jeff Bezos, watched Amazon's second headquarters rise from the ground on the outskirts of Washington, D.C., where rents soared so high that even Chasten Buttigieg, whose husband, Pete Buttigieg, was the U.S. secretary of transportation, complained the city was barely affordable. When "Mayor" Pete's boss, Joe Biden, secured the U.S. presidency in late 2020, the smart-cities industry began salivating at all the investment he might make in their sector. Two months after that, Nevada announced new legislation that would effectively allow tech companies to create their own governments, just as Larry Page had always wanted. One person exploring that state was Diapers.com founder Marc Lore, who in 2021 said he wanted to purchase 200,000 acres of land to build a city for five million people, launch a trust-like foundation to own the land, and funnel the money generated from its appreciation to pay for social services. Lore hired Bjarke Ingels as his chief architectural designer—six years after Google had hired Ingels's firm, alongside Vessel designer Heatherwick Studio, to build a dome-like structure at the Googleplex that was far less ambitious than the weather-beating hemispheres its Javelin team had once imagined. Ingels was also designing a thirty-five-storey tower just north of Kreuzberg to house Amazon offices, giving Berlin's Google opponents a fresh project at which to direct their scorn.

At least 150 new city-building projects were under way worldwide when Lore announced his plan, some designed by governments, others by wealthy private-sector businesspeople. "It's so seductive to say 'I'll start over' rather than just pay my taxes," said Sarah Moser, a McGill University professor studying these projects, when Lore announced his plans. "Then they present themselves as this beacon of hope for humanity." She and her research colleagues came to describe this philosophy as "unicorn planning."

Waterfront Toronto, at least, learned that buying into a billionaire's glossy dream might not be the best way to build a city. After taking some time to recover from the fallout of the Quayside project, the agency began putting together a new procurement contest for the site. The agency's staff and board spent nearly ten months trying to figure out what to ask for, with the consequences of the pandemic weighing on their minds. The new Quayside would focus on affordable housing, sustainability and even senior citizens, whose care homes were ravaged the world over in the early days of the pandemic. This time, Waterfront sought out developers, not an "innovation and funding partner." As Waterfront got the process rolling, Sidewalk Labs' old office at 307 Lake Shore got a new tenant that would make some urbanists cringe. It became a truck-rental depot.

It was a grey day in Toronto, the first dusting of November snow still sitting on the ground from the night before. Storefronts were on the cusp of closing again: the second wave of COVID-19 was well on its way. By that point in late 2020, the pandemic *had* begun reshaping the real estate market, but not in the way Sidewalk Labs had anticipated.

Home sales were on fire. Prices around Toronto shot up as people clamoured for more living space. So many were finding themselves priced out of cities, in fact, that markets across the country saw similar patterns. No longer tied to the office, tens of thousands of Torontonians moved to rural communities and farther-flung parts of Canada, from Yukon to Nova Scotia. Toronto's decades-long failure to plan for its growth was helping to make housing less affordable across the country, and Sidewalk wasn't around to pitch itself as a solution.

That November day, urbanists and technologists from across Toronto were staring at their computer monitors, watching Jim Balsillie, Stephen Diamond and former Barcelona smart-city leader Francesca Bria do a post-mortem of Waterfront Toronto's ill-fated Sidewalk partnership. There was no sparring. They mostly spoke in lengthy diatribes.

Reading from his notes, Balsillie seemed uncharacteristically lost, his eyes rarely focusing on the camera. But his words were as pointed and cynical as ever. He assailed Canada for falling victim to neoliberalism, becoming a country "where everything gets resolved in the marketplace." Skyscrapers need building codes and roads have speed limits; he insisted that governments need to similarly intervene when technology companies make intrusions into city life. "Technology is not governance," he said. "It needs to be governed."

Diamond spoke more calmly, sitting in his shirt sleeves, hands clasped. He went on a tear, arguing that cities were always afraid of change; even the Eiffel Tower, he pointed out, had frustrated Parisians more than a century ago. "Technological innovation is here to stay, and it's only going to increase its influence on our society," the Waterfront chair said. He agreed with Balsillie that governments needed to step up, and that they failed to do so with Sidewalk. They needed to act fast to prevent the future of corporate data collection from going unchecked. "You can't come at it backwards again," he said.

He offered up a quote by the Italian author Italo Calvino, from his 1972 novel *Invisible Cities*. "Cities, like dreams, are made of desires and fears," Diamond recited. He gave the quote his own addendum:

"We cannot allow our fears to take over our dreams." But in doing so, Diamond cut off the rest of the original quote. Calvino had more to say about cities: "The thread of their discourse is secret, their rules are absurd, their perspectives deceitful, and everything conceals something else."

EPILOGUE

Break It to Them Gently

NEARLY TWO YEARS into the COVID-19 pandemic, Google's fleet of legal staff were handed a decision that cracked the armour of their company's aggressive intellectual property strategy: the U.S. International Trade Commission ruled that the company had infringed a series of audio technology patents owned by the wireless smart-speaker company Sonos.

The battle forced Google to find a non-infringing workaround or stop importing some of its smart speakers, Pixel phones and other devices from overseas. The January 2022 decision was hailed as a sign that smaller companies and entities could go toe to toe with Google and win. Sonos CEO Patrick Spence had spent the past decade steeling the company for what he felt would be an inevitable battle with the giants of tech, and his preparation had paid off. "We built a business that allows us to be able to fight back," he says. "We probably would not have done it if I hadn't been there with Jim."

Balsillie. He was Spence's old boss at Research in Motion. Corporate lore goes that Spence sold the first BlackBerry; he was also one of the top contenders to replace Balsillie when his boss stepped down as co-CEO during the company's collapse. Instead, Spence joined the executive ranks at Sonos in 2012. He had watched RIM's nine-figure legal battle over patent infringement in the 2000s and vowed he would stand up for inventors in the future. Balsillie was one of the first people to call Spence after the Google patent news broke, telling his old protege he was proud of him for sticking through the fight.

While Balsillie found validation in his life's work as the pandemic dragged on, his nemesis did not. Dan Doctoroff's company was a shell of its former self. The company that just a few years earlier had wanted to remake cities "from the internet up" had been sapped of its grandstanding ambition. In the months after the only significant city-building project Sidewalk Labs ever attempted was cancelled, Doctoroff was trying to convince the world that everything was fine.

In May 2021, a conference called City Summit invited me to interview Doctoroff for a webcast. I asked if there would be parameters around the conversation. The conference organizers went silent for several days, then told me the day before we were scheduled to record that they'd picked someone else to do the interview.

When I finally watched the webcast, Doctoroff described Sidewalk as if he'd never set foot in Toronto. The CEO gave the same progressive windup he usually did—innovating in cities is hard, climate-positive building is good, car ownership shouldn't be necessary, affordable housing is important—but the pitch was different. "We've created about seven or eight different companies over the course of the last several years," he said. "Things as varied as a new approach to urban health care, or a new approach to urban data, mass timber buildings, literally just announcing a new approach to a sensor that can be deployed on the streets, manage parking assets, and a number of others."

Just over a year after Sidewalk Labs left Toronto, its CEO was presenting the company as a kind of incubator—something like the next Bell Labs, where a bunch of clever engineers were trying to solve a bunch of thorny problems. It certainly didn't sound like a company that would build a whole neighbourhood or city. The ideas were much smaller. It was still clinging to the idea of managing parking to minimize congestion. Five years after the *Guardian* had raised concerns about its app Flow, Sidewalk had just introduced Pebble, a parking space occupancy sensor that the company said was built using the concepts of Privacy by Design. Some of the Sidewalk-incubated companies that Doctoroff was describing predated his push into Canada. One, Cityblock Health, was now a genuinely ambitious company that worked with insurers to help low-income patients, and was worth more than $5 billion.

Start-ups often talk about failing fast to learn from those failures. It was clear that Sidewalk Labs had hand-picked the most financially feasible ideas from its work in Toronto as its new priorities—meaning that, if you ignore all the money and time that was lost, Quayside was, in some ways, worth it. Sidewalk had always done *some* incubating, and now that's how it was framing its best ideas from Toronto. There was more talk of tall "mass timber" buildings during the webcast, and Doctoroff at one point brought up Sidewalk's widely panned urban data concept. The idea of using algorithms to optimize neighbourhood design for things like sunlight access and density, first explored in depth in the Toronto draft master plan, had become a new Sidewalk product called Delve.

Not all of Sidewalk's spinoff companies worked out: Portland, Oregon, gave up on Replica, which used machine learning to develop "synthetic" privacy-protecting data to help cities learn how people move around. Portland officials alleged that Replica hadn't been transparent enough about its data—though it appeared that the city wanted more demographic data than Replica was willing to hand over, because Replica was trying to protect the privacy of the individuals who'd inspired its data set. Privacy had become one of the era's defining human-rights issues, but the world still struggled to agree on how best to protect it.

By mid-2021, Sidewalk was looking a lot like Google X after the collapse of Glass: a series of loosely connected, semi-independent companies whose products were lower-stakes than a flagship effort that had crashed and burned. Sidewalk had more in common with BlackBerry, too, than either Dan Doctoroff or Jim Balsillie would want to admit: when their marquee products failed, both companies scavenged the remains of their business to focus on a less flashy future. For BlackBerry, that was eventually cybersecurity. But by the time that company was reinvented, Balsillie was long gone. In mid-2021, Doctoroff was staying put.

If implemented carefully, many of Sidewalk Labs' individual ideas for Toronto could have changed countless lives for the better. Though few of its ideas were truly new, they had rarely been sold in such a lucrative-looking package, and could have set new standards around the world for energy-efficient buildings, home ownership models, garbage reduction and, eventually, maybe, neighbourhoods designed for self-driving cars. Sidewalk hired dozens of employees who brushed aside all the drama the company courted and really tried to make these things happen. But Sidewalk bid for a project from a relatively unknown public agency in a city it didn't know well that was asking for innovation it couldn't really define. And Dan Doctoroff built a company that couldn't stop sabotaging its own great ideas, repeatedly asking for more than anyone could offer it—against the advice of the agency he partnered with, the city he wanted to work in and even the people he courted for support. Sidewalk Labs wanted to win so badly that it just kept losing.

In the months after Sidewalk pulled out of Toronto, most of its local staff were laid off. Almost all of Doctoroff's top deputies quietly departed, too. Josh Sirefman, Rohit Aggarwala, Alyssa Harvey Dawson and Micah Lasher either moved to vague "advisory" roles or disappeared from the company altogether. Many of the people who put their reputations on the line for Sidewalk, who spent years of their lives trying to find ways to make cities better for everyone, were left with little to show for it but exhaustion. Sidewalk's $50 million Toronto plan and $300 million or more budget stalled more careers than it introduced revolutionary city ideas. A pattern was emerging. A decade and a half earlier, Doctoroff had convinced hundreds of people and organizations to hand over $35 million for an Olympic dream that didn't happen. And though Bloomberg reported in 2021 that Hudson Yards had delivered the city $663 million, the stadium-less development was still accused of leaving New Yorkers an unexpected $2.2 billion bill for a neighbourhood most of them wouldn't enjoy. It was becoming increasingly difficult to believe Doctoroff's self-branding as a city-builder when he so often struggled to succeed at what he first set out to do.

Doctoroff disagrees with this characterization. In his eyes, these were city-building successes, even if the results, each time, looked very different from what he had first hoped to achieve. He also objected to John Tory's description of Sidewalk's desperate final months in Toronto. Reflecting on the company at a digital version of the Collision conference in April 2021, the Toronto mayor said he felt Sidewalk was trying to get a "bargain basement price" for Quayside after the pandemic had put the real estate market in the lurch. Soon after, Doctoroff e-mailed the mayor, chastising him for describing negotiations this way. Even in matters of math, Doctoroff preferred his version of the truth.

At the City Summit webcast a month later, the Sidewalk CEO might not have mentioned Toronto, but if you really reached, you could almost wonder if he was sending coded messages about his experience there. "Cities are adaptive creatures. They learn to adapt—sometimes not as quickly as we'd like, but they do," he said. The bread crumbs kept piling up: "Hopefully cities will use the pandemic, and the impact, to really rethink the way they operate in the future." As he finished that thought, he flashed a smile. Doctoroff was suddenly his old self. "A crisis is, as we know, a terrible thing to waste."

He had two crises left to deal with.

Dan Doctoroff's father, Martin, died in 2002 of amyotrophic lateral sclerosis, better known as ALS, or Lou Gehrig's disease. To receive a diagnosis of ALS is to be sentenced to death: the body's motor neurons are gradually ravaged, and what begins as muscle weakness turns to clumsiness, a struggle to move and, eventually, the inability to eat or breathe. The younger Doctoroff knew the toll was devastating. His father had been limping before they attended the 2000 Olympic Games in Sydney together. There, the trudge between venues became too much; Martin soon needed a wheelchair. An ALS diagnosis came shortly thereafter. For the next two years, Doctoroff shuttled between New York and Detroit, spending time with his father before he died. The experience was seared in his brain.

ALS is genetic for at least one in every twenty sufferers. And a few years later, Doctoroff's uncle got a diagnosis. "Ever since then," Dan later recalled, "my family has been living with a spectre that was always vaguely present." He became determined to find out if he was doomed to the same fate, and underwent testing to see if he had any genetic mutations that could cause ALS. He learned that some of his family members carried one of those mutations, but he didn't. And it didn't appear that he had any other genetic mutations that could cause the disease. It seemed he might be free of the spectre, after all.

But around 2019, he started to feel some weakness in his hip. Then, on a trip to Iceland between waves of the pandemic two years later, he experienced more hip trouble. Breathing issues, too. Back in the United States, he began visiting ALS specialists in New York and Baltimore, trying to figure out if the disease that killed his father had found its way to him, too. He spent eight weeks buzzing between specialists. Once it was over, he granted me his first and only interview for this book. He did not disclose his medical troubles.

Six weeks after that, he announced that he believed he had ALS.

Genetic mutation or not, the symptoms were there, and the odds of them being from ALS, he said, "are very high." He had decided to dedicate the rest of his life to doing what he did best. Through his research foundation, he would influence people to hand over their money—this time $250 million—to defeat the disease. And so he was stepping down as the CEO of Sidewalk Labs.

Many of the growing ranks of ex-Sidewalkers found out at the same time the public did, through an early-morning blog post. Baffled, they tried to dissect the news, quickly noticing that an important piece of information was missing. Sidewalk would need to find another CEO, yet Doctoroff had no clear successor. None was announced. Many of the senior executives who had stood by him for years were long gone, and anyone from outside would see a company that had done little more than incubate a few start-ups after seven years of work.

There was no mention of a replacement. Instead, Doctoroff said Sidewalk would move four of its five main business lines into Google, including the latest version of its parking management system. Its tall-wood-building construction company, now named Canopy Buildings in homage to the dome Sidewalk never built, would be spun out as its own business. Two of the best executives who'd stuck with Sidewalk, Craig Nevill-Manning and Prem Ramaswami, would follow Sidewalk's business lines to Google.

I asked the company's PR team that day to confirm what was increasingly apparent each time I reread the blog post: that Sidewalk Labs would no longer exist. They didn't respond. The *New York Times*, meanwhile, wrote that Sidewalk was "planning to be subsumed inside Google's urban sustainability product efforts." Then the website Engadget said Sidewalk Labs was "winding down." When I asked about this, a spokesman finally got back to me, refusing to say the company would be no more. He even told me he was trying to get Engadget to change the article. A shutdown, he told me, was "not what we are saying today."

"Today." His response hung on a single word.

It turned out the company that had always struggled to generate good news couldn't even get the story of its own death straight. The next month, I even asked Kent Walker, Alphabet's chief legal officer and head of global affairs, if he had any idea whether Sidewalk still existed. His response was "I don't know." But many of the people Sidewalk Labs had disappointed since 2015 already knew the answer. The company would, in fact, be folding into Google. To say publicly that it was shutting down, though, would be to admit defeat—to acknowledge that tens if not hundreds of millions of dollars had disappeared along the way. That so many top executives had quit after its repeated failures that its CEO didn't have a successor. That there was nothing left to lead.

As Doctoroff cut his career short to grapple with a personal tragedy, the company he'd shaped in his image was trying to sell the world a story one last time. Sidewalk Labs didn't want to go out on a bad note.

Acknowledgements

THANK YOU FIRST to my sources, many of whom took professional and personal risks to help this book's readers better understand the Quayside saga because they care about the cities of tomorrow.

Thank you to Random House Canada for seeing the potential in this book. My editor, Craig Pyette, shared my excitement in untangling the story's many mysteries, and saved both me and readers from falling down too many rabbit holes. Further thanks to the rest of the team that helped make *Sideways* come together, including Sue Kuruvilla, Anne Collins, Linda Friedner, Sue Sumeraj, Matthew Flute, Tim Hilts, Gillian Watts, Deirdre Molina, Daniella D'Souza, Sonia Vaillant, Erin Kelly and Chalista Andadari. My agent, Eric Lupfer of Fletcher and Co., immediately saw the consequences of what happened with Quayside as extending well beyond Toronto, and I am grateful that he found this book such a great home. My friend and colleague Melissa Tait kindly took my author photo overlooking Quayside on a day that was probably way too cold to do so.

I grew up obsessed with cities, fascinated by the idea that you could fit the entire population of my hometown of Saint John into the SkyDome, home of the Toronto Blue Jays. It was journalism that brought me to Toronto, and the *Globe and Mail* that gave me a reason to stay—plus a desk from which I could actually see the SkyDome. This book is built on two and a half years of *Globe* stories. Former technology editor Claire Neary encouraged me to pry into the Quayside saga and always gave me the freedom to chase new leads.

Mark Stevenson, Mark Heinzl, Sarah Efron, Aron Yeomanson, Derek DeCloet and John Daly helped shape my stories, too. Gary Salewicz, Dennis Choquette, David Walmsley, Sinclair Stewart, Angela Pacienza and the rest of the *Globe*'s masthead graciously gave me time off during the wildest news cycle in generations to write this book, and always supported my coverage.

Many journalists helped the world understand the Quayside saga. At the *Globe*, they included Alex Bozikovic, Jeff Gray, Rachelle Younglai, Bill Curry, Tom Cardoso (who also served as an access-to-information adviser, Yellow Book Whisperer and friendly neighbour), Fred Lum, Oliver Moore, Stefanie Marotta, David Milstead, Laura Stone, Rick Cash and Stephanie Chambers. Elsewhere across Toronto, Tara Deschamps, John Lorinc, James McLeod, Amanda Roth, Donovan Vincent and the entire BetaKit team stayed on this story for the long haul, and I'm thankful for everything they unearthed.

This project may owe its biggest debt to Sean Silcoff, my longtime *Globe* reporting partner and a great friend. He cleared many important roadblocks for me to make this book happen—including doubling his workload while I took time off to write it.

Sean was one of this book's first readers, alongside Alex Bozikovic and Adrian Lee, each of whom helped me make sense of the manuscript's many tangled stories. Sheema Khan kindly reviewed segments about intellectual property to ensure I got the technicalities right. Any errors in this book, however, are mine and mine alone.

The Berlin segments of *Sideways* expand upon stories I reported from the *Handelsblatt* capital bureau during an Arthur F. Burns fellowship in 2019. I'm grateful that the International Center for Journalists and Internationale Journalisten-Programme selected me as a fellow (and I hope Dr. Frank-Dieter Freiling now understands why I didn't want to report from Munich). I returned to Berlin after COVID-19 vaccinations in 2021 to finish a crucial draft of this book; thank you to the staff of HeartSpace Coffee, the Black Lodge, Lass uns Freunde bleiben and Myxa for hosting endless hours of writing and revising.

For helpful conversations big and small as this book came together, I thank Jared Bland, Eliot Brown, Dennis Choquette, Janice Dickson, Robyn Doolittle, Robert Fife, Garfy, Jen Knoch, Manisha Krishnan, Bruce Livesey, Jacques Poitras, Jennie Russell, Oliver Sachgau, David Shipley, David Skok, and Max Tholl & Eva Björg Hafsteinsdóttir. Thanks, too, to the 2019 and 2021 Burns crews, Fizzy Water team, Brunch Club, Wolf Parade Subcommittee, Beckwith Street Alumni and the many Maritimers and non-Maritimers of the "activitis" diaspora.

This book was made possible by Kareen Sarhane, whose patience and kindness carried me through many highs and lows, and whose draft feedback was respectfully unvarnished. You make me a better person. Thank you for letting me be part of your life.

Finally, thanks to my family—especially Jon, Joey, Hilari, Mackenzie, Antoinette, Magdi, Dina, Simon, Henry and Izzy—for grounding me when journalism consumes too much of my brain. *Sideways* is dedicated to my parents, Joanne and Mick O'Kane. The curiosity and perseverance they've imbued in their sons has helped us build greater lives than we could have ever imagined.

A Note on Sources

THIS BOOK DRAWS from interviews with more than 150 people, the vast majority of whom had direct connections to the Quayside project and its founding partners. Many worked for or with Waterfront Toronto, Waterfront's three shareholder governments—the City of Toronto, the province of Ontario and Canada's federal government—or Sidewalk Labs, Google or Alphabet. This also includes the ranks of community leaders, activists, businesspeople and academics who were consulted for or carefully studied Sidewalk's plans for Toronto. *Sideways* is also built from a trove of both public and confidential documents I collected between 2018 and 2022. These range from Sidewalk's 2016 Yellow Book to Waterfront board meeting minutes to hundreds upon hundreds of pages of memos, speeches, briefings, slide decks and e-mail threads obtained through government access-to-information requests.

Conversations, meetings and other events that appear in this manuscript were reconstructed in one of two ways. Many are the results of recordings. I attended and personally taped many public Quayside meetings for *Globe and Mail* stories, while other events were broadcast and archived online. Those public recordings are cited in the book's endnotes. In some cases, dialogue has been condensed for brevity and clarity. Except for a handful of events that are described in the text as a specific person's recollection, all other events were described and verified to me by at least two people who were present or directly briefed about what occurred. Any dialogue that appears in these scenes is the product of people's memories, but not necessarily the memory of the person identified as speaking, since not everyone who attended each event agreed to an interview or granted enough interview time to reconstruct events.

Quotations that appear in the present tense come from interviews conducted during my reporting and serve as reflections or commentary on how the Quayside story unfolded. Commentary in the past tense comes from other sources, such as news stories, which are cited in the book's endnotes. Dialogue from meetings and other events in the past tense that appears without a citation in the endnotes comes from my own reporting. All dollar figures are presented in U.S. dollars, with which Alphabet does business, unless prefaced with a C, for Canadian dollars.

Many people who were interviewed for this book only agreed to do so "on background," requiring me not to identify the source of the information that was shared. In virtually all cases, this was because they were either contractually bound not to discuss certain matters publicly or faced reprisal for speaking up. Some individuals' on-the-record quotes are juxtaposed directly in the text with details shared with me by a different person or people who were only able to speak on background. In several cases, sources only agreed to be interviewed for fact-checking purposes, and while they are occasionally quoted to provide responses to situations in which they were involved, they were not necessarily the initial source of that information.

Waterfront Toronto eagerly participated in this endeavour, offering numerous on-the-record interviews with each executive who worked directly on the Quayside project, as well as its external subject-matter experts, lawyers and other staff. The agency also shared any documents I asked for that they were legally able to make public. Much credit for this openness goes to Andrew Tumilty, Kristina Verner and Stephen Diamond, who encouraged their colleagues to participate in a book about a project that some might rather forget.

Official government participation was hit-or-miss. Mayor John Tory granted a full interview, as did several Toronto councillors and members of parliament. The federal government department overseeing Waterfront agreed to respond to a series of written questions, and the Prime Minister's Office responded to several e-mailed queries. Representatives for Ontario premier Doug Ford and his cabinet minister who oversaw Waterfront during its most tumultuous time never returned my interview requests.

Representatives for Larry Page similarly did not respond to multiple interview requests, though Google did offer some clarifying details about the January 2017 call between Eric Schmidt and Justin Trudeau when I told the company I was aware of it. Four books helped shape my understanding of Page's thinking and work, on top of interviews with people close to him: Stephen Levy's *In the Plex*, Ken Auletta's *Googled*, Richard L. Brandt's *The Google Guys* and David A. Vise and Mark Malseed's *The Google Story*.

Sidewalk Labs granted me a one-hour interview with CEO Dan Doctoroff over Zoom in November 2021, after ten months of negotiations, just weeks before he revealed to the public that he believed he had ALS and was stepping down from the company. Sidewalk additionally provided various documents and made a handful of staff available to discuss technical details from the company's proposals. Thank you to Dan Levitan and Eric Jaffe for organizing this. I also interviewed Doctoroff five times in 2018 and 2019, and attended many of his public presentations. His thinking is represented in this book through our interviews, interviews with many people who worked closely with him, public remarks, news stories from his time as a New York deputy mayor and Olympic campaigner, and details he shared in his 2017 memoir about those years, *Greater Than Ever: New York's Big Comeback*.

As I wrote this book, it was back-channelled to me numerous times that Sidewalk Labs felt my *Globe and Mail* coverage was unfair. Those stories always sought to sift truth from a sea of self-mythologizing. This book seeks to do the same.

Notes

Prologue: After the Gold Rush

xiii **"Campus in Kreuzberg substation":** For this and subsequent blog post quotes, see Carolin Silbernagl, "#komminsHaus: Campus im Kreuzberger Umspannwerk wird Haus für soziales Engagement," Betterplace.org, October 24, 2018, archived via WayBack Machine, https://web.archive.org/web/20181207082901/betterplace.org/c/neues/haus.

xvii **"set a new standard":** Sidewalk Labs, Vision sections of RFP submission, made public October 17, 2017, 15, https://storage.googleapis.com/sidewalk-toronto-ca/wp-content/uploads/2017/10/13210553/Sidewalk-Labs-Vision-Sections-of-RFP-Submission.pdf.

xvii **buildings could have a "radical" mix:** Sidewalk Labs, Vision, 27.

xviii **double the Canadian government's budget:** William Francis Morneau, "Building a Strong Middle Class: #Budget2017," Government of Canada, March 22, 2017, https://www.budget.gc.ca/2017/docs/plan/budget-2017-en.pdf.

Chapter 1: Everything Now

1 **"healthy disregard for the impossible":** Larry Page, University of Michigan Commencement Address, May 2, 2009, https://googlepress.blogspot.com/2009/05/larry-pages-university-of-michigan.html.

1 **He happened to hate:** Lawrence D. Burns with Christopher Shulgan,

Autonomy: The Quest to Build the Driverless Car—and How It Will Reshape Our World (New York: Ecco, 2018), 4, 5.

1 **"It was a futuristic way":** Page, Commencement Address. When he deviates from the prepared text, the difference can be heard on this YouTube video: https://www.youtube.com/watch?v=qFb2rvmrahc.

2 **He'd brought the idea up:** Nathan Bomey, "Google's Larry Page Still Has Vision for Ann Arbor Transportation Changes," *Ann Arbor News*, October 9, 2009, http://www.annarbor.com/business-review/googles-larry-page-takes-interest-in-ann-arbors-transportation-future/.

2 **couldn't figure out why:** Steven Levy, *In the Plex: How Google Thinks, Works, and Shapes Our Lives* (New York: Simon & Schuster, 2011), 12.

2 **idolizing the brilliance of Nikola Tesla:** Levy, *In the Plex*, 13.

2 **didn't bother putting them back together:** Richard L. Brandt, *The Google Guys: Inside the Brilliant Minds of Google Founders Larry Page and Sergey Brin* (New York: Portfolio, 2011), 23.

3 **automated, robotic, low-cost taxis:** David A. Vise and Mark Malseed, *The Google Story: Inside the Hottest Business, Media, and Technology Success of Our Time*, rev. ed. (New York: Bantam, 2018), 39.

3 **city planning and zoning regulations:** Ken Auletta, *Googled: The End of the World as We Know It*, rev. ed. (New York: Penguin Press, 2010), 34.

3 **some of those annotations:** Levy, *In the Plex*, 16–18.

4 **indexed at least a billion pages:** Levy, 45.

4 **"hiding strategy":** Levy, 69.

5 **$347 million in revenue:** Levy, 70.

5 **they went to Burning Man:** Vise and Malseed, *Google Story*, 211.

5 **"It never occurred to me":** Levy, *In the Plex*, 171.

6 **a reported $50 million:** Daniel Roth, "Google's Open Source Android OS Will Free the Wireless Web," *Wired*, June 23, 2008, https://www.wired.com/2008/06/ff-android/.

6 **Page thought it would be smart:** Levy, *In the Plex*, 215.

6 **most popular operating system:** Michelle Meyers, "Android Inches Ahead of Windows as Most Popular OS," CNET, April 3, 2017,

https://www.cnet.com/tech/mobile/android-most-popular-os-beats-windows-statcounter/.

7 **a late-2000s attack:** Ellen Nakashima, "Chinese Hackers Who Breached Google Gained Access to Sensitive Data, U.S. Officials Say," *Washington Post*, May 20, 2013, https://www.washingtonpost.com/world/national-security/chinese-hackers-who-breached-google-gained-access-to-sensitive-data-us-officials-say/2013/05/20/51330428-be34-11e2-89c9-3be8095fe767_story.html.

7 **worried the company was losing its knack:** Liz Gannes and Kara Swisher, "Exclusive: Google CEO Larry Page Reorgs Staff, Anoints Sundar Pichai as New Product Czar," Recode, October 24, 2014, archived via WayBack Machine, https://web.archive.org/web/20141025054610/https://www.recode.net/2014/10/24/google-ceo-larry-page-reorgs-staff-anoints-sundar-pichai-as-new-product-czar/.

7 **He soon summoned Brin and Schmidt:** Nicholas Carlson, "The Untold Story of Larry Page's Incredible Comeback," Insider, April 24, 2014, https://www.businessinsider.com/larry-page-the-untold-story-2014-4.

7 **"gospel of 10x":** Steven Levy, "Big Ideas: Google's Larry Page and the Gospel of 10x," *Wired*, March 30, 2013, https://www.wired.co.uk/article/a-healthy-disregard-for-the-impossible.

7 **Google became a teenager:** Rolfe Winkler, "In New Structure, Google CEO Page Aims for 'Faster, Better Decisions,'" *Wall Street Journal*, October 27, 2014, https://blogs.wsj.com/digits/2014/10/27/in-new-structure-google-ceo-page-aims-for-faster-better-decisions/.

7 **brought a time machine** and **"A big part of my job":** Steven Levy, "Google's Larry Page on Why Moon Shots Matter," *Wired*, January 17, 2013, https://www.wired.com/2013/01/ff-qa-larry-page/.

8 **"One thing we're doing":** Richard Waters, "FT Interview with Google Co-founder and CEO Larry Page," *Financial Times*, October 31, 2014, https://www.ft.com/content/3173f19e-5fbc-11e4-8c27-00144feabdc0.

8 **just two inches deep:** Jason Hiner, "Google Fiber's Secret Weapon in Its Gigabit Comeback Has Failed," CNET, February 7, 2019, https://www.cnet.com/news/google-fibers-secret-weapon-in-its-gigabit-comeback-has-failed/.

8 **"No one wants to face":** Max Chafkin and Mark Bergen, "Google Makes So Much Money, It Never Had to Worry about Financial Discipline—Until Now," Bloomberg, December 8, 2016, https://www.bloomberg.com/news/features/2016-12-08/google-makes-so-much-money-it-never-had-to-worry-about-financial-discipline.

8 **He hated conflict:** Chafkin and Bergen, "Google Makes So Much Money."

8 **a *Vanity Fair* exposé:** Vanessa Grigoriadis, "O.K., Glass: Make Google Eyes," *Vanity Fair*, March 12, 2014, https://www.vanityfair.com/style/2014/04/sergey-brin-amanda-rosenberg-affair.

8 **Page summoned about a dozen:** Amir Efrati, "At Google, Larry Page's Dreams Keep Getting Bigger," The Information, September 17, 2014, https://www.theinformation.com/articles/at-google-ceo-page-s-dreams-keep-getting-bigger.

8 **"Today's big bets":** Eric Schmidt and Jonathan Rosenberg with Alan Eagle, *How Google Works* (New York: Grand Central Publishing, 2014), xiv.

9 **He wanted to know where someone was** and **ideas were even "creepier":** Efrati, "At Google."

9 **"He could produce something evil":** Ashlee Vance, *Elon Musk: Tesla, SpaceX, and the Quest for a Fantastic Future*, rev. ed. (New York: Ecco, 2017), 3.

9 **he'd gotten guests at his own wedding:** Vise and Malseed, *Google Story*, xiii.

10 **Cities everywhere have spent decades:** Some smart-city efforts from this section were explored at length by Anthony M. Townsend in his book *Smart Cities: Big Data, Civic Hackers, and the Quest for a New Utopia* (New York: W. W. Norton, 2013), and are bolstered here by news articles and proprietary information from each company.

10 **"Intelligent Operations Centers":** IBM Corp., "IBM Intelligent Operations Center for Smarter Cities," June 2013, https://www.ibm.com/downloads/cas/EMJY7VY4.

10 **hundreds of cameras and thirty agencies:** Natasha Singer, "Mission Control, Built for Cities," *New York Times*, March 3, 2012,

https://www.nytimes.com/2012/03/04/business/ibm-takes-smarter-cities-concept-to-rio-de-janeiro.html; and Greg Lindsay, "Building a Smarter Favela: IBM Signs Up Rio," *Fast Company*, December 27, 2010, https://www.fastcompany.com/1712443/building-smarter-favela-ibm-signs-rio.

11 **it would earn €6 billion:** Siemens AG, "Siemens Wants to Win Smart Grid Orders Worth over €6 Billion by 2014," press release, September 4, 2009, https://press.siemens.com/global/en/pressrelease/siemens-wants-win-smart-grid-orders-worth-over-eu6-billion-2014.

11 **"rebel with a cause":** Siemens AG, "Rebel with a Cause: How a Canadian Power Utility Together with Siemens Is Shaping the Smart Grid of the Future," July 5, 2019, https://new.siemens.com/global/en/company/stories/infrastructure/2019/shaping-the-smart-grid-of-the-future.html.

11 **"harness big data":** Siemens AG, "Siemens Digitalization Hub Factsheet," undated, https://assets.new.siemens.com/siemens/assets/api/uuid:60b62968-3745-4d19-8e76-aa6862b56669/factsheet-digitalization-hub-e.pdf.

11 ***the world's "smartest" city:*** Linda Poon, "Sleepy in Songdo, Korea's Smartest City," Bloomberg CityLab, June 22, 2018, https://www.bloomberg.com/news/articles/2018-06-22/songdo-south-korea-s-smartest-city-is-lonely.

12 **north of $40 billion:** Poon, "Sleepy in Songdo."

12 **"city in a box":** John Boudreau, "Cisco Wires 'City in a Box' for Fast-Growing Asia," *San Jose Mercury News*, June 8, 2010, archived via WayBack Machine, https://web.archive.org/web/20100619065543/https://www.newsobserver.com/2010/06/08/520176/cisco-wires-city-in-a-box-for.html.

12 **promised to spend $47 million:** Cisco Systems Inc., "Cisco and New Songdo International City Development Join Forces to Create One of the Most Technologically Advanced Smart+Connected Communities in the World," July 4, 2011, https://www.cisco.com/c/dam/en_us/about/ac78/docs/Cisco_and_New_Songdo.docx.

12 **Traffic patterns could be tracked:** Anthony M. Townsend, *Smart Cities: Big Data, Civic Hackers, and the Quest for a New Utopia* (New York: W. W. Norton, 2013), 24.

12 **"The city aims to do nothing less":** David McNeill, "New Songdo City: Atlantis of the Far East," *Independent*, June 22, 2009, https://www.independent.co.uk/news/world/asia/new-songdo-city-atlantis-of-the-far-east-1712252.html.

12 **$30 billion business opportunity:** Boudreau, "Cisco Wires."

12 **Only about 100,000:** Poon, "Sleepy in Songdo."

12 **less than half of the planned office space:** Peter Grant, "Developer Feuds with Korean Partner over Busted 'Smart' City," *Wall Street Journal*, June 11, 2019, https://www.wsj.com/articles/developer-feuds-with-korean-partner-over-busted-smart-city-11560261729.

12 **at least $1 billion:** Boudreau, "Cisco Wires."

13 **"Instead of Captain Kirk":** Bruce Gibney, "What Happened to the Future?," Founders Fund, 2011, https://foundersfund.com/the-future/#/. One thing I've never seen acknowledged about this widely shared statement is that it's not just an indictment of society's approach to innovation, but also of William Shatner's approach to his later career.

14 **funnelled half a million dollars:** Alexis Madrigal, "Peter Thiel Makes Down Payment on Libertarian Ocean Colonies," *Wired*, May 18, 2008, https://www.wired.com/2008/05/peter-thiel-makes-down-payment-on-libertarian-ocean-colonies/.

14 **"building floating societies":** The Seasteading Institute, home page, last updated 2021, https://www.seasteading.org/ (accessed March 31, 2022).

15 **according to an internal document:** Isaac Taylor did not provide me with any documents or written material, including those described here.

16 **"Larry's the worst person":** Levy, *In the Plex*, 234.

17 **"As long as their contributions":** Schmidt and Rosenberg, *How Google Works*, 51.

Chapter 2: Big Rings

18 **"Our city needs these Games":** As my access to Dan Doctoroff for this book was limited, this chapter is largely built upon news clippings

from his time as a New York deputy mayor, interviews with people who knew him, and his memoir, *Greater Than Ever: New York's Big Comeback* (New York: PublicAffairs, 2017), which provided helpful source material to describe significant events from his life, such as this one (pages 266–69).

18 **Since watching Italy oust Bulgaria:** Daniel L. Doctoroff, *Greater Than Ever: New York's Big Comeback* (New York: PublicAffairs, 2017), 7–9.

19 **"Give us a chance":** Doctoroff, 267.

19 **terrified of failure:** Doctoroff, 273.

19 **"The prospect of being shamed":** Doctoroff, 10.

20 **"alien city":** Doctoroff, xi.

20 **"run down":** Doctoroff, 1.

20 **came to admire the Giuliani administration's:** Doctoroff, 5, 6, 10.

20 **racist tactic that unfairly punishes:** Daniel A. Medina, "Stop and Frisk: NYPD's 'Broken Windows' Policing 'Criminalizes' Young Black Men," *Guardian*, February 2, 2015, https://www.theguardian.com/world/2015/feb/02/nypd-stop-and-frisk-keeshan-harley-young-black-men-targeted.

20 **"new hope":** Doctoroff, *Greater Than Ever*, 183.

20 **worth $11.25 million:** Katherine Clarke, "Former Bloomberg Number 2 Dan Doctoroff Sells Upper West Side Mansion for $11.25M," *New York Daily News*, August 11, 2014, https://www.nydailynews.com/life-style/real-estate/dan-doctoroff-sells-upper-west-side-mansion-11-25m-article-1.1899317.

21 **donated to the top four candidates:** Doctoroff, *Greater Than Ever*, 24.

21 **the Games would cost $1.9 billion:** Charles V. Bagli, "New York Olympic Stadium Plan Masks Shaky Political Coalition," *New York Times*, October 2, 2000, https://www.nytimes.com/2000/10/02/nyregion/new-york-olympic-stadium-plan-masks-shaky-political-coalition.html.

21 **a sort of outreach program:** Doctoroff, *Greater Than Ever*, 21.

21 **his then-teenage son, Jacob:** Doctoroff, 25.

22 **"ever more tightly embedded":** Scott Larson, *"Building Like Moses with Jacobs in Mind": Contemporary Planning in New York City* (Philadelphia: Temple University Press, 2013), 144.

22 **"There is an orientation here":** Jennifer Steinhauer, "Bloomberg Passes Hat, Aiming at Corporate Help," *New York Times*, February 6, 2002, https://www.nytimes.com/2002/02/06/nyregion/bloomberg-passes-hat-aiming-at-corporate-help.html.

23 **railed against the promotion of sugar:** Sewell Chan, "Why Snapple Deal Shrank," *New York Times*, March 10, 2006, https://www.nytimes.com/2006/03/10/nyregion/why-snapple-deal-shrank.html.

23 **potentially more profitable proposals:** Mike McIntire, "Comptroller Seeks to End Snapple Deal," *New York Times*, October 31, 2003, https://www.nytimes.com/2003/10/31/nyregion/comptroller-seeks-to-end-snapple-deal.html.

23 **"We were asking to build":** Doctoroff, *Greater Than Ever*, 165.

23 **one of the Olympic bid's biggest fundraisers:** Charles V. Bagli, "For City Official and Developer, Close Ties Mean Close Scrutiny," *New York Times*, November 19, 2004, https://www.nytimes.com/2004/11/19/nyregion/for-city-official-and-developer-close-ties-mean-close-scrutiny.html.

23 **a $3.2 million personal loan:** Charles V. Bagli and Robin Shulman, "Transforming Bronx Terminal Market, but at a Steep Price," *New York Times*, October 24, 2005, https://www.nytimes.com/2005/10/24/nyregion/transforming-bronx-terminal-market-but-at-a-steep-price.html.

23 **"When you look at the pattern"** and **"set a standard":** Bagli, "For City Official."

24 **it was he who called Ross:** Doctoroff, *Greater Than Ever*, 315.

24 **name Hudson Yards the Bloomberg Center:** Doctoroff, 250.

25 **real estate in the area:** Doctoroff, 97.

25 **more than a million new residents:** New York City Global Partners, "Best Practice: PlaNYC: NYC's Long-Term Sustainability Plan," July 21, 2010, https://www.nyc.gov/html/ia/gprb/downloads/pdf/NYC_Environment_PlaNYC.pdf.

25 **pony up $354 million:** Nicholas Confessore, "$8 Traffic Fee for Manhattan Gets Nowhere," *New York Times*, April 8, 2008, https://www.nytimes.com/2008/04/08/nyregion/08congest.html.

25 **administration approached rezoning:** Amy Armstrong, Vicki Been, Josiah Madar and Simon McDonnell, "How Have Recent Rezonings Affected the City's Ability to Grow?," Furman Center for Real Estate & Urban Policy, New York University, March 2010, https://furmancenter.org/files/publications/Rezonings_Furman_Center_Policy_Brief_March_2010.pdf.

26 **"perhaps the greatest concentration of hipsters":** Doctoroff, *Greater Than Ever*, xiv.

26 **more than 30 percent of their income:** Richard Florida, "Where New York Is Gentrifying and Where It Isn't," Bloomberg CityLab, May 12, 2016, https://www.bloomberg.com/news/articles/2016-05-12/nyu-s-furman-center-report-new-york-s-recent-history-of-gentrification.

26 **nearly a billion dollars an acre:** Carl Swanson, "The Only Man Who Could Build Oz," *New York*, February 18, 2019, https://nymag.com/intelligencer/2019/02/stephen-ross-hudson-yards.html.

26 **the biggest private development project:** Jessica Tyler and Aria Bendix, "Hudson Yards Is the Most Expensive Real-Estate Development in US History. Here's What It's Like Inside the $25 Billion Neighborhood," Insider, last updated March 15, 2019, https://www.businessinsider.com/hudson-yards-tour-of-most-expensive-development-in-us-history-2018-9?op=1.

26 **at least $1.2 billion in financing:** Kriston Capps, "The Hidden Horror of Hudson Yards Is How It Was Financed," Bloomberg CityLab, April 12, 2019, https://www.bloomberg.com/news/articles/2019-04-12/the-visa-program-that-helped-pay-for-hudson-yards.

26 **a claim to some of the rights:** Sopan Deb, "Following Outcry, Hudson Yards Tweaks Policy over Use of Vessel Pictures," *New York Times*, March 19, 2019, https://www.nytimes.com/2019/03/19/arts/design/hudson-yards-vessel-instagram.html.

26 **The mayor joked:** Doctoroff, *Greater Than Ever*, 337.

27 **Revenue jumped 67 percent:** Bloomberg LP, "Daniel L. Doctoroff Stepping Aside as CEO of Bloomberg LP at End of 2014," press release, September 3, 2014, https://www.bloomberg.com/company/press/daniel-l-doctoroff-stepping-aside-ceo-bloomberg-lp-end-2014/.

27 **visions for the money-losing news side:** Kevin Dugan, "Bloomberg Ousting His Company's CEO over Clash of Vision," *New York Post*, September 4, 2014, https://nypost.com/2014/09/04/ousting-of-bloomberg-lp-chief-signals-clash-in-vision/.

27 **"People only want you":** Doctoroff, *Greater Than Ever*, 250.

Chapter 3: Building a Mystery

29 ***You don't want to go out on a bad note*:** Daniel L. Doctoroff, *Greater Than Ever: New York's Big Comeback* (New York: PublicAffairs, 2017), 273.

29 **a budget of between $2 billion and $3 billion:** Doctoroff declined to confirm or comment on how much funding he sought from Page and Alphabet, except to say "it was going to require a lot of capital."

29 **Neom had a $500 billion budget:** Bill Bostock, "Everything We Know about Neom, a 'Mega-City' Project in Saudi Arabia with Plans for Flying Cars and Robot Dinosaurs," Insider, September 23, 2019, https://www.businessinsider.com/neom-what-we-know-saudi-arabia-500bn-mega-city-2019-9?op=1.

29 **"relatively modest":** Larry Page blog post announcing Sidewalk Labs in June 2015 on the since-deleted social network Google+, archived via Archive.is, https://archive.is/10qto (accessed March 10, 2022).

30 **"Larry's retirement letter":** Jessica E. Lessin and Amir Efrati, "What Alphabet Means," August 10, 2015, https://www.theinformation.com/articles/what-aphabet-means.

30 **cost him $1,500:** TRD Staff, "Former EDC Interim President Fined after Illegal Lobbying for Rail Yards," The Real Deal, September 29, 2009, https://therealdeal.com/2009/09/29/former-edc-interim-president-fined-after-illegal-lobbying-for-rail-yards/.

32 **"Many of you are reading":** Larry Page Google+ post, 2015.

33 **"We're calling this reimagining":** Google TechTalks, "Sidewalk Labs: Reimagining the City as a Digital Platform," posted March 22, 2016, https://www.youtube.com/watch?v=bPu8HvD7d9U.

33 **"urban-tech divide":** Daniel L. Doctoroff, "It's Time for Urbanists and Technologists to Start Talking," *Sidewalk Talk* (blog), April 4, 2016,

https://medium.com/sidewalk-talk/it-s-time-for-urbanists-and-technologists-to-start-talking-df1b57abfbd1#.ir30sxfco.

34 **Googley-sounding code name Project X:** Matt Patches, "Inside Walt Disney's Ambitious, Failed Plan to Build the City of Tomorrow," *Esquire*, May 20, 2015, https://www.esquire.com/entertainment/news/a35104/walt-disney-epcot-history-city-of-tomorrow/.

34 **"always be in a state of becoming":** Craig Dezern, "Back to the Future," *South Florida Sun-Sentinel*, January 3, 1994, https://www.sun-sentinel.com/news/fl-xpm-1994-01-03-9401040321-story.html.

35 **"There's a gigantic difference":** Patches, "Inside Walt Disney's."

35 **Two-thirds of the district's land:** Jason Garcia and Orlando Sentinel, "Disney's Reedy Creek Government Has Rare Board Vacancy, but Don't Bother Running," *Orlando Sentinel*, May 9, 2011, https://www.orlandosentinel.com/business/os-xpm-2011-05-09-os-disney-reedy-creek-election-20110509-story.html.

37 **more than $100,000 worth:** Josh O'Kane, "Sidewalk's End: How the Downfall of a Toronto 'Smart City' Plan Began Long before COVID-19," *Globe and Mail*, May 24, 2020, https://www.theglobeandmail.com/business/article-sidewalks-end-how-the-downfall-of-a-toronto-smart-city-plan-began/.

38 **"a bad listener with limited patience":** Doctoroff, *Greater Than Ever*, 38.

38 **Senator Chuck Schumer calling him out:** Doctoroff, *Greater Than Ever*, 167.

38 **"over the shared adversity":** For this and subsequent quotes and details about Doctoroff's shouting and anger, see Doctoroff, *Greater Than Ever*, 210–15.

40 **new software called Flow:** Laura Bliss, "Google Will Roll Out New Real-Time Transit Tech in an Underserved Neighbourhood," Bloomberg CityLab, March 17, 2016, https://www.bloomberg.com/news/articles/2016-03-17/google-s-sidewalk-labs-announces-urban-transit-data-platform-flow.

40 **"new superpowers to extend access":** Mark Harris, "Secretive Alphabet Division Funded by Google Aims to Fix Public Transit in US,"

Guardian, June 27, 2016, https://www.theguardian.com/technology/2016/jun/27/google-flow-sidewalk-labs-columbus-ohio-parking-transit.

40 **reduce traffic by as much as 30 percent:** Donald Shoup, "Cruising for Parking," *ACCESS*, Spring 2007, http://shoup.bol.ucla.edu/CruisingForParkingAccess.pdf.

41 **"optimised" with artificial intelligence algorithms:** Harris, "Secretive Alphabet Division."

41 **In 2012, the city began:** New York City Information Technology & Communications, "Pay Phones," undated, https://www1.nyc.gov/site/doitt/residents/pay-phones.page (accessed March 31, 2022).

41 **de Blasio announced that a consortium:** New York City Press Office, "De Blasio Administration Announces Winner of Competition to Replace Payphones with Five-Borough Wi-Fi Network," November 17, 2014, https://www1.nyc.gov/office-of-the-mayor/news/923-14/de-blasio-administration-winner-competition-replace-payphones-five-borough. (accessed March 31, 2022).

41 **The group began installing LinkNYC kiosks:** Michael del Castillo, "LinkNYC Starts Building Its $500M Wi-Fi Advertising Platform," *New York Business Journal*, December 29, 2015, https://www.bizjournals.com/newyork/news/2015/12/29/linknyc-starts-building-its-500m-wi-fi-advertising.html.

41 **using them to watch porn:** Janet Burns, "LinkNYC Drops Web Access from Kiosks after Some Users Watch Porn, Predictably," *Forbes*, September 16, 2016, https://www.forbes.com/sites/janetwburns/2016/09/16/linknyc-drops-web-browsing-from-its-kiosks-after-people-unsurprisingly-watch-porn/.

41 **"nearly limitless retention":** Shahid Buttar and Amul Kalia, "LinkNYC Improves Privacy Policy, Yet Problems Remain," Electronic Frontier Foundation, October 4, 2017, https://www.eff.org/deeplinks/2017/09/linknyc-improves-privacy-policy-yet-problems-remain.

42 **"one of the basic necessities":** Mariko Hirose and Johanna Miller, New York Civil Liberties Union letter to mayoral counsel Maya Wiley, March 15, 2016, https://www.nyclu.org/sites/default/files/releases/city%20wifi%20letter.pdf.

42 **putting sticky notes:** Steven Rosenbaum, "Privacy Battle Brewing: Are LinkNYC Kiosks Surveillance Devices?" HuffPost, December 7, 2017, https://www.huffpost.com/entry/privacy-battle-brewing-are-linknyc-kiosks-surveillance_b_5a284856e4b0650db4d40caf.

42 **had yet to be installed:** Reuven Blau and Gabriel Sandoval, "City Hall May Pull Plug on LinkNYC Owner over Missing Kiosks—and $75M Owed," The City, March 3, 2020, https://www.thecity.nyc/2020/3/3/21210474/city-hall-may-pull-plug-on-linknyc-owner-over-missing-kiosks-and-75m-owed.

42 **Doctoroff imagined the company:** Intersection, "Titan and Control Group Become Intersection," PRNewswire, September 16, 2015, https://www.prnewswire.com/news-releases/titan-and-control-group-become-intersection-300144002.html.

Chapter 4: Is Anybody Home?

44 **"street mesh":** Much of this chapter details ideas that Sidewalk Labs described in its confidential May 2016 Yellow Book, which I obtained a copy of.

45 **Alphabet's massive cash reserves:** This valuation comparison is of publicly available market capitalizations for Alphabet, Ford and General Motors from the first week of February 2016.

46 **"like any other product":** Daniel L. Doctoroff, *Greater Than Ever: New York's Big Comeback* (New York: PublicAffairs, 2017), 68.

47 **spending $800 to get one:** Doctoroff, 12.

47 **the dome was a crucial component:** Doctoroff said in an interview that "there were no expectations that we would build a dome," despite numerous references to a dome in Javelin documents and later in the Yellow Book, which Doctoroff signed off on.

47 **running joke inside the company:** Cory Weinberg, "Sidewalk Labs' Grand Vision Meets Reality in Toronto," The Information, November 29, 2018, https://www.theinformation.com/articles/sidewalk-labs-grand-vision-meets-reality-in-toronto; and *Saturday Night Live*, "The Bubble," posted November 20, 2016, https://www.youtube.com/watch?v=vKOb-kmOgpI.

47 **1935 Frank Lloyd Wright exhibition:** Jennifer Gray, "Reading Broadacre," *The Whirling Arrow* (blog), Frank Lloyd Wright Foundation, October 1, 2018, https://franklloydwright.org/reading-broadacre/.

49 **"cat and mouse game":** For this and subsequent quotes, see Eric Schmidt and Jared Cohen, *The New Digital Age: Reshaping the Future of People, Nations and Business* (New York: Alfred A. Knopf, 2013), 119.

50 **as much as $83 billion in revenue:** Weinberg, "Sidewalk Labs' Grand Vision."

55 **as much as $7.4 billion in commitments:** This is the sum of the maximum estimated values of individual capital expenditures for a "full build" of a Project Sidewalk community as outlined at the end of the Yellow Book.

55 **"unanimous excitement":** In an interview, Doctoroff repeatedly downplayed the significance of the Yellow Book, calling it "just a feasibility study," despite the extremely detailed plans it outlined, including cost analyses for building in Denver, Detroit and Alameda and extensive plans for how a bidding process for cities would work.

56 **buildings on wheels:** This idea was echoed in Cory Weinberg, "Sidewalk Labs Chose Toronto Despite Pushback from Larry Page," The Information, October 19, 2017, https://www.theinformation.com/articles/sidewalk-labs-chose-toronto-despite-pushback-from-larry-page.

57 **"My new CEO and I":** Ontario auditor general Bonnie Lysyk's 2018 annual report, December 5, 2018, 689, https://www.auditor.on.ca/en/content/annualreports/arreports/en18/v1_315en18.pdf.

Chapter 5: Shoreline

58 **error by an influential local historian:** Jeff Gray, "A Defining Moment for Tkaronto," *Globe and Mail*, October 17, 2003, https://www.theglobeandmail.com/news/national/a-defining-moment-for-tkaronto/article18432992/.

58 **the city's official version:** City of Toronto, "Natives and Newcomers, 1600–1793," undated, https://www.toronto.ca/explore-enjoy/history

-art-culture/museums/virtual-exhibits/history-of-toronto/natives-and-newcomers-1600-1793/ (accessed March 10, 2022).

58 **the cash and goods were an unrelated gift:** Mississaugas of the Credit First Nation, "Toronto Purchase Specific Claim: Arriving at an Agreement," undated pamphlet uploaded in April 2017, http://mncfn.ca/wp-content/uploads/2017/04/MNCFN-Toronto-Purchase-Specific-Claim-Arriving-at-an-Agreement.pdf.

58 **a C$145 million settlement:** Mississaugas of the Credit First Nation, "The Toronto Purchase Treaty No. 13 (1805)," May 28, 2017, http://mncfn.ca/torontopurchase/.

59 **drew its curtains on Sundays:** Carola Vyhnak, "Once Upon a City: When Tobogganing Was a Crime in Toronto," *Toronto Star*, November 26, 2015, https://www.thestar.com/yourtoronto/once-upon-a-city-archives/2015/11/26/once-upon-a-city-when-tobogganing-was-a-crime-in-toronto.html.

59 **could be joined by another half million:** T.O. Health Check, "Population Demographics," 2019, 7, https://www.toronto.ca/wp-content/uploads/2019/11/99b4-TOHealthCheck_2019Chapter1.pdf.

59 **a slogan lifted from Detroit:** Adrian Lee, "Toronto Has a Self-Esteem Problem," *Maclean's*, October 29, 2015, https://www.macleans.ca/society/toronto-vs-everybody-not-so-much/.

60 **With shipping activity on the decline:** Gene Desfor and Jennefer Laidley, "Introduction," in *Reshaping Toronto's Waterfront*, ed. Gene Desfor and Jennefer Laidley (Toronto: University of Toronto Press, 2011), 13.

60 **it amplified concerns:** Jennefer Laidley, "Creating an Environment for Change: The 'Ecosystem Approach' and the Olympics on Toronto's Waterfront," in *Reshaping Toronto's Waterfront*, ed. Gene Desfor and Jennefer Laidley (Toronto: University of Toronto Press, 2011), 208.

61 **The task force decided:** Robert A. Fung and the Toronto Waterfront Revitalization Task Force, "Our Toronto Waterfront: Gateway to the New Canada," 2000, https://www.toronto.ca/wp-content/uploads/2017/11/91f5-torontow.pdf.

61 **governments pledged c$1.5 billion:** Jennifer Lewington, "Waterfront to Get $1.5-billion," *Globe and Mail*, October 20, 2000, https://www.theglobeandmail.com/news/national/waterfront-to-get-15-billion/article25473588/.

61 **The agency had been modelled:** Gabriel Eidelman, "Who's in Charge? Jurisdictional Gridlock and the Genesis of Waterfront Toronto," in *Reshaping Toronto's Waterfront*, ed. Gene Desfor and Jennefer Laidley (Toronto: University of Toronto Press, 2011), 269, 277.

63 **"vastly superior":** Christopher Hume, "Waterfront Toronto's City-Building CEO Steps Down: Hume," *Toronto Star*, September 17, 2015, https://www.thestar.com/news/gta/2015/09/17/waterfront-torontos-city-building-ceo-steps-down-hume.html.

63 **could catch only glimpses:** "What Brought Renowned Urban Planner Will Fleissig to Toronto?" CBC News, *Metro Morning*, April 7, 2016, https://www.cbc.ca/news/canada/toronto/programs/metromorning/will-fleissig-1.3523833.

64 **"one percent of the one percent":** "What Brought Renowned."

64 **"The hair on the back of my neck":** "What Brought Renowned."

66 **a decade earlier for c$68 million:** Ontario auditor general Bonnie Lysyk's 2018 annual report, December 5, 2018, 662, https://www.auditor.on.ca/en/content/annualreports/arreports/en18/v1_315en18.pdf.

69 **front-page photo from 1971:** The photo from the May 24, 1971, *Telegram* front page was reproduced to accompany a story by Jamie Bradburn, "'World's Most Exciting Island': The Triumphs and Mishaps of Ontario Place's Opening," TVO, May 19, 2021, https://www.tvo.org/article/worlds-most-exciting-island-the-triumphs-and-mishaps-of-ontario-places-opening.

69 **city as a lab for solving issues:** Jillian D'Onfro, "Would Google Build a City from Scratch? 'Great Idea,'" Insider, April 5, 2016, https://www.businessinsider.com/google-dan-doctoroff-sidewalk-labs-building-a-city-from-scratch-2016-4.

69 **Smart-city experts warned:** Mark Harris, "Secretive Alphabet Division Funded by Google Aims to Fix Public Transit in US," *Guardian*, June 27, 2016, https://www.theguardian.com/technology

/2016/jun/27/google-flow-sidewalk-labs-columbus-ohio-parking-transit.

70 **"Google has purportedly told":** Lysyk, 2018 annual report, 706.

Chapter 6: Rose-Coloured Glasses

75 **Oxford didn't seem to be interested:** Oxford spokesperson Daniel O'Donnell told me by e-mail, "At that time, Oxford's focus was on building out our own development pipeline in Toronto, representing more than C$10-billion in capital requirements, and diversifying our global real estate holdings to increase our weighting outside of Canada. Adding greater exposure to Toronto via another substantial development project, a market in which we have tens of billions of dollars of capital already at work, just didn't make sense for our business."

76 **They'd talked to fifty-two organizations:** Kristina Verner, memo to Quayside Committee of Waterfront Board of Directors, December 7, 2017, Appendix A. https://quaysideto.ca/wp-content/uploads/2019/04/MEMO-Quayside-Outreach-and-Market-Sounding-December-7-2017-REDACTED.pdf. (accessed April 2, 2022.)

76 **Sidewalk was a no-show:** Verner, Appendix B.

76 **"innovation and funding partner":** Waterfront Toronto, "Request for Proposals: Innovation and Funding Partner for the Quayside Development Opportunity," March 17, 2017, https://quaysideto.ca/wp-content/uploads/2019/04/Waterfront-Toronto-Request-for-Proposals-March-17-2017.pdf. (accessed April 2, 2022.)

78 **"21st-century city":** Alex Bozikovic, "Waterfront of Dreams," *Globe and Mail*, March 17, 2017, https://www.theglobeandmail.com/news/toronto/waterfront-of-dreams-will-fleissig-wants-to-redefine-how-the-gta-approaches-lakeontario/article34340855/.

78 **information session for potential bidders:** Verner, memo to Quayside Committee, 2.

78 **putting aside, at least temporarily:** Cory Weinberg, "Sidewalk Labs Chose Toronto Despite Pushback from Larry Page," The Information, October 19, 2017, https://www.theinformation.com/articles/sidewalk-labs-chose-toronto-despite-pushback-from-larry-page.

79 **almost all of Sidewalk's resources:** Speaking to Canadian Parliament's Standing Committee on Access to Information, Privacy and Ethics on April 2, 2019, Dan Doctoroff said, "We decided to devote more or less the full resources of Sidewalk Labs toward our response" to the Quayside RFP.

80 **"smart disposal chain":** Sidewalk Labs, Vision sections of RFP submission, made public October 17, 2017, Appendix, 92, https://storage.googleapis.com/sidewalk-toronto-ca/wp-content/uploads/2017/10/13210553/Sidewalk-Labs-Vision-Sections-of-RFP-Submission.pdf.

81 **sought to make a regulatory case:** Sidewalk Labs, Vision, Appendix, 169.

81 **cost of living down 14 percent:** Sidewalk Labs, Vision, 32.

82 **not just a misalignment:** Of the controversies that ensued about the amount of land Sidewalk wanted, Dan Doctoroff told me in November 2021: "I think political pressure was built over time. . . . There was no misunderstanding at the very beginning. Otherwise, why would they have selected us when we mentioned [the eastern waterfront] in the RFP [response] three hundred times?"

83 **"What happens in Quayside":** Sidewalk Labs, Vision, 15.

83 **"The Eastern Waterfront is critically important":** After my November 2021 interview with Doctoroff, Sidewalk provided me with this portion of the RFP response that had not previously been made public.

85 **nearly C$400 million apiece:** Chris Fox, "Port Lands Flood Protection Plan Gets Nearly $1.2B Investment," *CTV News Toronto*, June 28, 2017, https://toronto.ctvnews.ca/port-lands-flood-protection-plan-gets-nearly-1-2b-investment-1.3480471. (John Tory's quote comes from the TV news story embedded in the article, reported by Paul Bliss.)

86 **until at least 2028:** Ontario auditor general Bonnie Lysyk's 2018 annual report, December 5, 2018, 649, https://www.auditor.on.ca/en/content/annualreports/arreports/en18/v1_315en18.pdf.

87 **leaked details of Sidewalk's application:** Mark Bergen, "Alphabet's

Sidewalk Labs Eyes Toronto for Its Digital City," Bloomberg News, May 8, 2017, https://www.bloomberg.com/news/articles/2017-05-08/alphabet-s-sidewalk-labs-eyes-toronto-for-its-digital-city.

88 **Waterfront began extensive:** These meetings were "commercially confidential," and though Waterfront staff described some details of each meeting in interviews, they did not disclose bidder names, which I found independently.

Chapter 7: Limelight

90 **years-long quest to minimize:** Tom Huddleston Jr., "Amazon Had to Pay Federal Income Taxes for the First Time Since 2016—Here's How Much," CNBC, February 4, 2020, https://www.cnbc.com/2020/02/04/amazon-had-to-pay-federal-income-taxes-for-the-first-time-since-2016.html.

91 **"quite far along":** Leslie Hook, "Alphabet Looks for Land to Build Experimental City," *Financial Times*, September 19, 2017, https://www.ft.com/content/22b45326-9d47-11e7-9a86-4d5a475ba4c5.

91 **He e-mailed colleagues:** These e-mail exchanges were obtained through a federal access-to-information request.

92 **"nearing a deal":** David George-Cosh and Eliot Brown, "Google Parent Nears Deal to Build Its Vision of a City in Toronto," *Wall Street Journal*, October 4, 2017, https://www.wsj.com/articles/alphabets-city-building-unit-nears-development-deal-in-toronto-1507142561.

92 **confirmed the scoop:** Alex Bozikovic and Jeff Gray, "Google's Sidewalk Labs Preferred Partner on Toronto Waterfront Development," *Globe and Mail*, October 4, 2017, https://www.theglobeandmail.com/news/toronto/googles-sidewalk-labs-preferred-partner-toronto-waterfront-development/article36491720/.

96 **lost money on a project only once:** Lisa Rochon, "A Rare Kind of Developer," *Globe and Mail,* December 7, 2006, https://www.theglobeandmail.com/arts/a-rare-kind-of-developer/article1329699/.

96 **"Imagine if you could create":** Transcript from the November 3, 2015, meeting of the Legislative Assembly of Ontario Standing Committee

on Government Agencies, https://www.ola.org/en/legislative-business/committees/government-agencies/parliament-41/transcripts/committee-transcript-2015-nov-03#P62_3811.

96 **getting high-level briefings:** Draft minutes of Waterfront Toronto's June 21, 2017, Investment and Real Estate Committee meeting, which are no longer online. In this meeting, Julie Di Lorenzo declared she had personal relationships with two members of one of the Quayside bids. I asked her about this, and she said she could only remember one: Peter Gilgan of Mattamy Group, who was part of the Made-in-Canada consortium and with whom she worked on a hospital fundraising initiative. Asked if she believed this influenced her opposition to the Sidewalk Labs bid, she told me, "Absolutely not. My reasons were outlined in any and all correspondence. I have never done business with Peter. It would be unfair and unfortunate to consider that."

98 **"urged—strongly":** Ontario auditor general Bonnie Lysyk's 2018 annual report, December 5, 2018, 691, https://www.auditor.on.ca/en/content/annualreports/arreports/en18/v1_315en18.pdf.

100 **called the meeting to order:** Details from this October 16, 2017, scene come from a mix of recollections from people who were present and draft minutes that Waterfront posted online, but deleted in early 2022 as part of a website migration. (I retained a copy.) Though Julie Di Lorenzo spoke to me for this book, she declined to share details from confidential events, including the in-camera portion of this meeting.

102 **vast holdings in a blind trust:** Robert Fife and Steven Chase, "Bill Morneau Didn't Place Assets in Blind Trust, Raising Conflict-of-Interest Risk," *Globe and Mail*, October 15, 2017, https://www.theglobeandmail.com/news/politics/finance-minister-bill-morneau-I-place-assets-in-blind-trust-raising-conflict-of-interest-risk/article36596635/.

103 **"Gord was my friend":** Charlie Smith, "Emotionally Shaken Justin Trudeau Delivers Tearful Tribute to Gord Downie," *Georgia Straight*, October 18, 2017, https://www.straight.com/music/982971/emotionally-shaken-justin-trudeau-delivers-tearful-tribute-gord-downie. (In what may have been a hedge, Trudeau followed this quote by saying "Gord was everyone's friend.")

103 **"What's going on up there ain't good":** "'Trained Our Entire Lives to Ignore': Gord Downie's Call to Action for Indigenous in the North," CBC News, August 21, 2016, https://www.cbc.ca/news/canada/north/gord-downie-praises-justin-trudeau-aboriginal-people-1.3729996.

103 **in court, denied responsibility:** Jorge Barrera, "Ottawa Says It's Not Liable for Cultural Damage Caused by Kamloops Residential School: Court Documents," CBC News, June 2, 2021, https://www.cbc.ca/news/indigenous/reparations-residential-school-1.6050501.

104 **deviate from the speech:** Speech obtained via access-to-information request.

104 **"Dan, I've been a big fan":** Sidewalk Labs, "Announcing Sidewalk Toronto: Press Conference Live Stream," streamed live on October 17, 2017, https://www.youtube.com/watch?v=A_yg_BsJy_o.

Chapter 8: Northern Touch

107 **On the final day of 2005:** This scene is adapted from Daniel L. Doctoroff, *Greater Than Ever: New York's Big Comeback* (New York: PublicAffairs, 2017), 286–89.

107 ***Failure is only failure*:** Doctoroff, 288.

107 **his $35 million Olympic bid:** Lynn Zinser, "London Wins 2012 Olympics; New York Lags," *New York Times*, July 7, 2005, https://www.nytimes.com/2005/07/07/sports/othersporontondon-wins-2012-olympics-new-york-lags.html.

108 **whether a country was "first world":** Doctoroff, *Greater Than Ever*, 213.

108 **hoovered up Peter C. Newman's 1975 book:** As Jim Balsillie's participation was limited for this book—he granted a roughly one-hour interview and answered fact-checking questions by e-mail—numerous details of his life and career prior to his exit from Research in Motion are adapted from Jacquie McNish and Sean Silcoff's *Losing the Signal: The Spectacular Rise and Fall of BlackBerry* (Toronto: HarperCollins, 2015). This is from page 12.

109 **"Jim believed everyone was out to kill us":** McNish and Silcoff, *Losing the Signal*, 43.

109 **"maximize adoption by minimizing complexity":** McNish and Silcoff, 50.

110 **RIM's revenue began doubling annually:** McNish and Silcoff, 53.

110 **"there is absolutely no shortage":** John A. McMahon, "Canada Has No Shortage of Internet Venture Capital," *Globe and Mail*, February 17, 2000, https://www.theglobeandmail.com/report-on-business/canada-has-no-shortage-of-internet-venture-capital/article766072/.

110 **fate of North Vancouver's ClipClop.com:** Derek DeCloet, "MoosePasture.com," *Canadian Business* 73, no. 5 (March 20, 2000): 57–58.

110 **"online mall":** July 2002 filing with the U.S. Securities and Exchange Commission, https://www.sec.gov/Archives/edgar/data/886184/000108503702000345/ww10qmar.htm.

111 **three-quarters of American internet traffic:** "Northern Light," *Forbes*, May 17, 1999, https://www.forbes.com/forbes/1999/0517/6310071a.html.

111 **by a factor of ten:** "Nortel Briefly Loses Title As Canada's Biggest Company," CBC News, August 13, 2001, https://www.cbc.ca/news/busiorontortel-briefly-loses-title-as-canada-s-biggest-company-1.269268.

111 **Nasdaq's telecom index fell 62 percent:** Elise A. Couper, John P. Hejkal and Alexander L. Wolman, "Boom and Bust in Telecommunications," Federal Reserve Bank of Richmond *Economic Quarterly* 89, no. 4 (Fall 2003), https://www.richmondfed.org/~/media/richmondfedorg/publications/research/economic_quarterly/2003/fall/pdf/wolman.pdf.

111 **compensation packages worth hundreds of millions:** James Bagnall, *100 Days: The Rush to Judgement That Killed Nortel* (Ottawa: Ottawa Citizen, 2013), 23.

111 **"one of the deepest and fastest downsizings":** Bagnall, 28.

111 **bounced back from its 75 percent share price collapse:** McNish and Silcoff, *Losing the Signal*, 89.

112 **didn't want to be "associated with him":** McNish and Silcoff, 159.

112 **more than $250 million a year:** McNish and Silcoff, 126.

112 **a $612.5 million payment to NTP:** Mark Heinzl and Amol Sharma, "RIM to Pay NTP $612.5 Million to Settle BlackBerry Patent Suit," *Wall Street Journal*, March 4, 2006, https://www.wsj.com/articles/SB114142276287788965.

113 **one of the biggest-ever sanctions:** McNish and Silcoff, *Losing the Signal*, 154.

113 **"flirted with valuations near a billion dollars":** Sean Silcoff and David Ebner, "With Offers Below Expectations, Hootsuite Abandons Auction," *Globe and Mail*, January 17, 2019, https://www.theglobeandmail.com/business/streetwise/article-with-offers-below-expectations-hootsuite-abandons-auction-process/.

114 **Tech grew faster than any other sector:** Matthew A. Winkler, "Trudeau Has Canada's Economy Humming," Bloomberg, October 17, 2019, https://www.bloomberg.com/opinion/articles/2019-10rontoudeau-has-canada-s-economy-humming.

Chapter 9: That Don't Impress Me Much

119 **"Trojan horse":** This scene is adapted from a recording and transcript of a round table called "Building Smarter Cities" on TVO's *The Agenda* from November 13, 2017, https://www.tvo.org/transcript/2472808/building-smarter-cities and https://www.tvo.org/video/building-smarter-cities.

120 **"cycle of ongoing improvement":** Sidewalk Labs, Vision sections of RFP submission, made public October 17, 2017, Appendix, 12, https://storage.googleapis.com/sidewalk-toronto-ca/wp-content/uploads/2017/10/13210553/Sidewalk-Labs-Vision-Sections-of-RFP-Submission.pdf. Bianca Wylie discussed this in a January 20, 2018 blog post: https://medium.com/@biancawylie/sidewalk-toronto-the-plan-for-r-d-with-our-civic-data-finally-comes-into-focus-f6aa3bd3e62.

126 **characterizing the Toronto project:** At the October 2017 project announcement, Dan Doctoroff said, "Our journey here . . . ends with us creating a place that creates a new model for urban life in the twenty-first century." A few minutes later, Will Fleissig said of their

proposed community: "Dan understands that it has to come out of the conversation [with Waterfront]—that the initial ideas that his team put together are just that."

126 **a newly public report:** Report by Deputy City Manager John Livey to Toronto's Executive Committee regarding Sidewalk Toronto, January 16, 2018, https://www.toronto.ca/legdocs/mmis/2018/ex/bgrd/backgroundfile-110745.pdf.

127 **"This is not an urban-planning project":** Toronto City Council, "Executive Committee—January 24, 2018—Part 2 of 2," streamed live on January 24, 2018, https://www.youtube.com/watch?v=y1SUwH6FoBw. The Quayside discussion begins at about the 5:56:00 mark.

129 **showed up in the *National Post* grimacing:** Natalie Alcoba, "Councillor Outraged after Waterfront Toronto 'Secretly' Spends $946,000 on Two Large Rocks and 36 Pink Umbrellas," *National Post*, January 24, 2015, https://nationalpost.com/orontoronto/councillor-outraged-after-waterfront-toronto-secretly-spends-946000-on-two-large-rocks-and-36-pink-umbrellas.

Chapter 10: In Undertow

131 **"We were very clear":** For this and the ensuing e-mail discussion, see Amanda Roth, "After Three High-Profile Resignations, Waterfront Toronto and Sidewalk Labs Attempt to Regain Control of Quayside Project," The Logic, August 1, 2018, https://thelogic.co/news/the-big-read/after-three-high-profile-resignations-waterfront-toronto-and-sidewalk-labs-attempt-to-regain-control-of-quayside-project/.

132 and 133 **"overplayed their hand a bit"** and **"answering a question that wasn't on the exam":** John Lorinc, "Lorinc: Sidewalk Labs Has Hit a Roadblock," *Spacing*, December 11, 2017, https://spacinorontoronto/2017/12/11/lorinc-sidewalk-labs-has-hit-a-roadblock/.

134 **"The solutions advanced":** "New York City: Sidewalk Labs & Grand Central Tech," *OnBoard* (Toronto Region Board of Trade magazine), Winter 2018, 24, https://wtctoronto.com/wp-content/uploads/2018/11/OnBoard-2018Winter_GoingGlobal_Web.pdf.

135 **co-wrote an op-ed:** Daniel L. Doctoroff and Will Fleissig, "The Neighbourhood of the Future Starts with Your Ideas," *Toronto Star*, November 1, 2017, https://www.thestar.com/opinion/commentary/2017/11/01/the-neighbourhood-of-the-future-starts-with-your-ideas.html.

135 **"It is not about doing cool things":** Ria Lupton, "#GoNorth17 AI Discussion Walks the Line between the Power of Technology and Its Social Impact," BetaKit, November 7, 2017, https://betakit.com/gonorth17-ai-discussion-walks-the-line-between-the-power-of-technology-and-its-social-impact/.

136 **"What is affordable for rich people?":** John Rieti, "Toronto Waterfront Won't Be a 'Tech Enclave,' Sidewalk Labs CEO Vows," CBC News, November 2, 2017, https://www.cbc.ca/orontaorontoronto/sidewalk-labs-meeting-1.4383359.

136 **bragged about lowballing:** Daniel L. Doctoroff, *Greater Than Ever: New York's Big Comeback* (New York: PublicAffairs, 2017), 75.

136 **there was even a line:** Nabeel Ahmed, "The City vs. Big Tech," *Briarpatch*, July 2, 2019, https://briarpatchmagazine.com/articles/view/the-city-vs.-big-tech.

136 **"the default platform":** Evgeny Morozov, "Google's Plan to Revolutionise Cities Is a Takeover in All but Name," *Guardian*, October 22, 2017, https://www.theguardian.com/technology/2017/oct/21/google-urban-cities-planning-data.

136 **"we're not going to gather up":** Mols Sauter, "Google's Guinea-Pig City," *Atlantic*, February 13, 2018, https://www.theatlantic.com/technology/archive/2018/02/googles-guinea-pig-city/552932/.

137 **"We have the right to walk away":** Rieti, "Toronto Waterfront."

137 **"harvested data on millions":** Harry Davies, "Ted Cruz Using Firm That Harvested Data on Millions of Unwitting Facebook Users," *Guardian*, December 11, 2015, https://www.theguardian.com/us-news/2015/dec/11/senator-ted-cruz-president-campaign-facebook-user-data.

138 **ties to a variety of personality quizzes:** McKenzie Funk, "Cambridge Analytica and the Secret Agenda of a Facebook Quiz," *New York*

Times, November 19, 2016, https://www.nytimes.com/2016/11/20/opiorontoeridge-analytica-facebook-quiz.html.

138 **"one of the largest data leaks":** Matthew Rosenberg, Nicholas Confessore and Carole Cadwalladr, "How Trump Consultants Exploited the Facebook Data of Millions," *New York Times*, March 17, 2018, https://www.nytimes.com/2018/03/17/us/poliorontoeridge-analytica-trump-campaign.html.

138 **fine Facebook $5 billion:** Makena Kelly, "FTC Hits Facebook with $5 Billion Fine and New Privacy Checks," The Verge, July 24, 2019, https://www.theverge.com/2019/7/24/20707013/ftc-facebook-settlement-data-cambridge-analytica-penalty-privacy-punishment-5-billion.

138 **since buying DoubleClick:** Liat Clark, "Google's Ad-Tracking Just Got More Intrusive. Here's How to Opt Out," *Wired*, October 24, 2016, https://www.wired.co.uk/article/google-ad-tracking.

Chapter 11: The Other Shoe

142 **speculation among project critics:** Mariana Valverde, "Mystery on the Waterfront: How the 'Smart City' Allure Led a Major Public Agency in Toronto into a Reckless Deal with Big Tech," Centre for Free Expression, Ryerson University, December 3, 2018, https://cfe.ryerson.ca/blog/2018/12/mystery-waterfront-how-smart-city-allure-led-major-public-agency-toronto-reckless-deal.

142 **"model in public-private partnerships":** UC Davis West Village home page, last updated July 10, 2021, https://westvillage.ucdavis.edu/.

145 **alleged that Lasher was "bigoted"** and **"I was part of an ugly campaign":** Michael Gartland, "Sharpton Fuming with Stringer Campaign Hire, 'Bigoted Strategist,' Says Rev.," *New York Daily News*, December 11, 2020, https://www.nydailynews.com/news/politics/new-york-elections-government/ny-nyc-mayoral-race-stringer-sharpton-lasher-20201211-lt73qmt2gva55hwtkunk2bpvje-story.html.

145 **At one 2018 event:** Brian Barth, "The fight against Google's smart city," *Washington Post*, August 8, 2018, https://www.washingtonpost.com/news/theworldpost/wp/2018/08/08/sidewalk-labs/.

145 **"NEVER" wanted taxation powers:** Tweet by Micah Lasher on July 3, 2019, https://twitter.com/MicahLasher/status/1146416391697944576. Sidewalk pointed out this tweet was about its 2019 draft master plan, but the definition of the word "never" includes the entirety of time.

147 **"the best mayor Toronto never had":** Sandra Martin, "Toronto Visionary David Pecaut Succumbs to Cancer," *Globe and Mail*, December 14, 2009, https://www.theglobeandmail.com/orontororontoronto-visionary-david-pecaut-succumbs-to-cancer/article1209546/.

149 **Even the mayor, John Tory, weighed in:** In an interview, Tory cautioned that he did not proactively tell the board to do anything, including about Will Fleissig's 360 review, because he respected the board's tri-governmental mandate. But he did not deny giving advice to the board, when asked, about working in Toronto's best interests, including around Fleissig's departure.

150 **"innovative vision":** "Media Statement—Issued on Behalf of Helen Burstyn, C.M., Chair of the Waterfront Toronto Board," Waterfront Toronto, July 4, 2018, archived via WayBack Machine, https://web.archive.org/web/20180725211845/https://www.waterfrontoronto.ca/nbe/portal/waterfront/Home/waterfronthome/newsroom/newsarchive/news/2018/july/media+statement+-+issued+on+behalf+of+helen+burstyn%2C+c.m.%2C+chair+of+the+waterfront+toronto+board.

Chapter 12: Left and Leaving

153 **"smart," data-hungry sensors:** Sidewalk Labs, Vision sections of RFP submission, made public October 17, 2017, Appendix, 97. https://storage.googleapis.com/sidewalk-toronto-ca/wp-content/uploads/2017/10/13210553/Sidewalk-Labs-Vision-Sections-of-RFP-Submission.pdf.

154 **examples of the kind of data:** Sidewalk Labs, "Sidewalk Toronto Responsible Data Use," June 7, 2018, https://quaysideto.ca/wp-content/uploads/2019/04/DSAP-Responsible-Data-Use-Presentation-June-7-2018.pdf. (accessed April 2, 2022.)

158 **"clear, consistent and coordinated":** "Plan Development Agreement

between Toronto Waterfront Revitalization Corporation and Sidewalk Labs LLC," July 31, 2018, 51, https://storage.googleapis.com/sidewalk-toronto-ca/wp-content/uploads/2019/06/13210552/Plan-Development-Agreement_July312018_Fully-Executed.pdf.

158 **"Circumstances have prevented me":** Julie Di Lorenzo, resignation letter, July 30, 2018, uploaded to Scribd by the *Toronto Star*, https://www.scribd.com/document/385327532/Resignation-Letter.

159 **"Any proposed options at scale":** "Plan Development Agreement," 32.

159 **"Both Parties recognize the value":** "Plan Development Agreement," 3.

160 **"ensure support" among "key constituents":** "Plan Development Agreement," 34.

160 **"goes sideways":** Jeff Gray and Josh O'Kane, "Waterfront Toronto, Sidewalk Labs Walk Back Plans in New Deal," *Globe and Mail*, July 31, 2018, https://www.theglobeandmail.com/canada/toronto/article-new-deal-between-waterfront-toronto-and-sidewalk-labs-walks-back-some/.

160 **"citizen-centered"** and **"foster monopolies":** "Plan Development Agreement," 47, 49.

160 **"Method for node ranking in a linked database":** Patent US6285999B1, first priority date January 10, 1997, https://patents.google.com/patent/US6285999B1/.

162 **"As it currently stands":** For this and quotes about subsequent July 2018 intellectual property concerns, see Josh O'Kane, "Sidewalk Labs' Toronto Deal Sparks Data, Innovation Concerns," *Globe and Mail*, August 1, 2018, https://www.theglobeandmail.com/business/technology/article-new-development-agreement-between-sidewalk-labs-and-waterfront-toronto/.

163 **"The taxpayer won't derive any benefit":** For this and subsequent quotes about the confidential contract, see Josh O'Kane and Alex Bozikovic, "Sidewalk Labs Taking Steps to Control Intellectual Property on Toronto's 'Smart City,' Document Shows," *Globe and Mail*, August 31, 2018, https://www.theglobeandmail.com/business/article-sidewalk-labs-taking-steps-to-control-intellectual-property-on-toronto/.

Chapter 13: Blue

167 **"It's way above my pay grade":** Fatima Syed, "Alphabet's Sidewalk Labs Was Secretly Considering Big Plans for Toronto Neighbourhood," Canada's National Observer, February 15, 2019, https://www.nationalobserver.com/2019/02/15/news/alphabets-sidewalk-labs-was-secretly-considering-big-plans-toronto-neighbourhood.

168 **exceeded eleven thousand:** Sidewalk Labs, "Sidewalk Toronto," last updated in 2021, https://www.sidewalklabs.com/toronto (accessed March 14, 2022).

171 **In one August 2018 discussion:** Details of this discussion did not come from Kurtis McBride, who spoke to me in more general terms about the future of urban technology and his time advising Waterfront Toronto.

173 **get his own work done more easily:** Daniel L. Doctoroff, *Greater Than Ever: New York's Big Comeback* (New York: PublicAffairs, 2017), 21.

173 **"As a result, the territorial jockeying":** Doctoroff, 156.

174 **"people would feel some ownership":** Doctoroff, 326.

174 **"As a thought leader in Toronto":** Bianca Wylie, "Sidewalk Toronto: A Brazen and Ongoing Corporate Hijack of Democratic Process," Medium, October 8, 2018, https://biancawylie.medium.com/sidewalk-toronto-a-brazen-and-ongoing-corporate-hijack-of-democratic-process-a96a1253fb2b.

174 **"Sidewalk Labs is using":** Wylie, "Sidewalk Toronto."

175 **called Toronto his "second home":** Sidewalk Labs provided Doctoroff's script to reporters, but did not allow us to attend the meeting.

175 **"parcels of land that have laid fallow":** Doctoroff publicly described the Port Lands as "fallow" several times, but this was not quite true. The urban planner Gil Meslin laid out why in a tweet thread in May 2021: https://twitter.com/g_meslin/status/1391812669846474753

Chapter 14: In Too Deep

177 **"cradle to grave":** Ann Cavoukian, "Privacy by Design: The 7 Foundational Principles," Privacy and Big Data Institute, Ryerson

University, undated, https://www.ryerson.ca/content/dam/pbdce/seven-foundational-principles/The-7-Foundational-Principles.pdf (accessed March 31, 2022).

178 **scan the footage:** Ali Al Shouk, "Dubai CCTV Cameras to Use AL, Face Recognition," *Gulf News*, January 27, 2018, https://gulfnews.com/uae/government/dubai-cctv-cameras-to-use-ai-face-recognition-1.2163726.

178 **more than twenty million images daily:** Robert Muggah and Greg Walton, "'Smart' Cities Are Surveilled Cities," *Foreign Policy*, April 17, 2021, https://foreignpolicy.com/2021/04/17/smart-cities-surveillance-privacy-digital-threats-internet-of-things-5g/.

178 **"smart cities of surveillance":** Ann Cavoukian repeated variations of this line to numerous media outlets, but this quote in particular comes from an interview for this book.

179 **pressed Sidewalk to add more details:** Sidewalk Toronto, Data Governance Working Group, summary of Meeting 1: May 8, 2018 (page 3), https://quaysideto.ca/wp-content/uploads/2019/05/Sidewalk-Toronto-Data-Governance-WG-Notes-Final.pdf. (accessed April 2, 2022.)

180 **only Google refused to comply** and **"honey pot":** Maria Godoy, "Google Records Subpoena Raises Privacy Fears," NPR, January 20, 2006, https://www.npr.org/templates/story/story.php?storyId=5165854.

180 **many concerns about privacy breaches:** Steven Levy, *In the Plex: How Google Thinks, Works, and Shapes Our Lives* (New York: Simon & Schuster, 2011), 337.

180 **"The question of whether Larry":** Richard L. Brandt, *The Google Guys: Inside the Brilliant Minds of Google Founders Larry Page and Sergey Brin* (New York: Portfolio, 2011), 14.

180 **outlining how to delete data:** "Data Deletion on Google Cloud Platform," Google Cloud Whitepaper, September 2018, https://services.google.com/fh/files/misc/data_deletion_on_gcp.pdf.

180 **"the option to 'delete' data":** Eric Schmidt and Jared Cohen, *The New Digital Age: Reshaping the Future of People, Nations and Business* (New York: Alfred A. Knopf, 2013), 54.

181 **tracking people's movements:** Ryan Nakashima, "AP Exclusive: Google Tracks Your Movements, Like It or Not," Associated Press, August 13, 2018, https://apnews.com/article/north-america-science-technology-business-ap-top-news-828aefab64d4411bac257a07c1afoecb.

181 **"data will never be connected":** Chris Stokel-Walker, "Why Google Consuming DeepMind Health Is Scaring Privacy Experts," *Wired*, November 14, 2018, https://www.wired.co.uk/article/google-deepmind-nhs-health-data.

182 **"Data trusts can play a role":** Bianca Wylie and Sean Martin McDonald, "What Is a Data Trust?," Centre for International Governance Innovation, October 9, 2018, https://www.cigionline.org/articles/what-data-trust/.

182 **"Do people even know":** Marco Chown Oved, "Sidewalk Labs Use of Cellphone Data in Proposed U.S. Deal Raises Concern in Toronto," *Toronto Star*, October 12, 2018, https://www.thestar.com/news/gta/2018/10/12/sidewalk-labs-use-of-cellphone-data-in-proposed-us-deal-raises-concern-in-toronto.html.

183 **"Sidewalk Toronto has only one beneficiary":** Jim Balsillie, "Sidewalk Toronto Has Only One Beneficiary, and It Is Not Toronto," *Globe and Mail*, October 5, 2018, https://www.theglobeandmail.com/opinion/article-sidewalk-toronto-is-not-a-smart-city/.

183 **"arrogance and gas-lighting":** Bianca Wylie, August 19, 2018, post on Medium, https://biancawylie.medium.com/debrief-on-sidewalk-toronto-public-meeting-3-a-master-class-in-gaslighting-and-arrogance-c1c5dd918c16.

185 **"civic data trust":** Sidewalk Labs, "Digital Governance Proposals for DSAP Consultation," October 2018, uploaded to Scribd by Sameer Chhabra, https://www.scribd.com/document/390927208/Sidewalk-Toronto-Digital-Governance-Proposals-for-DSAP-Consultation.

186 **warned Harvey Dawson and her team:** Sidewalk Toronto, Data Governance Working Group, summary of Meeting 3: October 12, 2018 (page 11), https://quaysideto.ca/wp-content/uploads/2019/05/Sidewalk-Toronto-Data-Governance-WG-Notes-Final.pdf. (accessed April 2, 2022.)

189 **demanding more substance:** Josh O'Kane, "Waterfront Toronto Advisers Threaten Resignations Ahead of Key Sidewalk Labs Project Meeting," *Globe and Mail*, October 16, 2018, https://www.theglobeandmail.com/business/article-waterfront-advisers-threaten-resignations-ahead-of-key-sidewalk-labs/.

190 **"I think, to some extent":** Josh O'Kane, "Sidewalk Labs CEO Calls Some Criticisms of Toronto Quayside Project 'Unfair'," *Globe and Mail*, October 17, 2018, https://www.theglobeandmail.com/business/article-sidewalk-labs-ceo-calls-some-criticisms-of-toronto-quayside-project/.

191 **a "real loss" for the panel:** A third person, Darin Graham, quietly left the digital panel in late summer because of time constraints.

193 **hours before Cavoukian had planned:** Amanda Roth, "Privacy Expert Ann Cavoukian Resigns As Adviser to Sidewalk Labs," The Logic, October 19, 2018, https://thelogic.co/news/privacy-expert-ann-cavoukian-resigns-as-adviser-to-sidewalk-labs/.

193 **she tried to convey to them:** The story of Lauren Reid's resignation was recounted to me in an interview with Kristina Verner. Reid did not respond to multiple interview requests.

Chapter 15: Prelude to the Feud

195 **a photograph about a hundred pages in:** The Invisible Committee, *To Our Friends*, trans. Robert Hurley (South Pasadena: Semiotext(e), 2015), 100.

195 **Google had begun chartering:** Cari Spivack, "Worth the Drive," Google Blog, September 13, 2004, archived via WayBack Machine, https://web.archive.org/web/20060323060827/https://googleblog.blogspot.com/2004/09/worth-drive.html.

195 **more than a thousand private commuter buses:** Zara Stone, "Inside a Secretive $250 Million Private Transit System Just for Techies," *OneZero*, February 25, 2020, https://onezero.medium.com/only-the-elite-have-nice-commutes-in-silicon-valley-8b2761863925.

195 **accused of taking up space:** Douglas Rushkoff, *Throwing Rocks at the Google Bus: How Growth Became the Enemy of Prosperity* (New York: Portfolio, 2017), 2.

195 **donated $6.8 million:** "Google to Fund San Francisco's Free Muni for Youth Program," KQED News, February 27, 2014, https://www.kqed.org/news/127970/google-to-fund-san-franciscos-free-muni-for-youth-program.

197 **fastest-rising home prices:** Knight Frank Global Residential Cities Index, Q4 2017, https://content.knightfrank.com/research/1026/documents/en/global-residential-cities-index-q4-2017-5413.pdf.

197 **average rent doubled:** Emily Schultheis, "Berlin's Radical Plan to Stop Rocketing Rents," BBC Worklife, February 25, 2019, https://www.bbc.com/worklife/article/20190226-berlins-radical-plan-to-stop-rocketing-rents.

197 **built in 1926** and **fifteen-year lease:** "Case Study—Umspannwerk Kreuzberg," Avignon Capital, June 19, 2018, https://avignoncapital.com/news/case-study-umspannwerk-kreuzberg/.

199 **Grassroots opposition coalesced:** David Goldblatt, "People Have Always Hated the Olympics," *Time*, August 4, 2016, https://time.com/4419436/olympics-history-of-protests/.

199 **more than 275 buildings:** Paul Hockenos, *Berlin Calling: A Story of Anarchy, Music, the Wall, and the Birth of the New Berlin* (New York: New Press, 2017), 100–101.

202 **surging 77 percent:** Jimmy Im, "If You Bought a House in San Francisco 10 Years Ago, Here's How Much It Could Be Worth Now," *USA Today*, July 2, 2019, https://www.usatoday.com/story/money/2019/07/02/san-francisco-housing-market-how-much-prices-have-risen-since-2009/1632567001/.

202 **"likely contributing":** Chris Pangilinan, "Learning More About How Our Roads Are Used Today," Uber Under the Hood, August 5, 2019, https://medium.com/uber-under-the-hood/learning-more-about-how-our-roads-are-used-today-bde9e352e92c.

203 **supposed to catch errors:** Alex Davies and Aarian Marshall, "Feds Pin Uber Crash on Human Operator, Call for Better Rules," *Wired*, November 19, 2019, https://www.wired.com/story/feds-blame-uber-crash-on-human-driver-call-for-better-rules/.

203 **in historically Black communities:** Sean Collins, "How Tech Companies Are Driving Gentrification in Pittsburgh," Blavity, March 30, 2017, https://blavity.com/how-tech-companies-are-driving-gentrification-in-pittsburgh.

203 **probably erased thirty-one thousand units:** Tom Cardoso and Matt Lundy, "Airbnb Likely Removed 31,000 Homes from Canada's Rental Market, Study Finds," *Globe and Mail*, June 20, 2019, https://www.theglobeandmail.com/canada/article-airbnb-likely-removed-31000-homes-from-canadas-rental-market-study/.

203 **listings for furnished condos:** Matt Lundy, "Toronto Sees 52% Spike in Furnished Condo Listings, Pointing to Airbnb Conversions," *Globe and Mail*, July 15, 2020, https://www.theglobeandmail.com/business/article-surge-in-furnished-greater-toronto-area-condo-listings-suggests/.

203 **City and state governments offered:** Nolan Hicks and Max Jaeger, "Cuomo: $3B Giveaway to Amazon 'Costs Us Nothing,'" *New York Post*, November 13, 2018, https://nypost.com/2018/11/13/cuomo-3-billion-giveaway-to-amazon-costs-us-nothing/.

203 **out of jealousy of Elon Musk:** Spencer Soper, Matt Day and Henry Goldman, "Behind Amazon's HQ2 Fiasco: Jeff Bezos Was Jealous of Elon Musk," Bloomberg, February 3, 2020, https://www.bloomberg.com/news/articles/2020-02-03/amazon-s-hq2-fiasco-was-driven-by-bezos-envy-of-elon-musk.

203–4 **no U.S. federal income taxes:** Christopher Ingraham, "Amazon Paid No Federal Taxes on $11.2 Billion in Profits Last Year," *Washington Post*, February 16, 2019, https://www.washingtonpost.com/us-policy/2019/02/16/amazon-paid-no-federal-taxes-billion-profits-last-year/.

204 **half a billion dollars in incentives:** Ben Casselman, "A $2 Billion Question: Did New York and Virginia Overpay for Amazon?" *New York Times*, November 13, 2018, https://www.nytimes.com/2018/11/13/business/economy/amazon-hq2-va-long-island-city-incentives.html.

204 **"This was about crowdsourcing data":** Jason Silverstein, "How Will Amazon Use the Data It Got from Cities Bidding On Its HQ2?,"

CBS News, November 15, 2018, https://www.cbsnews.com/news/amazon-new-hq2-bidding-process-gave-the-company-priceless-data-on-cities-how-will-it-be-used/.

Chapter 16: Twist My Arm

210 **In the talking points:** Josh O'Kane, "Federal Staff Were Advised to Press Sidewalk Labs CEO on IP Concerns at October Dinner," *Globe and Mail*, January 17, 2019, https://www.theglobeandmail.com/business/article-federal-staff-advised-to-press-sidewalk-labs-toronto-project-at/.

210 **"fatal" for the German capital:** Christopher F. Schuetze, "Google Retreats from Berlin Plan Opposed by Local Groups," *New York Times*, October 25, 2018, https://www.nytimes.com/2018/10/25/world/europe/google-berlin-kreuzberg-campus.html.

211 **"epidemic proportions":** James C. Scott, *Seeing Like a State: How Certain Schemes to Improve the Human Condition Have Failed* (New Haven: Yale University Press, 1998), as excerpted by the *New York Times*: https://archive.nytimes.com/www.nytimes.com/books/first/s/scott-state.html.

211 **drew a line between "myopic":** Ben Green, *The Smart Enough City: Putting Technology in Its Place to Reclaim Our Urban Future* (Cambridge, MA: MIT Press, 2019), 145.

211 **"If we are building":** Shawn Micallef, "Ghosts of Spadina Expressway Haunt Us Still: Micallef," *Toronto Star*, October 6, 2017, https://www.thestar.com/news/gta/2017/10/06/ghosts-of-spadina-expressway-haunt-us-still-micallef.html.

211 **"market victories justify":** Vincent Mosco, *The Smart City in a Digital World* (Bingley, UK: Emerald Publishing, 2019), 20.

212 **"tech goggles":** Green, *Smart Enough City*, 1–4.

212 **"relentlessly targeted":** Dhruv Mehrotra for Gizmodo, Surya Mattu, Annie Gilbertson and Aaron Sankin, "How We Determined Crime Prediction Software Disproportionately Targeted Low-Income, Black, and Latino Neighborhoods," The Markup, December 2, 2021, https://themarkup.org/show-your-work/2021/12/02/how-we-determined

-crime-prediction-software-disproportionately-targeted-low-income-black-and-latino-neighborhoods.

212 **If cities want to invest:** Green, *Smart Enough City*, 77.

212 **"lay the foundations":** Daniel L. Doctoroff, "Reimagining Cities from the Internet Up," Sidewalk Talk, November 30, 2016, https://medium.com/sidewalk-talk/reimagining-cities-from-the-internet-up-5923d6be63ba.

Chapter 17: It's My Way

216 **of the 268 audits:** Amanda Roth, "Ontario Auditor General Conducting Audit of Waterfront Toronto," The Logic, October 4, 2018, https://thelogic.co/news/exclusive/ontario-auditor-general-conducting-audit-of-waterfront-toronto/.

217 **"Waterfront Toronto's communications":** Ontario auditor general Bonnie Lysyk's 2018 annual report, December 5, 2018, 648, https://www.auditor.on.ca/en/content/annualreports/arreports/en18/v1_315en18.pdf. The following details in the paragraphs that follow also come from Lysyk's report: **powerful enough:** 666; **the autonomy it did have:** 649; **Ontario government take the lead:** 695; **"an unfair and unequal advantage":** 689–90; **six weeks:** 690; **weren't "adequately" consulted:** 652; **how little time:** 690–91; **expand across the Port Lands:** 690; **bring in revenue:** 680; **exactly that:** 688; **information they'd shared:** 689; **"internal Waterfront Toronto email":** 689.

219 **did not make public:** When I asked the auditor general's office for more details about her investigation shortly after her report was published, a spokesperson told me that "we rarely provide details beyond those which have already appeared in our annual or special reports." As I researched this book several years later, I sought clarity on Waterfront Toronto executives' insistence that her office began intensely interviewing Quayside opponents and changing the focus of its questions late in the audit. I was told: "Information about Sidewalk Labs was provided to our Office in response to initial information requests made in early in 2018. More details were sought through additional information requests and numerous stakeholder interviews

to gain further understanding throughout the summer and well into the fall of that year."

219 **"headline hunting"**: Martin Regg Cohn, "What's Going On with Ontario's Auditor General? People Are Asking," *Toronto Star*, December 2, 2020, https://www.thestar.com/politics/political-opinion/2020/12/02/whats-going-on-with-ontarios-auditor-general-im-not-the-only-one-asking.html.

220 **"decisive action"**: Josh O'Kane, "Ontario Auditor-General Warns Waterfront Toronto to Slow Down Project with Google-Affiliate Sidewalk Labs," *Globe and Mail*, December 5, 2018, https://www.theglobeandmail.com/business/article-ontario-auditor-general-warns-waterfront-toronto-to-slow-down-project/.

221 **"the world's largest Ferris wheel"**: "Doug Ford's Waterfront Includes a Monorail, Ferris Wheel," CTV News, August 30, 2011, https://toronto.ctvnews.ca/doug-ford-s-waterfront-includes-a-monorail-ferris-wheel-1.690321.

221 **"We had fifteen people in the room"**: CBC Radio *Metro Morning* interview quoted in David Rider and Daniel Dale, "Doug Ford's Dream Waterfront? Ferris Wheel, Monorail and a Boat-In Hotel," *Toronto Star*, August 30, 2011, https://www.thestar.com/news/gta/2011/08/30/doug_fords_dream_waterfront_ferris_wheel_monorail_and_a_boatin_hotel.html.

222 **a mayor who wanted to run the province:** I cannot take credit for this observation, which I first heard through the urban affairs writer Shawn Micallef in a December 1, 2018, tweet: https://twitter.com/shawnmicallef/status/1068882094418468864. It is, however, worth mentioning that at numerous points in Robyn Doolittle's book *Crazy Town: The Rob Ford Story* (Toronto: Viking Canada, 2014), she reports that Doug Ford wanted to be premier.

222 **"six-million dollar man"**: Patricia Best, "Power Outage: Inside the Epic Battle between Doug Ford and Hydro One," *Globe and Mail*, *Report on Business* magazine, January 23, 2019, https://www.theglobeandmail.com/business/rob-magazine/article-power-outage-inside-the-epic-battle-between-doug-ford-and-hydro-one/.

222 **cost as much as C$9 million:** David Milstead, "Retiring Hydro One CEO Can Get $9-Million Compensation Package, Globe Analysis Reveals," *Globe and Mail*, July 12, 2018, https://www.theglobeandmail.com/business/article-retiring-hydro-one-ceo-can-get-8-million-to-compensate-for-stock/.

223 **He said yes:** Monte McNaughton's representatives did not respond to an interview request for this book.

224 **"The staff and the board members":** Josh O'Kane, "Ontario Government to Fire Three Waterfront Toronto Directors over Sidewalk Labs Partnership," *Globe and Mail*, December 6, 2018, https://www.theglobeandmail.com/canada/toronto/article-ontario-government-to-fire-three-waterfront-toronto-directors-over/.

225 **"I'm determined to remain":** Josh O'Kane, "Fired from Waterfront Toronto Board, Acting CEO Michael Nobrega Opens Up," *Globe and Mail*, December 10, 2018, https://www.theglobeandmail.com/business/commentary/article-michael-nobrega-the-acting-ceo-fired-from-waterfront-toronto-board/.

225 **"has never won the argument":** James McLeod, "Premier Ford Could 'Sabotage' Toronto Waterfront Plans Based On Sidewalk Labs Concerns: MP," *Financial Post*, January 3, 2019, https://financialpost.com/technology/premier-ford-could-sabotage-toronto-waterfront-plans-based-on-sidewalk-labs-concerns-mp.

226 **bridge seemingly impossible divides:** Catherine McIntyre, "The Man Who Stood Up to Sidewalk," The Logic, October 17, 2019, https://thelogic.co/news/the-big-read/the-man-who-stood-up-to-sidewalk/.

226 **was a mogul:** "Case Study: Building a Legacy: Lessons from Cadillac Fairview's First Leader," Rotman School of Management, University of Toronto, 2013, https://www.rotman.utoronto.ca/-/media/Files/Programs-and-Areas/CanadianBusinessHistory/CadillacFairview_Case_9%2027%2013_rev.pdf.

227 **said he wanted to resign:** Josh O'Kane and Rachelle Younglai, "Amid Smart-City Tumult, Tory Intervened to Prevent Another Waterfront Toronto Resignation," *Globe and Mail*, December 19, 2018,

https://www.theglobeandmail.com/business/article-amid-smart-city-tumult-mayor-tory-intervened-to-prevent-another/.

227 **"You cannot step off"** and **"Not only do I see":** McIntyre, "Man Who Stood Up." The rest of the quotes and details for this scene come from my own interviews, and I thank Catherine McIntyre for providing a jumping-off point to help recreate the conversation.

Chapter 18: Raise a Little Hell

228 **As Doctoroff has told the story:** For this and subsequent quotes, see Daniel L. Doctoroff, *Greater Than Ever: New York's Big Comeback* (New York: PublicAffairs, 2017), 159–62.

229 **"essentially pays for itself":** Doctoroff, 162.

229 **analysis by New School researchers:** Bridget Fisher and Flávia Leite, "The Cost of New York City's Hudson Yards Redevelopment Project," Schwartz Center for Economic Policy Analysis and Department of Economics, The New School for Social Research, Working Paper Series 2018-2, https://www.economicpolicyresearch.org/images/docs/research/political_economy/Cost_of_Hudson_Yards_WP_11.5.18.pdf.

229 **That month, he and his team:** Marco Chown Oved, "Google's Sidewalk Labs Plans Massive Expansion to Waterfront Vision," *Toronto Star*, February 14, 2019, https://www.thestar.com/news/gta/2019/02/14/googles-sidewalk-labs-plans-massive-expansion-to-waterfront-vision.html. I later acquired the slide deck described in the story.

230 **"only had an introductory conversation":** The Canada Infrastructure Bank confirmed this to my *Globe* colleague Bill Curry in March 2019, though the detail was trimmed from the story we later wrote: Josh O'Kane and Bill Curry, "Sidewalk Labs in Talks with Investors for Toronto Smart-City Infrastructure, Document Shows," *Globe and Mail*, March 11, 2019, https://www.theglobeandmail.com/business/article-sidewalk-labs-finished-seed-financing-now-in-preliminary-agreements/.

230 **$6 billion over thirty years:** Oved, "Google's Sidewalk Labs Plans."

231 **National Observer got copies:** Fatima Syed, "Alphabet's Sidewalk Labs Was Secretly Considering Big Plans for Toronto

Neighbourhood," Canada's National Observer, February 15, 2019, https://www.nationalobserver.com/2019/02/15/news/alphabets-sidewalk-labs-was-secretly-considering-big-plans-toronto-neighbourhood.

231 **"We don't think that 12 acres":** Oved, "Google's Sidewalk Labs Plans."

231 **"I can tell you":** Sidewalk was extensively lobbying numerous members of Waterfront's three shareholder governments at the time, and registering those many meetings on public databases, but given Gillis's reaction—found via an access-to-information request—the meetings weren't always with key decision-makers.

232 **the city should pull out:** David Rider, Marco Chown Oved, Robert Benzie and Alex Boutilier, "Politicians React with Shock, Anger to Google's Sweeping Vision for Port Lands," *Toronto Star*, February 15, 2019, https://www.thestar.com/news/gta/2019/02/15/google-vision-for-port-lands-a-no-go-with-ontario-government-source-says.html.

232 **"there is no way":** Rider et al., "Politicians React."

233 **devoted a few pages:** Shoshana Zuboff, *The Age of Surveillance Capitalism: The Fight for a Human Future at the New Frontier of Power* (New York: PublicAffairs, 2019), 228–32.

233 **more than 102,000 families:** City of Toronto, Social Housing Waiting List Reports, Q1, 2019, https://www.toronto.ca/city-government/data-research-maps/research-reports/housing-and-homelessness-research-and-reports/social-housing-waiting-list-reports/.

233 **once called "strong":** Margi Murphy, "Sidewalk Labs: How Google's Plan for a Sensor-Packed, Data-Hoovering 'Smart Neighbourhood' Has Divided a City," *Telegraph*, May 28, 2019, https://www.telegraph.co.uk/technology/2019/05/28/googles-plan-sensor-packed-data-hoovering-smart-neighbourhood/.

235 **A hundred and fifty people:** Bruce DeMara, "Citizens Pack Public Meeting Opposing Sidewalk Labs Plan," *Toronto Star*, April 17, 2019, https://www.thestar.com/news/gta/2019/04/17/citizens-pack-public-meeting-opposing-sidewalk-labs-plan.html.

235 **"They really raised expectations":** Josh O'Kane, "Opponents of Sidewalk Labs Get Advice from German Tech Protesters," *Globe and Mail*, November 24, 2019, https://www.theglobeandmail.com

/business/article-opponents-of-sidewalk-labs-get-advice-from -german-tech-protesters/.

236 **"reverse Reaganism":** Lee Greenberg, "Ontario to 'Invest' up to $2B in Business," *National Post*, May 5, 2009, https://www.pressreader.com/canada/national-post-latest-edition/20090505/282595963860144.

236 **dreams of running the province:** Leah McLaren, "Michael Bryant's Very Bad Year: His Life on Bail, How He Got Off, and His Surprise Comeback," *Toronto Life*, September 16, 2010, https://torontolife.com /city/michael-bryants-very-bad-year-his-life-on-bail-how-he-got-off -and-his-surprise-comeback/.

236 **needed to be challenged:** Sean Fine, "Michael Bryant's Second Life: Once He Wielded the Power of the State, Now He Challenges It," *Globe and Mail*, January 6, 2019, https://www.theglobeandmail.com /canada/article-michael-bryants-second-life-once-he-wielded-the -power-of-the-state/.

238 **"potential" ownership of the data:** "Plan Development Agreement between Toronto Waterfront Revitalization Corporation and Sidewalk Labs LLC," July 31, 2018, 48, https://storage.googleapis.com /sidewalk-toronto-ca/wp-content/uploads/2019/06/13210552/Plan -Development-Agreement_July312018_Fully-Executed.pdf.

238 **"effect historically unprecedented":** The CCLA and Lester Brown's Application for Judicial Review regarding Waterfront Toronto and its parent governments, Ontario Superior Court of Justice Divisional Court, Court File No. 211/19, April 16, 2019, https://ccla.org/wp-content/uploads/2021/06/Amended-Notice-of -Application.pdf.

239 **"reset" the project:** Michael Bryant and Brenda McPhail, "Open Letter from CCLA: Calling for a Reset on Waterfront Toronto," Canadian Civil Liberties Association, March 5, 2019, https://ccla.org /open-letter-ccla-calling-reset-waterfront-toronto/.

244 **One of Diamond's first actions:** Catherine McIntyre, "The Man Who Stood Up to Sidewalk," The Logic, October 17, 2019, https://thelogic.co /news/the-big-read/the-man-who-stood-up-to-sidewalk/. Some people close to the project, but outside of Sidewalk, believed that the

November 2018 slide deck presentation to Alphabet executives was leaked to the media by Sidewalk. My reporting suggests otherwise. But Diamond's core frustration was functionally the same: he was angry that Waterfront found out about the deck *through* the media, for whom Sidewalk verified the slide deck's details.

Chapter 19: Shine a Light

245 **"It's not surprising":** Transcript of the February 21, 2019, meeting of the Parliament of Canada's Standing Committee on Access to Information, Privacy and Ethics. All dialogue and details from the opening scene of this chapter come from this transcript and its embedded videos. Portions of dialogue were cut for brevity. https://www.ourcommons.ca/DocumentViewer/en/42-1/ETHI/meeting-137/evidence.

249 **even telling some people:** Jim Balsillie told me this in an interview.

249 **significant disagreements with OMERS:** Josh O'Kane and Sean Silcoff, "A Truck Nearly Killed Him, but Venture Capitalist John Ruffolo Has Kept Moving Forward," *Globe and Mail*, March 19, 2021, https://www.theglobeandmail.com/business/article-a-truck-nearly-killed-him-but-venture-capitalist-john-ruffolo-has-kept/.

250 **"close" to a rival Quayside bid:** John Sewell, "Controversial Sidewalk Labs Plan May Actually Be Good for Toronto," Streets of Toronto, July 5, 2019, https://streetsoftoronto.com/controversial-sidewalk-labs-plan-may-be-good-for-toronto/.

250 **Bianca Wylie tried to figure out:** Since-deleted tweet quoted in Rosemary Frei, "Plan to Re-imagine Toronto's Waterfront: How Much Does Public Know About It?" Rabble.ca, December 6, 2018, https://rabble.ca/news/2018/12/plan-re-imagine-toronto%E2%80%99s-waterfront-how-much-does-public-know-about-it.

250 **"Because Google has more money":** Sarah Fulford, "Is Google's Smart City Stupid?" *Toronto Life*, September 9, 2019, https://torontolife.com/city/is-googles-smart-city-stupid/.

250 **"Majority Still Backs":** Toronto Region Board of Trade, "Environics Research Poll: Majority Still Backs Sidewalk Labs/Quayside Project,"

May 23, 2019, https://www.bot.com/Portals/0/NewsDocuments/5232019NEWS%20RELEASE%20-%20Quayside%20Poll%20May_FINAL.pdf.

250 **"Fewer than one-half of GTA residents":** Environics Research, "Quayside GTA Survey Report," May 16, 2019, https://www.bot.com/Portals/0/NewsDocuments/Environics%20-%20Board%20of%20Trade%20-%20Polling%20FGTA2019%20-%20Report.pdf.

251 **"Alternatively, the main takeaway":** Tweet by Brian Kelcey, May 23, 2019, https://twitter.com/stateofthecity/status/1131598205014216704.

252 **included the Canada Infrastructure Bank:** Sidewalk Labs and John Brodhead have consistently maintained that he followed all ethics laws when leaving the federal government for Sidewalk Labs.

252 **"I'm sure you'll forgive":** Transcript of the April 2, 2019, meeting of the Parliament of Canada's Standing Committee on Access to Information, Privacy and Ethics. Dialogue and details from the closing scene of this chapter come from this transcript and its embedded videos. Portions of dialogue were cut for brevity. https://www.ourcommons.ca/DocumentViewer/en/42-1/ETHI/meeting-141/evidence.

253 **"it may be beneficial":** Waterfront Toronto, "Request for Proposals: Innovation and Funding Partner for the Quayside Development Opportunity," March 17, 2017, 6, https://quaysideto.ca/wp-content/uploads/2019/04/Waterfront-Toronto-Request-for-Proposals-March-17-2017.pdf. (accessed April 2, 2022.)

253 **sent the committee a letter:** Julie Di Lorenzo's letter to the ethics committee, which I have seen, largely argued that Waterfront had mischaracterized the number of meetings her committee had before the vote. This was a longstanding dispute between Di Lorenzo and Waterfront.

255 **someone was putting pressure:** The word "pressure" was Charlie Angus's. The auditor general did not use it in her 2018 annual report (https://www.auditor.on.ca/en/content/annualreports/arreports/en18/v1_315en18.pdf), but did say on page 691: "We found internal Waterfront Toronto emails indicating that the Board felt it was being 'urged—strongly' by the federal and provincial governments to approve

and authorize the Framework Agreement with Sidewalk Labs as soon as possible." Angus's line of questioning did not bring up the provincial government.

Chapter 20: Maybe Tomorrow

256 **"ironically," Doctoroff later said:** Many details from this chapter come from the text of Sidewalk Labs' draft Master Innovation and Development Plan (MIDP), released in June 2019 and titled *Toronto Tomorrow*, which consisted of four books, numbered 0 through 3. Citations from the plan will begin with the book number, followed by the page number; "ironically" can be found in Doctoroff's foreword, MIDP 0-26, 0-27. As of March 16, 2022, digital copies of these books were available at https://quaysideto.ca/document-library/.

257 **Months later, he would tell:** Donovan Vincent, "Why the Much-Touted 'Raincoats' Sidewalk Labs Wanted to Install on Toronto's Waterfront Were Rejected," *Toronto Star*, March 4, 2020, https://www.thestar.com/news/gta/2020/03/04/why-the-much-touted-raincoats-sidewalks-labs-wanted-to-install-on-torontos-waterfront-were-rejected.html.

258 **was nowhere near ready:** Asked why Sidewalk went forward with the draft master plan when Waterfront was still pushing back on details, Dan Doctoroff said, "I don't think that's fair at all. We shared literally every line of the MIDP with Waterfront Toronto. That's not to say that we always started in the same place."

258 **13 percent more energy:** MIDP 1-395.

258 **If all 2.6 million square feet:** MIDP 2-210.

258 **to just 15 percent:** MIDP 1-69; **"ultra" efficiency:** MIDP 0-109; **"library" of prebuilt parts:** MIDP 0-172, 0-183; **generated geothermally:** MIDP 2-343, 2-336; **whole blocks might be dreamed up:** MIDP 2-142.

259 **"largest climate-positive district":** MIDP 0-176.

259 **Extending Sidewalk's reach onto Villiers:** MIDP 0-91; **grant Sidewalk significant privileges:** MIDP 0-92; **"the scale of Quayside alone":** MIDP 0-224.

260 **at least twenty-one thousand:** MIDP 0-66; **streets that focused on pedestrians:** MIDP 2-109, 2-196, 2-291.

260 **"not trying to develop the Port Lands":** MIDP 3-22, 3-23; **share only 10 percent of profits:** MIDP 3-127.

261 **call an elementary school an "innovation":** MIDP 0-113; **planting trees a "technical system":** MIDP 1-99.

261 **the term still appeared:** MIDP 0-114 (and many other pages). Asked why Sidewalk Labs kept pressing forward with the urban data concept well after Waterfront and many of the privacy experts Sidewalk consulted had said it wouldn't work in the Canadian context, Doctoroff said: "This was uncharted territory. No jurisdiction that I was aware of ever had developed a framework for managing data in public space. I'm not sitting here saying the ideas we put forth were all acceptable right away—we didn't know what was going to be acceptable. I think the key thing is, is that we ultimately got to a place that satisfied everybody—ourselves, and our government partners and their constituents." In fact, Sidewalk's government partners, their constituents and even Sidewalk's own former advisers would disagree. When I pressed about Waterfront's rejection of Sidewalk's data terms for a second time, Doctoroff changed tack: "In the fall of 2019, we reached a common point of view. And that is what you really have to look at."

262 **Toronto wanted to dedicate C$800,000:** Report for Action for the City of Toronto's Executive Committee about Quayside by Gregg Lintern and David Stonehouse, June 3, 2019, https://www.toronto.ca/legdocs/mmis/2019/ex/bgrd/backgroundfile-133867.pdf.

262 **"potential" sources of revenue:** MIDP 3-179; **"performance payments":** MIDP 3-177; **tax increment financing re-emerged:** MIDP 2-40; **"limited set" of new technologies":** MIDP 3-123.

263 **would lay out C$900 million:** MIDP 3-152.

263 **Quayside was worth about C$590 million:** Waterfront Toronto, "Overview of Realignment of MIDP Threshold Issues," October 29, 2019, https://quaysideto.ca/wp-content/uploads/2019/10/Overview-of-Thresold-Issue-Resolution-Oct-29.pdf. (accessed April 2, 2022.)

263 **worth nearly c$400 million:** Alex Bozikovic, "The $500-Million Discount at the Heart of Sidewalk Labs' Smart-City Proposal for Toronto," *Globe and Mail*, September 16, 2019, https://www.theglobeandmail.com/arts/art-and-architecture/article-sidewalk-labs-proposal-for-toronto-is-complicated-and-so-far-poorly/. At the time, Sidewalk disputed Bozikovic's analysis, and Dan Doctoroff insisted in our November 2021 interview that the company was not seeking a discount.

263 **set up a public administrator:** MIDP 3-62 to 3-70; **change laws and regulations:** MIDP 3-224 to 3-226.

265 **weighed 18 pounds:** Tweet by the tech news website BetaKit, June 24, 2019, https://twitter.com/BetaKit/status/1143298855376687105.

265 **"democracy grenade":** Jordan Pearson, "Sidewalk Labs' 1,500-Page Plan for Toronto Is a Democracy Grenade," Motherboard: Tech by Vice, June 24, 2019, https://www.vice.com/en/article/vb9nd4/sidewalk-labs-midp-plan-for-toronto-quayside-is-a-democracy-grenade.

265 **"seems to be about pushing back":** John Lorinc, "Lorinc: Sidewalk Labs and the Problem of Smart City Governance," *Spacing*, June 25, 2019, https://spacing.ca/toronto/2019/06/25/lorinc-sidewalk-labs-and-the-problem-of-smart-city-governance/.

265 **"I'm not sure that Sidewalk Labs":** Blayne Haggart, "No Longer Liveblogging Sidewalk Labs' MIDP, Entry 44: The Digital Strategy Advisory Panel's Preliminary Commentary and Questions on the MIDP," Blayne Haggart's Orangespace, September 12, 2019, https://blaynehaggart.com/2019/09/12/no-longer-liveblogging-sidewalk-labs-midp-entry-44-the-digital-strategy-advisory-panels-preliminary-commentary-and-questions-on-the-midp/.

266 **"not up for sale":** Josh O'Kane and Jeff Gray, "Sidewalk Labs Unveils $1.3-Billion Plan for Toronto's Waterfront, Revealing a Vision Much Larger Than Initially Proposed," *Globe and Mail*, June 24, 2019, https://www.theglobeandmail.com/business/article-sidewalk-labs-unveils-13-billion-plan-for-torontos-waterfront/.

Chapter 21: Underwhelmed

267 **Elliott took a photo:** Tweet by Matt Elliott, July 3, 2019, https://twitter.com/GraphicMatt/status/1146527078839443456/photo/1.

267 ***What happens on your iPhone*:** N. Ingraham, "Apple Took Out a CES Ad to Troll Its Competitors over Privacy," Engadget, January 5, 2019, https://www.engadget.com/2019-01-05-apple-ces-2019-privacy-advertising.html.

267 ***Privacy is King*:** Tweet by Josh McConnell, June 28, 2019, https://twitter.com/joshmcconnell/status/1144691437247827970.

268 **Sidewalk, Google's little sister:** If, like all its other deliberately located ads, Apple's Lake Ontario billboard really was directed at Sidewalk, its subtle accusation was admittedly a stretch. Privacy did not loom as large in the critical response to Sidewalk's draft master plan as other issues, and the company had long taken pains to separate phone-based data collection from what it wanted to do at Quayside.

268 **"phalanx" of lobbyists:** Matt Elliott, "Forget the Crystal Ball: Here's What the Lobbying Registry Says Will Be the Big Issues at City Hall This Year," *Toronto Star*, January 5, 2021, https://www.thestar.com/opinion/contributors/2021/01/05/forget-the-crystal-ball-heres-what-the-lobbying-registry-says-will-be-the-big-issues-at-city-hall-this-year.html.

269 **claimed about needing a discount:** Sidewalk Labs' draft Master Innovation and Development Plan (MIDP), *Toronto Tomorrow*, June 2019, 3-158, 3-159, available at https://quaysideto.ca/document-library/.

269 **Sidewalk and Waterfront set a deadline:** Tara Deschamps, "Deal Allows Waterfront Toronto to Block Controversial Sidewalk Labs Project by Fall If Key Issues Not Resolved," *Globe and Mail*, August 2, 2019, https://www.theglobeandmail.com/business/article-deal-allows-waterfront-toronto-to-block-controversial-sidewalk-labs/.

269 **deliberately lowball the amount** and **"leaving a little juice":** Daniel L. Doctoroff, *Greater Than Ever: New York's Big Comeback* (New York: PublicAffairs, 2017), 75.

271 **"technology-enabled infrastructure":** MIDP 3-147.

271 **Alphabet and Teachers invested $400 million:** U.S. Securities and Exchange Commission filing by Sidewalk Infrastructure Partners, September 3, 2019, https://www.sec.gov/Archives/edgar/data/0001768502/000176850219000001/xslFormDX01/primary_doc.xml.

271 **commit $100 million to a start-up:** Adele Peters, "A $100 Million Investment Will Fund the Largest 'Virtual Power Plant' in the U.S.," *Fast Company*, December 7, 2020, https://www.fastcompany.com/90582902/a-100-million-investment-will-fund-the-largest-virtual-powerplant-in-the-u-s.

271 **roads designed for self-driving cars:** Jonathan Shieber, "Starting with Michigan, Sidewalk Infrastructure Is Looking to Build Roads Specifically for Autonomous Cars," TechCrunch, August 13, 2020, https://techcrunch.com/2020/08/13/starting-with-michigan-sidewalk-infrastructure-is-looking-to-build-roads-specifically-for-autonomous-cars.

272 **a kind of "IP grab":** Catherine McIntyre, "Sidewalk Labs Ramps Up Patenting of Technologies Conceived for Now-Defunct Toronto Smart City," The Logic, April 6, 2021, https://thelogic.co/news/sidewalk-labs-ramps-up-patenting-of-technologies-conceived-for-now-defunct-toronto-smart-city/.

273 **Brian Beamish, sent a letter:** Letter from Brian Beamish to Stephen Diamond, "Re: Sidewalk Labs' Proposal," September 24, 2019, https://www.ipc.on.ca/wp-content/uploads/2019/09/2019-09-24-ltr-stephen-diamond-waterfront_toronto-residewalk-proposal.pdf.

274 **"This resulted in a grossly misleading":** I received a copy of this letter, which is quoted in Josh O'Kane, "Indigenous Group Speaks Out over 'Grossly Misleading' Sidewalk Labs Consultation," *Globe and Mail*, October 25, 2019, https://www.theglobeandmail.com/business/article-indigenous-leaders-speak-out-over-grossly-misleading-sidewalk-labs/.

276 **"You need to be frank":** This meeting was corroborated on the record by three people in the room, but a fourth person with knowledge of how it unfolded suggested that the Quayside project's future was not so dramatically on the line. That person did not go on the record. A

spokesperson for Dan Doctoroff declined to comment on the meeting.

278 **"Any disposition of land":** October 21, 2019, letter from Tracey Cook and Brian Johnston to George Zegarac, in Waterfront Toronto, "Threshold Issues Resolution Documents," 12–13, https://quaysideto.ca/wp-content/uploads/2019/10/Threshold-Issues-Resolution-Documents-October-29-2019.pdf. (accessed April 2, 2022.)

280 **"the Jane Jacobs of the Smart Cities Age":** Laura Bliss, "Meet the Jane Jacobs of the Smart Cities Age," Bloomberg CityLab, December 21, 2018, https://www.bloomberg.com/news/articles/2018-12-21/toronto-privacy-advocate-bianca-wylie-v-sidewalk-labs.

280 **"a modern-day Robert Moses":** E.B. Solomont, "The Closing: Dan Doctoroff," *Real Deal*, March 1, 2015, https://therealdeal.com/issues_articles/the-closing-dan-doctoroff/.

280 **a website called Some Thoughts:** Some Thoughts, October 2019, https://www.some-thoughts.org/ (accessed March 18, 2022).

281 **"That timing change":** October 30, 2019, letter from Joshua J. Sirefman to George Zegarac, in Waterfront Toronto, "Threshold Issues Resolution Documents," 14, https://quaysideto.ca/wp-content/uploads/2019/10/Threshold-Issues-Resolution-Documents-October-29-2019.pdf. (accessed April 2, 2022.)

283 **"The vast majority of the planning work":** This interview was excerpted in Josh O'Kane, "Waterfront Toronto Moving Forward on Sidewalk Labs's Smart City, but with Limits on Scale, Data Collection," *Globe and Mail*, October 31, 2019, https://www.theglobeandmail.com/business/article-waterfront-toronto-votes-to-move-forward-with-sidewalk-labs/.

283 **"becomes possible only when considered":** MIDP 1-349.

Chapter 22: Make and Break Harbour

285 **widely publicized sexual-harassment crisis:** Daisuke Wakabayashi and Katie Benner, "How Google Protected Andy Rubin, the 'Father of Android,'" *New York Times*, October 25, 2018, https://www.nytimes.com/2018/10/25/technology/google-sexual-harassment-andy-rubin.html.

285 **nowhere to be found:** Mark Bergen and Austin Carr, "Where in the World Is Larry Page?" *Bloomberg Businessweek*, September 13, 2018, https://www.bloomberg.com/news/features/2018-09-13/larry-page-is-a-no-show-with-google-under-a-harsh-spotlight.

285 **"to leave the roost":** Larry Page and Sergey Brin, "A Letter from Larry and Sergey," *The Keyword* (blog), December 3, 2019, https://blog.google/alphabet/letter-from-larry-and-sergey/.

286 **"While we take a long-term view":** Adam Lashinsky, "The Conversation: Google and Alphabet CEO Sundar Pichai on Managing a Tech Giant's Growing Pains," *Fortune*, January 22, 2020, https://fortune.com/longform/sundar-pichai-google-alphabet-ceo-conversation/.

286 **"Why can't this be bigger?":** Conor Dougherty, "How Larry Page's Obsessions Became Google's Business," *New York Times*, January 22, 2016, https://www.nytimes.com/2016/01/24/technology/larry-page-google-founder-is-still-innovator-in-chief.html.

286 **$659 million that year:** Alphabet Inc. annual report, 2019, 80, https://abc.xyz/investor/static/pdf/2019_alphabet_annual_report.pdf?cache=c3a4858.

288 **On the rooftop patio:** Josh O'Kane, "Sidewalk Labs Courting Local Venture Capitalists to Back Smart-City Project on Toronto Waterfront," *Globe and Mail*, October 20, 2019, https://www.theglobeandmail.com/business/article-sidewalk-labs-courting-local-venture-capitalists-to-back-smart-city/.

288 **found its partner:** Amanda Roth and Catherine McIntyre, "Sidewalk Labs and Plaza Ventures Planning Toronto-Based Venture Fund Focused on Smart-City Technology," The Logic, November 21, 2019, https://thelogic.co/news/exclusive/sidewalk-labs-and-plaza-ventures-planning-toronto-based-venture-fund-focused-on-smart-city-technology/.

288 **"wherever possible":** Sidewalk Labs, "Master Innovation & Development Plan Digital Innovation Appendix," November 14, 2019, 45, https://storage.googleapis.com/sidewalk-toronto-ca/wp-content/uploads/2019/11/15093613/Sidewalk-Labs-Digital-Innovation-Appendix.pdf.

289 **different financial scenarios:** Sidewalk Labs, "Digital Innovation Appendix," 133–34.

289 **"I don't know":** James McLeod, "Sidewalk Labs Digital Update Brings New Details, But Many Questions Remain," *Financial Post*, November 15, 2019, https://financialpost.com/technology/sidewalk-labs-digital-update-brings-new-details-but-many-questions-remain.

291 **"We're still in the midst":** Interview conducted for Josh O'Kane, "Google Plans to Triple Canadian Work Force to As Many As 5,000 Employees," *Globe and Mail*, February 6, 2020, https://www.theglobeandmail.com/business/technology/article-google-plans-to-triple-canadian-work-force-across-toronto-montreal/.

293 **willing to lose a significant amount of money:** When asked about Waterfront staff's recollection of this, Sidewalk Labs did not deny the event happened, but said the "characterization is incorrect." The company pointed to a draft master plan excerpt that said it expected a "reasonable return," though the Waterfront staff who described the event said it took place well before the plan was published.

294 **effectively uninhabitable:** Josh O'Kane, "Waterfront Toronto Endorses Bulk of Sidewalk Labs Smart-City Innovations, but Still Uncertain of Risks," *Globe and Mail*, February 18, 2020, https://www.theglobeandmail.com/business/article-waterfront-toronto-endorses-bulk-of-sidewalk-labs-smart-city/.

294 **"speculative" and "premature":** Josh O'Kane, "Waterfront Toronto Tries to Block Legal Attempt to Shut Down Sidewalk Labs Smart-City Project," *Globe and Mail*, January 30, 2020, https://www.theglobeandmail.com/business/article-waterfront-toronto-tries-to-block-legal-attempt-to-shut-down-sidewalk/.

295 **At least three others:** Eric Levenson, "After Latest Suicide, the Vessel in New York City's Hudson Yards Ponders Its Future," CNN, August 7, 2021, https://edition.cnn.com/2021/08/07/us/vessel-hudson-yards-suicide-wellness/index.html.

296 **Manhattan health-care worker:** "Manhattan Woman, 39, Is NYC's First COVID-19 Case; Husband's Test Results Are Pending," NBC New York, March 1, 2020, https://www.nbcnewyork.com

/news/coronavirus/person-in-nyc-tests-positive-for-covid-19-officials/2308155/.

296 **bailed on negotiations:** Cory Weinberg, "Alphabet Leads Tech Retreat on Real Estate Deals," The Information, April 21, 2020, https://www.theinformation.com/articles/alphabet-leads-tech-retreat-on-real-estate-deals.

296 **prices could plummet by 18 percent:** Tess Kalinowski, "CMHC Says House Prices Could Decline By 18%—and Might Not Recover until 2023," *Toronto Star*, May 27, 2020, https://www.thestar.com/business/2020/05/27/cmhcs-worst-case-scenario-sees-housing-recovery-needing-until-2023.html.

296 **Dan Doctoroff wrote an op-ed:** Daniel L. Doctoroff, "I Helped New York Rebound from 9/11. Here's How to Recover after the Pandemic." *New York Times*, April 15, 2020, https://www.nytimes.com/2020/04/15/opinion/coronavirus-new-york-economy.html.

297 **The RFP was only for a plan:** Sidewalk Labs claims that "it was always understood that the RFP was a step toward the disposition of the Quayside site," though the document did not explicitly say that the partnership would include a land sale.

299 **Ruth Porat had already been agitating:** Vipal Monga and Rob Copeland, "Google Parent Alphabet Drops Controversial 'Smart City' Project," *Wall Street Journal*, May 7, 2020, https://www.wsj.com/articles/alphabet-subsidiary-sidewalk-labs-abandons-toronto-smart-city-project-11588867545.

300 **had raised $400 million:** Jeff John Roberts, "Exclusive: Alphabet Vets Raise $400 Million to Remake America's Infrastructure," *Fortune*, May 7, 2020, https://fortune.com/2020/05/07/alphabet-vets-infrastructure-funding-400-million/.

300 **eight-month-old:** U.S. Securities and Exchange Commission filing by Sidewalk Infrastructure Partners, September 3, 2019, https://www.sec.gov/Archives/edgar/data/0001768502/000176850219000001/xslFormDX01/primary_doc.xml.

300 **"Failure is only failure":** Daniel L. Doctoroff, *Greater Than Ever: New York's Big Comeback* (New York: PublicAffairs, 2017), 288.

301 **one controversy too many:** Josh O'Kane, Chris Hannay and Bill Curry, "From Sunny Ways to Icy Reception: How the Liberals Are Handling Issues Involving Big Tech Firms," *Globe and Mail*, December 23, 2020, https://www.theglobeandmail.com/politics/article-from-sunny-ways-to-deep-freeze-how-the-liberals-are-handling-issues/.

304 **tried to buy a village:** Nancy Keates and Mark Maremont, "Elon Musk's SpaceX Is Buying Up a Texas Village. Homeowners Cry Foul," *Wall Street Journal*, May 7, 2021, https://www.wsj.com/articles/elon-musk-spacex-rocket-boca-chica-texas-starbase-11620353687.

304 **"war on regulators":** Susan Pulliam, Rebecca Elliott and Ben Foldy, "Elon Musk's War on Regulators," *Wall Street Journal*, April 28, 2021, https://www.wsj.com/articles/elon-musk-tesla-spacex-regulators-crash-11619624227.

304 **complained the city was barely affordable:** Oliver O'Connell, "Chasten Buttigieg Criticised for Calling Unaffordable DC 'Almost Unaffordable'," *Independent*, July 27, 2021, https://www.independent.co.uk/news/world/americas/buttigieg-chasten-mocked-washington-rent-b1891573.html.

304 **he wanted to purchase 200,000 acres:** Joshua Brustein, "The Diapers.com Guy Wants to Build a Utopian Megalopolis," *Bloomberg Businessweek*, September 1, 2021, https://www.bloomberg.com/news/features/2021-09-01/how-diapers-com-founder-marc-lore-plans-to-build-utopian-city-telosa.

304 **Google had hired Ingels's firm:** Oliver Wainwright, "Google's New Headquarters: An Upgradable, Futuristic Greenhouse," *Guardian*, February 27, 2015, https://www.theguardian.com/artanddesign/2015/feb/27/googles-new-headquarters-upgradable-futuristic-greenhouse.

305 **At least 150 new city-building projects** and **"It's so seductive to say":** Brustein, "Diapers.com Guy."

307 **"The thread of their discourse":** Italo Calvino (translation by William Weaver), *Invisible Cities* (London: Vintage, 1997), 38.

Epilogue: Break It to Them Gently

309 **"We built a business":** This quote is from an interview conducted for Josh O'Kane and Sean Silcoff, "The BlackBerry veteran who took on Google and won," *Globe and Mail*, *Report on Business* magazine, March 21, 2022, https://www.theglobeandmail.com/business/rob-magazine/article-the-blackberry-veteran-who-took-on-google-and-won/.

311 **Worth more than $5 billion:** Erin Brodwin, "Cityblock Health Raises Another Mega-Round of Funding, Tipping Its Valuation over $5 Billion," STAT, September 3, 2021, https://www.statnews.com/2021/09/03/cityblock-health-series-d-valuation/.

311 **Portland officials alleged:** Kate Kaye, "Portland Ditches Google's Smart City Tech Sibling Replica," RedTail, February 20, 2021, https://redtailmedia.org/2021/02/20/portland-ditches-googles-smart-city-tech-sibling-replica/.

312 **hundreds of people and organizations:** Russell Winer, Jinsong Du, Rebecca Horan, Katsura Kikuzawa, Dana Meir, Esilda Seng and Juliana Wu, "NYC2012" case study, Leonard N. Stern School of Business, New York University, Case Number: MKT04-02, December 2004, https://people.stern.nyu.edu/rwiner/NYC2012%20Case.pdf.

313 **Hudson Yards had delivered the city $663 million:** Martin Z. Braun, "NYC Reaping Hudson Yards Dividends After Luxury Building Boom," Bloomberg, October 18, 2021, https://www.bloomberg.com/news/articles/2021-10-18/nyc-reaping-hudson-yards-dividends-after-luxury-building-boom.

313 **accused of leaving New Yorkers:** Bridget Fisher and Flávia Leite, "The Cost of New York City's Hudson Yards Redevelopment Project," Schwartz Center for Economic Policy Analysis and Department of Economics, The New School for Social Research, Working Paper Series 2018-2, https://www.economicpolicyresearch.org/images/docs/research/political_economy/Cost_of_Hudson_Yards_WP_11.5.18.pdf.

313 **Doctoroff disagrees:** "If you actually look at my investment track record, over the course of time, the reason people have backed me is

because I have made people an extraordinary amount of money," Doctoroff told me in an interview. A spokesperson later sent a link to an analysis by Mitchell Moss, director of the Rudin Center for Transportation Policy & Management at New York University, which his team said "demonstrates that the investment made in NYC's Olympic bid shaped development and transformed major areas of New York City." The spokesperson also shared a Bloomberg story from October 2021 that reported that Hudson Yards "has transferred about $663 million in surplus property tax and other revenue to the city since 2017," and also said that the city will get far more than $2.2 billion in income from the development in the coming years. The Sidewalk spokesperson also said that describing the $50 million Sidewalk spent preparing the draft master plan as a failed investment "is refuted by the enormous value generated to date by Sidewalk's broad research and development efforts." As an example, the company cited the success of spinout Cityblock Health. But Sidewalk and Waterfront consistently said the $50 million was to focus specifically on the draft master plan in Toronto, and Cityblock Health had long been treated as its own business line.

313 **"bargain basement price":** Isabelle Kirkwood and Josh Scott, "Toronto Mayor Believes Sidewalk Labs Wanted 'Bargain Basement Price' for Quayside, Hootsuite CEO Pro Social Regulation (Collision 2021)," BetaKit, April 21, 2021, https://betakit.com/toronto-mayor-believes-sidewalk-labs-wanted-bargain-basement-price-for-quayside-hootsuite-ceo-pro-social-regulation-collision-2021/.

313 **died in 2002:** Daniel L. Doctoroff, "My Next Chapter: Fighting ALS," *Sidewalk Talk* (blog), December 16, 2021, https://medium.com/sidewalk-talk/my-next-chapter-fighting-als-207ce7ca69c8/.

314 **His father had been limping:** Daniel L. Doctoroff, *Greater Than Ever: New York's Big Comeback* (New York: PublicAffairs, 2017), 23.

314 **"Ever since then":** Doctoroff, "My Next Chapter."

315 **"planning to be subsumed":** Andrew Ross Sorkin, Jason Karaian, Sarah Kessler, Stephen Gandel, Lauren Hirsch and Ephrat Livni, "Dealbook Newsletter: 'I Want to Try to Enjoy Every Day,'"

New York Times, December 16, 2021, https://www.nytimes.com/2021/12/16/business/dealbook/doctoroff-als.html.

315 **"winding down"**: K. Holt, "Sidewalk Labs Products Will Be Folded into Google Proper," Engadget, December 16, 2021, https://www.engadget.com/sidewalk-labs-products-google-alphabet-151740698.html.

Index

JOSH O'KANE has been a reporter with the *Globe and Mail*, Canada's largest national newspaper, since 2011. He won Germany's 2019 Arthur F. Burns Award for transatlantic political and cultural reporting for his coverage of that country's broad pushback against Big Tech. O'Kane's reporting has also won numerous Best in Business Canada awards from the Society for Advancing Business Editing and Writing. His first book, *Nowhere with You*, was a Canadian bestseller. He lives in Toronto.